188

2

NON-LINEAR PROBLEMS
IN STRESS ANALYSIS

Papers presented at the 1977 Annual Conference of the Stress Analysis Group of the Institute of Physics held at the University of Durham, England, 20–22 September, 1977.

NON-LINEAR PROBLEMS
IN STRESS ANALYSIS

Edited by

P. STANLEY

DEPARTMENT OF MECHANICAL ENGINEERING
UNIVERSITY OF NOTTINGHAM, ENGLAND

APPLIED SCIENCE PUBLISHERS LTD
LONDON

APPLIED SCIENCE PUBLISHERS LTD
RIPPLE ROAD, BARKING, ESSEX, ENGLAND

British Library Cataloguing in Publication Data

Institute of Physics. Stress Analysis Group.
Annual Conference, University of Durham, 1977
Non-linear problems in stress analysis.
1. Strains and stresses—Congresses
I. Stanley, Peter
620.1'123   TA407

ISBN 0-85334-780-8

WITH 27 TABLES AND 197 ILLUSTRATIONS

Printed in Great Britain by Galliard (Printers) Ltd, Great Yarmouth, Norfolk

Preface

The non-linear behaviour of structures and components, and of structural materials, offers a wide range of challenging and important problems to the stress analyst. Twenty-two papers on these problems, presented at the 1977 Annual Conference of the Stress Analysis Group, are brought together in this volume.

Two particularly important aspects of non-linear material behaviour received much attention, viz. creep and the effects of cyclic plastic loading. Five papers deal with creep. The creep which develops in a body subjected to cyclic plastic straining is studied, both in cyclically hardening and in cyclically softening materials. The plastic flow and creep behaviour of anisotropic aluminium have been examined and extensions of existing theories are proposed to cover the observations. A design approach based on the steady-state creep under redistributed stresses is proposed and the properties of a lead–antimony–arsenic alloy used for the modelling of structural creep are described. The fifth paper presents an analysis of the creep collapse of a non-uniformly heated, pressurised tube. In creep-related investigations, the effects of gross plastic cycling on the toughness of mild steel are evaluated in terms of the J contour integral and related to the cyclic creep phenomenon; cyclic-hardening effects in several irons and steels are also studied and interpreted in the light of the associated cyclic creep. Two papers are concerned with strain accumulation due to cyclic thermal loading, that is, ratcheting. The first uses finite-element analysis to study shakedown in thick tubes and illustrates the effects of stress concentrations on ratcheting. The second assesses the influence of material characteristics on the incremental deformation of a pressurised cylinder.

A paper on crack-tip deformation fields considers in detail the validity of using crack opening displacement and the J contour integral as characterising parameters in cases of large-scale yielding.

Finite-element studies are prominent. An introductory paper on the application of finite-element methods with non-linear constitutive laws is included and several finite-element analyses are described. The energy and extended energy methods are applied in a turbine blade creep problem and hardening effects are assessed. Degenerate isoparametric elements are used in a study of the effects of large displacements and material non-linearities in shells; stability characteristics can also be examined. Finite-element methods are applied in a further study of the buckling behaviour of shells of revolution. Detailed stress analyses are presented for an austenitic steel weld and the exhaust valve and piston crown of a typical medium-speed diesel engine; the need for suitable temperature-dependent material data is emphasised in the latter study. The plastic strains in the threads of a steam turbine flange bolt and around the tip of a crack in a flat plate are also analysed.

An analytical approach to the structural buckling problem, relevant in the study of wind-loading on cooling towers, is described. The non-linear behaviour of polyhedral sandwich shell structures is outlined, with emphasis on the time-dependent nature of the structural deformations. Further analytical investigations deal with the loading and unloading of an elastic half-space by a rigid punch with interfacial friction and the non-linear deflection of imperfect annular plates.

The Stress Analysis Group Committee are grateful to the authorities of the University of Durham and Trevelyan College for the excellent facilities and accommodation provided, and, in particular, to Dr P. Braiden for his tireless efforts in taking care of all local arrangements.

P. STANLEY

List of Contributors

W. S. BLACKBURN
C. A. Parsons & Co. Ltd, Heaton Works, Newcastle upon Tyne NE6 2YL.

M. R. BRIGHT
Department of Combined Engineering, Lanchester Polytechnic, Lower Hillmorton Road, Rugby, Warwickshire CV21 3TG.

D. CHRISTOPHERSON
Vice-Chancellor and Warden, University of Durham, Old Shire Hall, Durham DH1 3HP.

J. DANKS
Department of Mechanical Engineering, Lanchester Polytechnic, Priory Street, Coventry CV1 5FB.

D. J. F. EWING
Research Division, Central Electricity Research Laboratories, Central Electricity Generating Board, Kelvin Avenue, Leatherhead, Surrey KT22 7SE.

M. K. EXETER
Department of Mechanical Engineering, University of Nottingham, University Park, Nottingham NG7 2RD.

H. FESSLER
Department of Mechanical Engineering, University of Nottingham, University Park, Nottingham NG7 2RD.

A. M. GOODMAN
   Research Department, Berkeley Nuclear Laboratories, Central Electricity Generating Board, Berkeley, Gloucestershire, GL13 9PB.

S. J. HARVEY
   Department of Combined Engineering, Lanchester Polytechnic, Lower Hillmorton Road, Rugby, Warwickshire CV21 3TG.

T. K. HELLEN
   Research Department, Berkeley Nuclear Laboratories, Central Electricity Generating Board, Berkeley, Gloucestershire GL13 9PB.

R. D. HENSHELL
   PAFEC Ltd, PAFEC House, 40 Broadgate, Beeston, Nottingham NG9 2FW.

D. HITCHINGS
   Department of Aeronautics, Imperial College of Science and Technology, Exhibition Road, London SW7 2AZ.

J. S. HOLT
   Department of Mechanical Engineering, University of Leeds, Leeds LS2 9JT.

T. H. HYDE
   Department of Mechanical Engineering, University of Nottingham, University Park, Nottingham NG7 2RD.

A. D. JACKSON
   C. A. Parsons & Co. Ltd, Heaton Works, Newcastle upon Tyne NE6 2YL.

A. G. JAMES
   Scientific Services Centre (Midlands Region), Central Electricity Generating Board, Ratcliffe-on-Soar, Nottingham NG11 0EE.

B. KRÅKELAND
   Engineering Department, AS Computas, Veritasveien 1, P.O. Box 310, 1322 Høvik, Norway.

A. B. LOMAX
Faculty of Mechanical Engineering, Derby College of Art and Technology, Kedleston Road, Derby DE3 1GB.

S. B. MATHUR
School of Mechanical, Aeronautical and Production Engineering, Kingston Polytechnic, Canbury Park Centre, Canbury Park Road, Kingston upon Thames KT2 6LA.

O. MO
Engineering Department, AS Computas, Veritasveien 1, P.O. Box 310, 1322 Høvik, Norway.

H. A. MONEY
Scientific Services Centre (Midlands Region), Central Electricity Generating Board, Ratcliffe-on-Soar, Nottingham NG11 0EE.

P. W. J. OLDROYD
Department of Mechanical Engineering, Polytechnic of Central London, 115 New Cavendish Street, London W1M 8JS.

B. PARSONS
Department of Mechanical Engineering, University of Leeds, Leeds LS2 9JT.

G. M. PARTON
Department of Engineering Science, University of Durham, Science Laboratories, South Road, Durham DH1 3LE.

C. PATTERSON
Department of Mechanical Engineering, University of Sheffield, Mappin Street, Sheffield S1 3JD.

R. PILKINGTON
Department of Metallurgy, University of Manchester, Grosvenor Street, Manchester M1 7HS.

J. C. RADON
Department of Mechanical Engineering, Imperial College of Science and Technology, Exhibition Road, London SW7 2AZ.

D. W. A. REES
  *School of Mechanical, Aeronautical and Production Engineering,
  Kingston Polytechnic, Canbury Park Centre, Canbury Park Road,
  Kingston upon Thames KT2 6LA.*

V. SAGAR DWIVEDI
  *Department   of   Mechanical   Engineering,   University   of
  Nottingham, University Park, Nottingham NG7 2RD.*

E. SMITH
  *Department of Metallurgy, University of Manchester, Grosvenor
  Street, Manchester M1 7HS.*

I. SMITH
  *Department of Engineering Science, University of Durham,
  Science Laboratories, South Road, Durham DH1 3LE.*

J. R. TURNER
  *Department of Engineering Science, University of Oxford, Parks
  Road, Oxford OX1 3PJ.*

G. J. TURVEY
  *Department of Engineering, University of Lancaster, Bailrigg,
  Lancaster LA1 4YR.*

J. J. WEBSTER
  *Department   of   Mechanical   Engineering,   University   of
  Nottingham, University Park, Nottingham NG7 2RD.*

Contents

1

The Characterisation of Crack-Tip Deformation Fields in Non-Linear Materials

E. SMITH AND R. PILKINGTON

University of Manchester

SUMMARY

This paper critically reviews the current position regarding the characterisation of the crack-tip deformation field in a non-linear material, for which the non-linearity arises from time-independent plastic deformation. The main characterising procedures are appraised, linear elastic material behaviour being used as a basis for the discussion. With linear elastic material, the crack-tip deformation field is characterised by the linear elastic stress intensity factor K_1; this representation is also applicable to a plastic–elastic material, provided that the plastic zone size is small. When the extent of the spread of plasticity is large, however, alternative procedures must be employed. The paper discusses the usefulness, ranges of applicability and relative merits of the J integral and crack opening displacement as characterising parameters for the crack-tip deformation field. The relation of these parameters to the linear elastic stress intensity factor is also considered and their relevance to fracture initiation criteria is explored in detail.

INTRODUCTION

The last twenty years have seen the development of fracture mechanics procedures as a rational basis for design against catastrophic fracture in engineering structures, with most attention being concentrated on the prevention of unstable fracture initiation from a pre-existing crack rather than the subsequent arrest of a propagating crack. The basis of these procedures is the behaviour of linear elastic material, for which the crack-tip deformation field can be charac-

1

terised by a single parameter K_I (assuming plane strain Mode I crack opening); K_I is referred to as the stress intensity factor and is a scaling factor for the local stresses and strains, which have a square-root singularity. The shape and size of the solid and magnitude of the applied loads do not affect the angular variation of the crack-tip deformation field, but govern its absolute magnitude through K_I.

This concept is readily extended to a material that is not perfectly elastic, but for which there is a limited amount of plastic deformation in the vicinity of a crack-tip, i.e. the plastic zone is small compared with the crack size and any other characteristic dimension of the solid. For this situation, which is commonly referred to as 'small-scale yielding', the shape and size of the solid and the magnitude of the applied loads again influence the crack-tip deformation behaviour only through the value of K_I (which is obtained by assuming the material to be perfectly elastic) and again the crack-tip deformation field is characterised by the single parameter K_I. Assuming that fracture initiates in the vicinity of a crack-tip when the material near the tip attains some critical state of deformation, the fracture initiation criterion is $K_I = K_{IC}$, where K_{IC} has a value that reflects the difficulty of initiating fracture by plastic processes and is material-dependent. There is ample experimental evidence, particularly for high yield strength materials, to support this viewpoint, and such behaviour is referred to as being covered by the linear elastic fracture mechanics (LEFM) approach.

In using the LEFM approach to predict the condition for fracture initiation at a crack-tip in an engineering structure, the first objective is to determine the critical value of K_I, i.e. K_{IC}, at which fracture occurs in a laboratory specimen, by measuring the applied stress required to initiate fracture at the tip of a crack of known size. A linear elastic stress analysis is then conducted for the structure and the values of K_I are determined for cracks that are assumed to be present within the structure. A comparison of these K_I values with K_{IC} gives an estimate of the size of defect that can be tolerated within the structure without fracture being initiated and this critical defect size can be compared with actual defect sizes, as revealed by appropriate inspection procedures. The integrity of the structure, with regard to fracture initiation, can then be assessed.

The plastic zone size at fracture initiation is proportional to $(K_{IC}/\sigma_y)^2$, where σ_y is the yield strength, and the general behaviour trend is that K_{IC} increases as σ_y decreases [1]. This means that with many materials, such as the low- and medium-strength steels that are used in many engineering structures that operate around room temperature, plastic deformation is far-reaching and it becomes impossible to measure K_{IC} in a laboratory test without using very large specimens containing very deep cracks. These circumstances focus attention on the crucial question of whether it is possible, as it is with small-scale yielding, to characterise the crack-tip deformation field in the large-scale yield situation in terms of a single parameter. If this were so, it could be assumed that fracture initiates when this parameter attains a critical value. If such a characterising parameter exists, it should be possible to determine its critical value from laboratory specimens and use the results to predict the behaviour of engineering structures. Considerable attention has been given to this particular aspect of the fracture problem during the last few years and there have been two main streams of development: (i) the *J* integral approach, developed primarily in the USA and (ii) the crack opening displacement (COD) approach, developed primarily in the UK.

The primary objective in this paper is to discuss the validity of both these approaches and their relation to each other and to the LEFM approach, and thereby to assess their respective merits. To provide a logical background for the discussion, both perfectly elastic and small-scale yielding behaviour are initially considered. The emphasis throughout is on physical principles and not detailed mathematical analyses or measurement techniques, both of which have been admirably considered in detailed publications in the literature.

THE CRACK-TIP DEFORMATION FIELD FOR A LINEAR ELASTIC MATERIAL

Irrespective of both the geometry of the solid and the magnitude of the applied load, the stress in the immediate vicinity of a crack-tip (see Fig. 1) in a linear elastic material takes the form [2]:

$$p_{ij} = \frac{K_I}{\sqrt{2\pi r}} f_{ij}(\theta) \qquad (1)$$

for a crack that opens by Mode I deformation (r and θ are polar coordinates measured with respect to the crack tip). The applied loads and geometry affect the local stress field only through the stress intensity factor K_I, and $f_{ij}(\theta)$ are known functions of θ; the crack-tip

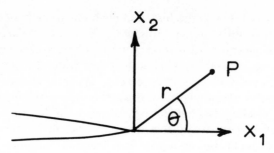

FIG.1. Coordinate system appropriate to a crack opening by Mode I deformation.

deformation field is therefore characterised by the single parameter K_I.

If the Griffith energetics approach [3] is pursued, the condition for fracture initiation at the crack tip is

$$K_I = \sqrt{\frac{2E\gamma}{(1 - \nu^2)}} \qquad (2)$$

where γ is the surface energy of the material, E is Young's modulus and ν is Poisson's ratio. The criterion has the same form if initiation is assumed to occur when a critical tensile stress is attained at a critical distance ahead of the crack-tip. As an example, for the important special case where an infinite solid contains a crack of length $2c$ in the x_1 direction and of infinite length in the x_3 direction (normal to the plane of the section shown in Fig. 1) and is loaded by a uniform applied tensile stress $p_{22} = \sigma$ normal to the plane of the crack, $K_I = \sigma(\pi c)^{1/2}$ and the fracture initiation condition represented in eqn (2) becomes the well-known Griffith relation

$$\sigma = \sqrt{\frac{2E\gamma}{\pi(1 - \nu^2)c}} \qquad (3)$$

It should be emphasised that the preceding considerations refer to a linear elastic continuum model and departures from material linearity are incorporated into the discussion only in a very indirect manner via the surface energy term γ, which is a measure of the work required to separate adjacent atomic planes, a process that clearly does involve non-linearity of material behaviour on the atomic scale.

THE CRACK-TIP DEFORMATION FIELD FOR A MATERIAL WITH LIMITED PLASTIC DEFORMATION IN THE VICINITY OF THE CRACK-TIP

For linear elastic materials, two bodies having different geometries, with cracks of different sizes and with different loading systems, will have the same crack-tip deformation fields if the crack-tip stress intensity factors are equal. This conclusion is also valid for a material which deforms plastically, provided that the applied loads are low enough for the plastic zone at a crack-tip to be surrounded by elastic material and to be small compared with characteristic dimensions of the solid such as crack length, uncracked width, etc. (see Fig. 2). In this situation, usually referred to as 'small-scale yielding', a boundary layer formulation [4] gives the near-tip deformation field within both

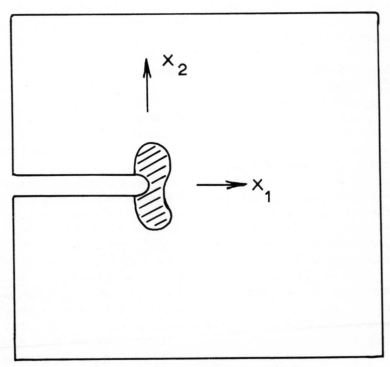

FIG. 2. Small-scale yielding near a crack tip.

the plastic region and the surrounding elastic region; the stresses can
be expressed as

$$p_{ij} = K_1 g_{ij}(r, \theta) \tag{4}$$

where $g_{ij}(r, \theta)$ are functions of the polar coordinates r and θ and also
of the material's flow properties, but assume different forms for the
plastic and elastic regions. $g_{ij}(r, \theta)$ are independent of both the applied
loading system and the geometry of the solid, whose effects on the
crack-tip deformation behaviour are manifested through K_I, the
crack-tip intensity factor as determined for a linear elastic material.
The linear elastic stress intensity factor, K_I, therefore uniquely
characterises the crack tip deformation field, even when there is
yielding, provided that this is small-scale. Proceeding on the basis
of the reasonable assumption that fracture initiates at a crack-tip
when the material near the crack-tip attains some critical state of
deformation, the fracture initiation criterion can be viewed in terms
of K_I attaining a critical value K_{IC} (i.e. $K_I = K_{IC}$) which is referred to
as the material's plane strain fracture toughness corresponding to
Mode I crack opening. K_{IC} depends on the flow and microstructural
properties of the material through their effects on the local fracture
mechanism, but is independent of the applied loads and the geometry
of the solid.

The power of these conclusions, which form the basis of what is
commonly referred to as linear elastic fracture mechanics (LEFM)
(even though there is a limited amount of plasticity), resides in the
fact that K_{IC} may be determined in a laboratory test, by measuring the
applied stress required to initiate fracture at the tip of a crack of a
known size. There is a vast amount of experimental evidence to
support the view that, for small-scale yielding, fracture initiates at a
critical K_I value. An important feature of the LEFM approach is that
the experimental measurements are made away from the crack-tip—
for example, at the outer boundary of a test specimen. Consequently,
experimental difficulties associated with making measurements in the
immediate vicinity of a crack-tip are avoided. The integrity of an
engineering structure with regard to fracture initiation is then assessed by conducting an elastic stress analysis for the structure, using
the design loadings as input, and K_I is determined for cracks which
are assumed to be present at various regions of the structure. With
the knowledge that fracture initiates when $K_I = K_{IC}$, it is therefore
possible to estimate the size of crack that the structure can tolerate

without fracture being initiated, and this critical size can be compared with actual defect sizes, as determined by inspection procedures.

THE CRACK-TIP DEFORMATION FIELD FOR A MATERIAL WITH EXTENSIVE PLASTIC DEFORMATION IN THE VICINITY OF THE CRACK-TIP

The small-scale yielding behaviour pattern is applicable for a wide variety of materials subjected to a wide range of testing conditions and this is an important reason why LEFM procedures are so useful when designing against fracture initiation in engineering structures. However, there are many materials—for example, the low and medium yield strength steels used in a variety of engineering components that operate around room temperature—for which the plastic zone at a crack-tip is large. As indicated in the introductory section, with such materials it is impossible, without using very large specimens containing very deep cracks, to measure K_{IC} in a laboratory test. Alternative procedures must therefore be developed, and attention is focused on the crucial question of whether the crack-tip deformation field in the large-scale yielding situation can be characterised by a single parameter, as is the case for the small-scale yielding. If such a unique characterisation is possible, it is reasonable to assume that fracture initiates when the characterising parameter attains a critical value; consequently, one should, in principle, be able to determine this critical value in laboratory specimens and use the results to predict the behaviour of engineering structures. This problem area has received considerable attention during the last few years, prompted particularly by arguments concerned with the integrity of nuclear reactor steel pressure vessels, and there have been two main streams of development: (i) the J integral approach developed primarily in the USA and (ii) the crack opening displacement (COD) approach developed primarily in the UK. There has been considerable discussion of the validity and relative merits of both these approaches and their relation to each other and to the LEFM approach; the key features of these considerations will be highlighted in this section.

The most significant development in recent years has been an increasing awareness, due largely to Rice's efforts [4], of the importance in crack-tip analyses of the J integral, which was first

discovered by Eshelby [5] in a very general discussion of elastic
singularities and referred to as the 'energy–momentum tensor'. For the
plane strain Mode I opening of a crack lying along the x_1 axis, the J
integral is defined [4] as

$$J = \int_\Gamma \left[W \, dx_1 - n_j p_{ij} \frac{\partial u_i}{\partial x_2} \, ds \right] \tag{5}$$

where Γ is a contour which surrounds the crack-tip (see Fig. 3),
starting from the lower flat crack surface and ending on the upper
surface, ds is an element of arc length, n_j is the outward normal to
the contour, while u_i and W are the displacement function and stored
energy function respectively. The usefulness of the J integral stems
from its path-independence; it has the same value irrespective of the
choice of contour Γ, although it should be emphasised that the
path-independence property is strictly valid only when the material
flows according to a 'deformation theory' of plasticity and not when it
flows according to an 'incremental theory' [4].

A consideration of plastic–elastic models [4, 6, 7] of work-hardening
materials has shown that the J integral gives the deformation field
within the plastic region in the immediate vicinity of a crack-tip,
irrespective of the applied loads, geometrical effects and plastic zone

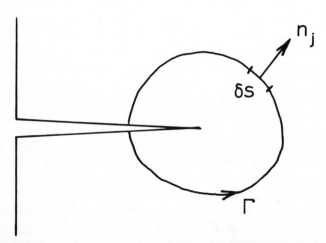

FIG. 3. A contour Γ around the tip of a crack. The magnitude of the J integral given
by eqn (5) is independent of the choice of Γ; it can pass through the purely elastic
region, the plastic region, or both plastic and elastic regions.

size. As an example, a material is considered which deforms in accordance with the Ramberg–Osgood relation, i.e.

$$\bar{\sigma} = \bar{\sigma}_1(\bar{\epsilon}_p)^n \qquad (6)$$

where $\bar{\sigma}$ is the equivalent stress, $\bar{\epsilon}_p$ the equivalent plastic strain, $\bar{\sigma}_1$ a constant and n the strain-hardening exponent. The stresses and strains within the plastic region in the immediate vicinity of a crack-tip [8] are:

$$p_{ij} = \bar{\sigma}_1 \left[\frac{J}{\bar{\sigma}_1 I_n} \right]^{(n/n+1)} \frac{1}{r^{(n/n+1)}} \tilde{p}_{ij}(\theta) \qquad (7)$$

$$e_{ij} = \left[\frac{J}{\bar{\sigma}_1 I_n} \right]^{(1/n+1)} \frac{1}{r^{(1/n+1)}} \tilde{e}_{ij}(\theta) \qquad (8)$$

where I_n is a function of n, and $\tilde{p}_{ij}(\theta)$ and $\tilde{e}_{ij}(\theta)$ are functions of θ and also of the material's flow behaviour. The crack-tip stresses and strains within the plastic region in the immediate vicinity of the crack-tip are singular in r and the J integral is a scaling factor. Equations (7) and (8) are analogous to the crack-tip stress and strain relations for a linear elastic material (see eqn 1) where the linear elastic stress intensity factor K_I was the appropriate scaling factor. Indeed, by considering the linear elastic situation and choosing an appropriate contour around the crack-tip, it is easily shown that J and K_I are related by the expression

$$J = \frac{(1 - \nu^2)K_I^2}{E} \qquad (9)$$

It also follows that the deformation field in the immediate vicinity of a crack-tip in the small-scale yielding situation is characterised by the J integral (see eqns 4 and 9).

For a material in which there is only small-scale yielding prior to fracture initiation, it has been assumed that fracture occurs when K_I attains the critical value K_{IC}. A logical extension for the case of large-scale yielding is the assumption that fracture initiates when J attains a critical value J_{IC}, using an obvious terminology. Regarding the determination of J, its path independence allows its value to be measured around a contour that is far removed from the crack-tip; indeed, its value may be obtained from load–displacement curves for laboratory-scale specimens. Consequently, by observing when fracture initiation occurs, the critical value J_{IC} can be determined; recent experimental evidence supports this approach [9–11]. The determination of J_{IC} for large-scale yielding by measuring load and displacement is analogous to determining K_{IC} in the case of small-scale yielding by measuring the applied stress required to extend a crack of known length; in both cases the response of the solid at points far removed from the crack-tip is being measured.

Having shown that the deformation field within the plastic region in the immediate vicinity of a crack-tip can be characterised by the J integral (see eqns 7 and 8), it follows that the crack-tip profile is also characterised by J. (It should be noted that present considerations have been based on small geometry change theories. These give the crack-tip profile only very approximately, but do allow the boundary conditions for some highly localised region to be specified, thereby enabling a large geometry change theory [12] to give the tip profile accurately. It is in this sense that the J integral characterises the crack-tip profile). If the crack-tip profile is such that a crack opening displacement can be satisfactorily defined (see Fig. 4), it can be expressed in terms of J and the material's flow behaviour expressed, for example, in terms of the parameters $\bar{\sigma}$ and n of the Ramberg–Osgood relation [7].

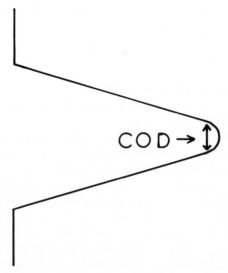

FIG. 4. An example of a situation where the COD can be satisfactorily defined.

Assuming that a crack-tip opening displacement can be defined and measured experimentally (the last fifteen years have seen extensive efforts made to attain this goal), then because the J integral and the COD are related in the manner discussed in the preceding paragraph, it follows that fracture should initiate when the COD attains a critical value, say Φ_C. This is the basis of the COD criterion for fracture initiation in materials where there is large-scale yielding at a crack-tip. In contrast with the requirements of the J integral approach, experimental measurements of the COD must be made at the crack-tip. This is clearly inconvenient and crack-mouth opening measurements have been correlated with crack-tip opening displacements, using relations that have been developed for specific materials and prescribed specimen geometries [13]. These correlations are not general, however, and in this respect the J integral approach has decided advantages over the COD approach.

THE USE OF CRITICAL J OR CRITICAL COD VALUES TO PREDICT THE BEHAVIOUR OF LARGE ENGINEERING STRUCTURES

The design stress levels for many engineering structures are appreciably less than the yield stress. Consequently, if there is serious concern about a structure failing owing to fracture initiation at some defect, it is important to know the critical defect size at which fracture will initiate at the design stress level; this critical size can then be compared with actual defect sizes as given by appropriate inspection procedures and the integrity of the structure assessed. Since it is being assumed that the design stresses are less than the yield stress, small-scale yielding conditions apply. One therefore performs an elastic stress analysis for the structure, assuming the existence of cracks in various regions of the structure, and determines the K_I values for these cracks. The critical defect size is then obtained by arguing that fracture initiates when $K_I = K_{IC}$. It is therefore essential to know K_{IC} for the material from which the structure is built. As indicated earlier in this paper, for a material where fracture initiates in laboratory specimens under small-scale yielding conditions, K_{IC} can be measured directly. With a tougher material where fracture initiates in laboratory-scale test specimens only after extensive plastic yielding, perhaps even after general yielding, this relatively simple procedure becomes impossible. However, the arguments of the preceding section may be used. Remembering that fracture initiates when J attains the critical value J_{IC}, which can be measured from load–displacement curves for laboratory-scale test specimens, eqn (9) shows that

$$K_{IC} = \left[\frac{EJ_{IC}}{(1 - \nu^2)} \right]^{1/2} \tag{10}$$

the important point being that this relation is independent of the material's flow characteristics.

In principle, therefore, it is possible to proceed directly from a critical J value for fracture initiation to a K_{IC} value. On the other hand, even when it is possible to satisfactorily define and experimentally measure a critical COD corresponding to fracture initiation in a laboratory specimen, it is essential to know the detailed plastic flow characteristics of the material and to know how to use this information to obtain a K_{IC} value. (This can be seen by examination of eqns 8 and 10). In other words, although the relationship between J_{IC} and K_{IC} is independent of the plastic flow properties of the material, any relationship between the critical COD value Φ_C and K_{IC} must depend on the plastic flow properties; this point is not always appreciated. Having emphasised this, however, it must be noted that although the correlation between critical COD values and K_{IC} values depend on the plastic flow properties, an approximate correlation, independent of detailed flow properties, may exist within a limited range of these properties. For example, Robinson and Tetelman [14] have shown that critical COD measurements can be satisfactorily correlated with K_{IC} values within limited ranges of testing and material variables by means of the expression $K_{IC} \approx \alpha(E\sigma_y\Phi_C)^{1/2}$, α being a constant and σ_y the yield stress. The COD approach therefore has some merits, although, in the light of this section's comments, great care must be exercised when rating dissimilar materials for their plane strain fracture toughness K_{IC}, solely in terms of their critical crack opening displacements.

If the loads applied to an engineering structure are so large that there are regions where the stress is an appreciable fraction of the yield stress, or may even exceed it, the K_{IC} value cannot be used in conjunction with an elastic stress analysis to give the critical size of defect which the structure can tolerate, since this procedure overestimates the critical defect size. In such situations there is no alternative but to conduct a plastic–elastic stress analysis for the structure for the applied loadings expected in service and with cracks situated in various regions, and to determine the magnitude of either the *J* integral or the COD associated with crack; the former is preferred, since it can be determined using a contour that need not be close to the crack-tip. The structure's integrity with respect to fracture in-itiation from a defect can then be assessed by comparing these *J* or COD values with J_{IC} and Φ_C, by arguing that fracture will not initiate if $J < J_{IC}$ or $COD < \Phi_C$.

DISCUSSION

This paper has described, in essentially physical terms, how the deformation field in the vicinity of a crack-tip may be characterised and how such a characterisation automatically leads to criteria for the initiation of fracture. It has been shown how the behaviour of linear elastic material and that of material with a limited amount of plastic deformation near a crack-tip can be characterised by the linear elastic stress intensification factor K_I, with fracture initiating when K_I attains a critical value K_{IC}, which depends on the material's flow charac-teristics and also on the fracture processes that are operative on a microstructural scale.

With the behaviour of these materials as a basis, the next section concentrated on the situation where there is extensive plastic defor-mation in the vicinity of a crack-tip and the usefulness of the *J* integral as a characterising parameter for crack-tip deformation was emphasised. Furthermore, the validity of fracture initiation criteria based on either the COD or the *J* integral attaining a critical value was discussed in detail, particularly with regard to their relation with the K_{IC} approach and their relative usefulness when assessing the integrity of engineering structures. Taking all factors into account, the *J* integral approach would seem to have distinct advantages when compared with the COD approach.

A point worthy of special emphasis is that the discussion has been based on consideration of realistic macroscopic plastic–elastic models. With regard to the characterisation of crack-tip deformation fields and fracture initiation criteria, conclusions drawn from an examination of more simplified models must be viewed with a high degree of caution [15]. For example, misleading conclusions can arise if work-hardening is neglected or if the crack-tip plastic relaxation is represented by a finite number of slip lines containing dislocations that have been emitted from the crack-tip.

It should also be emphasised that, although the J integral characterises the crack-tip deformation behaviour of a work-hardening material and although fracture should, in general, occur when J attains a critical value, J_{IC}, the latter may not always be the case. For example, when cleavage fracture in steels is initiated by the cracking of carbide particles, the critical event in the fracture initiation process occurs at a finite distance from the crack-tip [16], where the stresses and the strains given by the large-scale yielding eqns (7) and (8) may not be applicable. On the other hand, if this initiation mechanism operates when the yielding is small-scale, the initiation should follow a $K_I = K_{IC}$ criterion, since the deformation throughout the plastic region is characterised by the parameter K_I. This is a contributory factor to the widespread success of the linear elastic fracture mechanics approach.

Finally, it should be noted that the discussions have been concerned entirely with crack-tip deformation under plane strain conditions. Furthermore, they have been concerned with fracture initiation or, more generally, with materials where the amount of stable crack growth prior to unstable fracture is so limited that it has a negligible effect on the overall fracture process. The phenomenon of stable crack growth is currently under consideration, and if it can be satisfactorily incorporated within the J integral approach, thereby making it less conservative, it will indeed become an extremely powerful approach for characterising plastic deformation at a crack-tip and, more importantly, for correlating the fracture behaviour of laboratory-scale specimens and large engineering structures.

REFERENCES

1. SMITH, E., COOK, T. S. and RAU, C A., Flow localisation and the fracture toughness of high strength materials, *Fracture* 1977, Proc. ICF-4, Vol. 1, University of Waterloo Press, pp. 215–236.
2. SIH, G. C. and LIEBOWITZ, H., Mathematical theories of brittle fracture, *Fracture*, Vol. 2 (Ed. H. Liebowitz), Academic Press, 1968, pp. 67–190.
3. GRIFFITH, A. A., The phenomena of rupture and flow in solids, *Phil. Trans. Roy. Soc.*, **A221**, 1921, 163–198.

4. RICE, J. R., Mathematical analyses in the mechanics of fracture, *Fracture*, Vol. 2 (Ed. H. Liebowitz), Academic Press, 1968, pp. 191–311.
5. ESHELBY, J. D., The force on an elastic singularity, *Phil. Trans. Roy. Soc.*, **A244**, 1951, 87–112.
6. HUTCHINSON, J. W., Singular behaviour at the end of a tensile crack in a hardening material, *J. Mech. Phys. Solids*, **16**, 1968, 13–31.
7. RICE, J. R. and ROSENGREN, G. F., Plane strain deformation near a crack tip in a power law hardening material, *J. Mech. Phys. Solids*, **16**, 1968, 1–12.
8. McCLINTOCK, F. A., Plasticity aspects of fracture, *Fracture*, Vol. 3 (Ed. H. Liebowitz), Academic Press, 1968, pp. 47–225.
9. BEGLEY, J. A. and LANDES, J. D., The *J* integral as a fracture criterion, *Fracture Toughness*, ASTM STP514, American Society for Testing and Materials, 1972, pp. 1–20.
10. LANDES, J. D. and BEGLEY, J. A., The effect of specimen geometry on J_{IC}, *Fracture Toughness*, ASTM STP514, American Society for Testing and Materials, 1972, pp. 24–39.
11. LANDES, J. D. and BEGLEY, J. A., *Fracture Analysis*, ASTM STP560, American Society for Testing and Materials, 1974, p. 170.
12. RICE, J. R. and JOHNSON, M. A., The role of large crack tip geometry changes in plane strain fracture, *Inelastic Behaviour of Solids*, (Ed. M. F. Kanninen *et al.*), McGraw-Hill, New York, 1970, pp. 641–672.
13. HAYES, D. J. and TURNER, C. E., An application of finite element techniques to post yield analysis of proposed standard 3-point bend fracture testpieces, *Int. J. Fracture Mech.*, **10**, 1974, 17–32.
14. ROBINSON, J. N. and TETELMAN, A. S., The relationship between crack tip opening displacement, local strain and specimen geometry, *Int. J. Fracture Mech.*, **11**, 1975, 453–468.
15. SMITH, E. (to be published).
16. RITCHIE, R. O., KNOTT, J. F. and RICE, J. R., On the relationship between critical tensile stress and fracture toughness in mild steel, *J. Mech. Phys. Solids*, **21**, 1973, 395–410.

2

A Variational Solution of the Frictional Unloading Problem

J. R. TURNER

University of Oxford

SUMMARY

This paper deals with a stress analysis problem in which the non-linearity enters through frictional boundary conditions. A variational formulation is obtained for the problem of loading and unloading a linear elastic half-space by a rigid circular cylindrical punch, when conditions of Coulomb friction act over the interface throughout. The loading problem can be solved by usual integral equation techniques because the way in which the contact area is divided into regions over which differential inward and outward sliding occur can be predicted in advance. But this is not so for the unloading problem. In a variational formulation it can be shown that the solution minimises the complementary energy, which can be expressed as a quadratic functional of the surface stresses alone, the frictional conditions acting as linear constraints on the minimisation.

A numerical solution is obtained using the finite-element technique. By limiting the surface stresses to a finite-dimensional subspace, the quadratic minimisation is reduced to matrix form without further approximation. This gives a quadratic programming problem, which is solved using the quadratic simplex algorithm.

The surface radial stress distribution is obtained for the discrete solution, giving information about the expected fracture of brittle materials on loading and unloading.

NOTATION

$[A_{11}], [A_{12}], [A_{22}]$ compliance matrices
E Young's modulus

J, J_1, J_2, J_3	complementary energies
$J_0(\alpha), J_1(\alpha)$	Bessel functions of first kind
$\mathbf{L}(r), \mathbf{M}(r), \mathbf{N}(r)$	integrals of $l(\rho)n(\rho)$
S	surface of the half space, $z = 0$
S_0	region of surface with stress boundary conditions $r \geqslant 1, z = 0$
S_1	region of surface with frictional/displacement conditions $r \leqslant 1, z = 0$
$\mathbf{S}_p(r), \mathbf{S}_q(r)$	vectors of finite-element spline functions
$\forall$	volume of the half space, $r \leqslant \infty, -\infty \leqslant z \leqslant 0$
c	adhesion radius
$\mathbf{c}, \mathbf{b}$	integrals of spline function
$g(r)$	known normal stress frictional constraint
i, j, k, l	tensor subscripts
$\mathbf{l}(\rho), \mathbf{n}(\rho)$	Hankel transforms of $\mathbf{S}p(r), \mathbf{S}q(r)$
$\mathbf{p}, \mathbf{q}$	vectors of nodal values
$p(r)$	normal surface traction $-\sigma_z(r, \theta, 0)$
$q(r)$	shear surface traction $\sigma_{rz}(r, \theta, 0)$
$\bar{p}(\rho), \bar{q}(\rho)$	Hankel transforms of $p(r), q(r)$
r	radial coordinate $= (x_1^2 + x_2^2)^{1/2}$
$u_i(r, z)$	Cartesian components of displacement
$u_r(r, z), u_\theta(r, z), u_z(r, z)$	polar components of displacement
$u_1(r)$	differential inward slip
$u_2(r)$	differential outward slip
$u(r) = u_r(r, 0), w(r) = -u_z(r, 0)$	surface values of displacement
$u_0(r), w_0(r)$	prescribed values for surface displacement
x_i	Cartesian coordinates, $i = 1, 2, 3$
z	polar coordinate $= x_3$
γ	$(1 - 2\nu)/(2 - \nu)$
$\delta(\ )$	variation of $(\ )$
δ_{ij}	Kronecker delta
ϵ_{ij}	Cartesian components of strain
θ	angular coordinate $= \tan^{-1} x_2/x_1$
λ	unloading parameter = present load/maximum load
λ_0	specific value of λ
μ	shear modulus
ν	Poisson's ratio
ρ	coefficient of Coulomb friction
ρ	independent variable in Hankel transforms
σ_{ij}	Cartesian components of stress

$\sigma_r, \sigma_\theta, \sigma_{r\theta}, \sigma_z, \sigma_{z\theta}, \sigma_{rz}$ polar components

$\sigma_1, \sigma_2, \sigma_3$ stress terms

INTRODUCTION

The problem of pressing two dissimilar elastic bodies together, when conditions of Coulomb friction act over the interface, has been of both theoretical and experimental interest for some time [1, 2].

The related idealised problem of contact between a rigid, circular-cylindrical indentor and an isotropic, homogeneous, linear-elastic half-space is considered in this paper. After non-dimensionalisation, the above problem is equivalent to a mixed boundary value problem, in which the elastic body occupies the half-space $z \le 0$, the normal stress, $\sigma_z(r, \theta, 0)(= p(r))$, and the shear stress, $\sigma_{rz}(r, \theta, 0)(= q(r))$, are specified zero for $r > 1$, and the normal displacement $-u_z(r, \theta, 0)(= w(r))$ is specified as unity for $r \le 1$. (A system of cylindrical polar coordinates, with the z coordinate normal to the half-space, has been adopted and the axisymmetric assumption, in which there is no shear stress, displacement or variation in the θ direction (i.e. $\sigma_{r\theta} = \sigma_{z\theta} = u_\theta = \partial/\partial\theta = 0$) made.) To completely specify the problem, it remains to define some relation between the horizontal displacement, $u_r(r, \theta, 0)(= u(r))$, and the shear, $q(r)$, in the contact region.

The simplest boundary conditions of frictionless contact $(q(r) \equiv 0)$ and adhesive contact $(u(r) \equiv 0)$, which preserve the linear and reversible nature of the equations, were solved by Hertz [3] in 1881 and by Mossakovski [4] in 1954. However, the experimental evidence is that the boundary conditions are not linear and do not exhibit reversible behaviour on loading and unloading.

Johnson *et al.* [2] report experiments performed on the indentation of a glass plate by a steel ball. It was observed that, on loading, the glass plate fractured a short distance out from the contact area when the normal force reached a certain critical value, corresponding to the radial stress reaching the fracture stress at that radius. However, if the plate was loaded short of the failure load, it was observed that, on unloading, the glass plate might still fracture, but at the edge of the region of contact and at a load less than the maximum load. This behaviour can be explained by introducing Coulomb friction in the contact region.

Physically, during any small change in the normal displacement, the shear stress, $q(r)$, can never be greater than some proportion, ρ, of the normal stress, $p(r)$ (ρ is the coefficient of Coulomb friction). If $q(r)$ is less than this proportion, the half-space will adhere to the indentor, and if it is equal, the half-space will slide in a direction opposite to the shear. This can be expressed mathematically as follows. The half-space occupies some initial stress state, $w(r) = 1 - d\lambda$ and $u(r) = u_0(r)$, prescribed for $r \leq 1$, and is loaded ($d\lambda > 0$) or unloaded ($d\lambda < 0$) to $w(r) = 1$. The final horizontal displacement is given by

$$u(r) = u_0(r) - u_1(r) + u_2(r) \tag{1}$$

where $u_1(r)$ is the inward slip and $u_2(r)$ is the outward slip. These two slip terms obey the conditions

$$\left.
\begin{array}{ll}
q(r) \leq \rho g(r), & q(r) \geq -\rho g(r) \\
u_1(r) \geq 0, & u_2(r) \geq 0 \\
u_1(r)(q(r) - \rho g(r)) = 0, & u_2(r)(q(r) + \rho g(r)) = 0
\end{array}
\right\} \tag{2}$$

where $g(r)$ is some normal stress. True Coulomb friction is given by $g(r) \equiv p(r)$, but the mathematical formulation of the next section requires that $g(r)$ be a known function. In a general problem $p(r)$ will not be known in advance, but it is shown below that it will be known if the half-space is incompressible, since the problems for solving for $p(r)$ and $q(r)$ decouple in that case. The complementary energy principle is derived below for the boundary conditions represented in eqns (2) above, where $g(r)$ is a known function, and it is then shown how $g(r)$ can be replaced by the known normal pressure in the incompressible limit.

The above conditions are unilateral. They maintain the linear character of the equation while $w(r)$ is monotonically increasing and so the problem can be solved by standard techniques [1]. In the solution obtained, the contact region ($r \leq 1$) splits into an inner region of adhesion ($u_1(r) \equiv u_2(r) \equiv 0$ for $r \leq c$) surrounded by an annulus of inward slip ($q(r) \equiv \rho p(r)$, $u_2(r) \equiv 0$ for $c \leq r \leq 1$). The radius of adhesion, c, depends on ρ and Poisson's ratio, ν, but is independent of the magnitude of the normal displacement, $w(r)(= \lambda)$ for $r \leq 1$. Thus, the problem can be scaled with respect to the loading parameter, λ.

Because the conditions represented in eqns (2) are irreversible, on unloading (i.e. with $w(r)$ decreasing), the problem becomes non-linear, leading to the behaviour observed experimentally.

In the next section a complementary energy principle will be derived for the above problem. It will then be shown that this reduces to a simple analytic form for an incompressible axisymmetric half-space. A finite-element approximation is introduced, and numerical results for the normal and shear stresses presented. Finally, it is shown how these results can be used to calculate the surface radial stress, and the corresponding distribution is given for frictional loading and unloading.

THE COMPLEMENTARY ENERGY PRINCIPLE

The posed problem is to solve the elasticity equations, namely

$$\frac{1}{2}\left(\frac{\partial u_i}{\partial x_j} + \frac{\partial u_j}{\partial x_i}\right) = \epsilon_{ij} \tag{3}$$

$$\epsilon_{ij} = -\frac{\nu}{E}\sigma_{kk}\delta_{ij} + \frac{1+\nu}{E}\sigma_{ij} \tag{4}$$

$$-\frac{\partial \sigma_{ij}}{\partial x_j} = 0 \tag{5}$$

for (r, z) contained in $\mathbf{V} = \{x_i | x_3 = z < 0\}$

where u_i, ϵ_{ij}, σ_{ij} are the Cartesian components of displacement, strain and stress, respectively; ν is Poisson's ratio; and E is Young's modulus. The relevant displacement boundary conditions are

$$\left.\begin{array}{l} -u_z(r, 0) = w(r) = w_0(r) \\ u_r(r, 0) = u(r) = u_0(r) - u_1(r) + u_2(r) \end{array}\right\} \tag{6}$$

for (r, z) contained in $S_1 = \{r, z | r \leq 1, z = 0\}$

where $w_0(r)$, $u_0(r)$ are prescribed and $u_1(r)$, $u_2(r)$ obey the conditions

$$\left.\begin{array}{l} u_1(r) \geq 0, g(r) \leq \rho g(r), u_1(r)(q(r) - \rho g(r)) = 0 \\ u_2(r) \geq 0, q(r) \geq -\rho g(r), u_2(r)(q(r) + \rho g(r)) = 0 \end{array}\right\} \tag{7}$$

with $g(r)$ a prescribed function. The relevant stress-free boundary conditions are

$$-\sigma_z(r, 0) = p(r) = 0, \sigma_{rz}(r, 0) = q(r) = 0 \tag{8}$$

for (r, z) contained in $S_0 = \{r, z | r > 1, z = 0\}$

The Cartesian components are used in eqns (3)–(5) to emphasise that the principle below has wider applicability, but the axisymmetric form is used in the boundary conditions because this greatly simplifies the ensuing equation.

The derivation of the complementary energy principle relies upon Green's theorem, which is the adjoint property relating the differential operators of eqns (3) and (5):

$$\int_V \left[\frac{1}{2}\left(\frac{\partial u_i}{\partial x_j} + \frac{\partial u_j}{\partial x_i}\right)\sigma_{ij} + u_i \frac{\partial \sigma_{ij}}{\partial x_j}\right] dV = \int_{\delta V} \sigma_{ij}n_j u_i \, da$$

$$= \int_0^\infty (u(r)q(r) + w(r)p(r))r \, dr \quad (9)$$

In this equation a 2π has been absorbed into the displacements $u(r)$, $w(r)$.

Consider now the complementary energy, J, given by the equation

$$J = \frac{1}{2}\int_V \sigma_{ij}\left(-\frac{\nu}{E}\sigma_{kk}\delta_{ij} + \frac{1+\nu}{2E}\sigma_{ij}\right)dV - \int_0^1 (u_0(r)q(r) + w_0(r)p(r))r \, dr$$

$$(10)$$

If the conditions on the components of the elasticity tensor are such that the quadratic term of J is positive definite (these conditions for the axisymmetric case are given by Turner [5]), then the solution of eqns (3)–(8) minimises J within the set of admissible stress states, i.e. those stress fields obeying eqns (5), (7) and (8). To prove this, it is only necessary to show that at the solution the first variation of J is non-negative, $\delta J \geq 0$, within the set of admissible stress states, and minimisation follows from positive definiteness of the quadratic part and, hence, the second derivative.

Consider the first variation, δJ, at the solution

$$\delta J = \int_V \left(-\frac{\nu}{E}\sigma_{kk}\delta_{ij} + \frac{1+\nu}{2E}\sigma_{ij}\right)\delta\sigma_{ij} \, dV$$

$$- \int_0^1 (u_0(r)\delta q(r) + w_0(r)\delta p(r))r \, dr$$

$$= \int_V \left(\frac{\partial u_i}{\partial x_j} + \frac{\partial u_j}{\partial x_i}\right)\delta\sigma_{ij} \, dV - \int_0^1 (u_0(r)\delta q(r) + w_0(r)\delta p(r))r \, dr \quad (11)$$

$$= \int_V u_i\delta\left(\frac{\partial \sigma_{ij}}{\partial x_j}\right)dV + \int_0^1 [(u(r) - u_0(r))\delta q(r) + (w(r) - w_0(r))\delta p(r)]r \, dr$$

$$+ \int_1^\infty (u(r)\delta q(r) + w(r)\delta p(r))r \, dr$$

using eqns (4), (3) and (9), respectively. Within the subset of admissible stress states $\delta(\partial\sigma_{ij}/\partial x_j) = 0$ for (r, z) contained in $\forall$ and $\delta p(r) = \delta q(r) = 0$ for $r > 1$, and at the solution $u(r) - u_0(r) = -u_1(r) + u_2(r)$ and $w(r) - w_0(r) = 0$ for $r \leqslant 1$. Also, if $g(r)$ is known, $\delta g(r) = 0$ and eqn (11) can be written as

$$\delta J = \int_0^1 [-u_1(r)\delta(q(r) - \rho g(r)) + u_2(r)\delta(q(r) + \rho g(r))] r \, dr \quad (12)$$

But at the solution either $u_1(r) = 0$ or $u_1(r) < 0$, in which case $\delta(q(r) - \rho g(r)) \leqslant 0$. Hence, $u_1(r)\delta(q(r) - \rho g(r)) \geqslant 0$ and similarly $u_2(r)\delta(q(r) + \rho g(r)) \geqslant 0$; thus, $\delta J \geqslant 0$, proving the result.

Thus, the solution to the posed problem is given by that stress state which minimises J given by eqn (10) subject to the conditions:

$$\frac{\partial\sigma_{ij}}{\partial x_j} = 0 \quad (13)$$

$$-\rho g(r) \leqslant q(r) \leqslant \rho g(r) \quad (r \leqslant 1) \quad (14)$$

$$q(r) = p(r) = 0 \quad (r > 1) \quad (15)$$

THE AXISYMMETRIC INCOMPRESSIBLE LIMIT

It can be shown that a stress function which satisfies eqn (13) will satisfy non-negativity of δJ if and only if it obeys the Beltrami–Mitchel equations for (r, z) contained in $\forall$. Introducing the stress function into eqn (10), integrating by parts and using the fact that it obeys the Beltrami–Mitchel equations, the quadratic term can be expressed in terms of the unknown surface stresses alone. In the axisymmetric case, with $w_0(r) \equiv 1$, the complementary energy becomes

$$J = \frac{1}{2}\int_0^\infty (\bar{p}^2(l) - 2\gamma\bar{p}(\rho)\bar{q}(\rho) + \bar{q}^2(\rho)) \, d\rho - \int_0^1 (u_0(r)q(r) + p(r)) r \, dr \quad (16)$$

where

$$\bar{p}(\rho) = \int_0^1 p(r)J_0(\rho r) r \, dr \quad (17)$$

and

$$\bar{q}(\rho) = \int_0^1 q(r)J_1(\rho r) r \, dr \quad (18)$$

the Hankel transforms, and $\gamma = (1 - 2\nu)/(2 - \nu)$. In the above, the transformations $p(r) \to \mu p(r)$, $q(r) \to \mu q(r)$, $u(r) \to (1 - \nu)u(r)$ and $w(r) \to (1 - \nu)w(r)$ have been made, and J has been scaled by a constant term involving ν, μ and π.

In the incompressible limit, $\nu \to \frac{1}{2}$, then $\gamma \to 0$ and the problems for $p(r)$ and $q(r)$ decouple. The normal stress is given as that function which minimises the quantity

$$J_1 = \frac{1}{2} \int_0^\infty \bar{p}^2(\rho) \, d\rho - \int_0^1 p(r)r \, dr \qquad (19)$$

This is the Hertzian solution, and is

$$p(r) = (2/\pi)(1 - r^2)^{-1/2} \qquad (20)$$

For the loading problem $u_0(r) \equiv 0$; therefore, to the first order $q(r)$ is given by that function which minimises

$$J_2 = \int_0^\infty \bar{q}^2(\rho) \, d\rho \qquad (21)$$

subject to the constraint, eqn (10). This function is trivially zero, as are the resulting displacements $u_1(r)$ and $u_2(r)$, and, hence, the initial displacement for the unloading problem. Thus, in the incompressible limit, the loading and unloading problems have to first order the trivial solution $\bar{q}(r) = 0$. However, if the scalings $q(r) \to \gamma q(r)$, $u(r) \to \gamma u(r)$ and $\rho \to \gamma \rho$, are made, the solution for $q(r)$ is that function which minimises

$$J_2 = \frac{1}{2} \int_0^\infty \bar{q}^2(\rho) \, d\rho - \int_0^\infty \bar{q}(\rho)\bar{p}(\rho) \, d\rho - \int_0^1 u_0(r)q(r)r \, dr \qquad (22)$$

subject to the conditions

$$-\rho p(r) \leq q(r) \leq \rho p(r) \qquad (23)$$

where $p(r)$ is given by eqn (20). The solution to $O(\gamma)$ is thus not trivially zero.

The solution to the adhesive problem is given by $u_0(r) \equiv 0$, $\rho \to \infty$, and that for the frictional loading problem by $u_0(r) \equiv 0$, ρ finite. It can be seen that the linear character of the problem is maintained in the loading problem, because if a normal pressure $p(r)$ gives a solution for the shear $q(r)$, then $\alpha p(r)$ gives $\alpha q(r)$. However, this is not so for the unloading problem, $u_0(r) \not\equiv 0$.

The problem solved below is that in which the normal displacement is increased monotonically to $w(r) = 1$, and then decreased monotonically back to zero. If at any point of the loading or unloading the normal displacement is $w(r) = \lambda$ (λ will be called the loading or unloading parameter), the corresponding normal stress is given by a suitable scaling of eqn (20) in the form:

$$p(r) = (\partial\lambda/\pi)(1 - r^2)^{-1/2} \tag{24}$$

The loading problem is solved directly with $\lambda = 1$ and $u_0(r) = 0$, and the slip, which is entirely inward, is found as the first Fréchet derivative of J_2 at the solution

$$-u_1(r) = \frac{\partial J_2}{\partial q} = \int_0^\infty (\bar{q}(\rho) - \bar{p}(\rho))J_1(\rho r)\, d\rho \tag{25}$$

This then becomes the initial displacement for the first step of the unloading, $u_0^1(r) = -u_1(r)$. The unloading problem is solved in a series of discrete steps. During the nth step the normal displacement is reduced from $w(r) = \lambda_0^{n-1}$ to $w(r) = \lambda_0^n$ ($0 < \lambda_0 < 1$); the initial horizontal displacement for each step is given by that from the step before and the slip occurring in that step, i.e.

$$u_0^{n+1}(r) = u_0^n(r) - u_1^n(r) + u_2^n(r) \qquad (n = 1, \infty) \tag{26}$$

Further, the slip is given by the Fréchet derivative of J_2 at the solution

$$-u_1^n(r) + u_2^n(r) = \frac{\partial J_2}{\partial q}$$

$$= \int_0^\infty (\bar{q}(\rho) - \bar{p}(\rho))J_1(\rho r)\, d\rho - u_0^n(r) \tag{27}$$

Taking discrete steps involves some approximation, the true solution being obtained as $\lambda \to 1$, but it was found that the errors introduced by the discretisation of the stress space were greater than that by the 'time' steps, and quite accurate results could be obtained with fairly large steps, i.e. λ_0 of the order of 0·8.

A numerical solution was obtained for the unloading problem using the finite-element technique. Numerical results were also obtained for the three loading problems, viz. frictionless, adhesive and frictional loading, to check the numerical method and to provide data for later steps.

NUMERICAL IMPLEMENTATION

It is not the intention to give here a full description of the numerical technique used to solve these problems (that can be found in the author's thesis [5]), but to show how the finite-element technique can be used to find an approximate solution to the above infinite-dimensional quadratic programming problems, and, hence, solve the stress problems described. The philosophy is to approximate the elements of the stress spaces $\{p(r)\}, \{q(r)\}$ by the elements of finite-dimensional subspaces. The finite-dimensional spaces are spanned by a basis of splines and, in this way, the approximate functions are expressed in terms of a discrete number of nodal values, $\mathbf{p}$, $\mathbf{q}$. Writing the two vectors of splines as $\mathbf{S}_p(r)$ and $\mathbf{S}_q(r)$, respectively, the approximate functions can be expressed as

$$p(r) = \mathbf{S}_p^T(r)\mathbf{p}; \qquad q(r) = \mathbf{S}_q^T(r)\mathbf{q} \qquad (28)$$

which, when substituted into eqns (17) and (18), give

$$\bar{p}(\rho) = \mathbf{l}^T(\rho)\mathbf{p}; \qquad \bar{q}(\rho) = \mathbf{n}^T(\rho)\mathbf{q} \qquad (29)$$

where

$$\mathbf{l}(\rho) = \int_0^1 \mathbf{S}_p(r)J_0(\rho r)r\,\mathrm{d}r; \qquad \mathbf{n}(\rho) = \int_0^1 \mathbf{S}_q(r)J_1(\rho r)r\,\mathrm{d}r \qquad (30)$$

An approximate discrete solution can then be found to the loading and unloading problems as follows.

The approximate solution for the pressure is given by that vector which minimises

$$J_1 = \mathbf{p}^T[A_{11}]\mathbf{p} - 2\mathbf{b}^T\mathbf{p} \qquad (31)$$

where

$$[A_{11}] = \int_0^\infty \mathbf{l}(\rho)\mathbf{l}^T(\rho)\,\mathrm{d}\rho \quad \text{and} \quad \mathbf{b} = \int_0^1 \mathbf{S}_p(r)r\,\mathrm{d}r$$

Because $[A_{11}]$ is positive definite and symmetric, the solution is obtained by simple calculus as $\mathbf{p} = [A_{11}]^{-1}$. This can be found without further approximation, since exact expressions exist for the integrals in eqns (30)–(32) below, for the splines used. The resulting pressure distribution was obtained using a FORTRAN program and is plotted in Fig. 1 against the Hertzian solution (eqn 20).

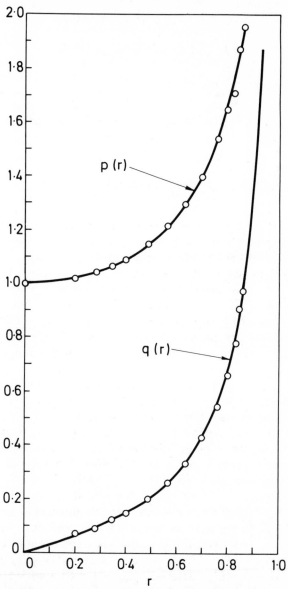

FIG. 1. Theoretical curves for normal stress, $p(r)$ (from eqn (20)), and shear stress, $q(r)$ (ref. [6]), with computed numerical values.

The approximate solution for the shear at any step of the loading or unloading is given by that vector which minimises

$$J_2 = \mathbf{q}^T[A_{22}]\mathbf{q} - 2\mathbf{p}^T[A_{12}]\mathbf{q} - 2\mathbf{c}^T\mathbf{q} \qquad (32)$$

subject to $-\rho\mathbf{p} \leq \mathbf{q} \leq \rho\mathbf{p}$, where

$$[A_{22}] = \int_0^\infty \mathbf{n}(\rho)\mathbf{n}^T(\rho)\,d\rho, \quad [A_{12}] = \int_0^\infty \mathbf{l}(\rho)\mathbf{n}^T(\rho)\,d\rho$$

and

$$\mathbf{c} = \int_0^1 u_0(r)\mathbf{S}_q(r)r\,dr$$

The pressure p is found as the solution to eqn (31). (It was found consistently that better results were obtained using the solution to eqn 31 rather than the nodal values of eqn 20.)

The approximate solution to the adhesive problem is given by $\rho \to \infty$, $c = 0$, and again follows from simple calculus as $\mathbf{q} = [A_{22}]^{-1}[A_{12}]^T\mathbf{p}$. This is also plotted in Fig. 1 against the theoretical solution obtained by Mossakovski [4] and given by Spence [6]. The linear problems are presented here because the close fit obtained lends some credence to the results obtained for the non-linear problems below.

The approximation to the frictional loading and unloading problems leads to quadratic programming problems, which can be solved by the quadratic simplex algorithm [7]. The loading profile, $\theta(r) = q(r)/\rho p(r)$, is plotted in Fig. 2 against the theoretical curve [6] for a value of $\rho = 0.6652$, chosen to give a theoretical adhesion radius of $c = 0.7$. The unloading profiles are plotted in Figs. 3–6 for values of the unloading parameter $\lambda = 0.8179$, $\lambda = 0.5472$, $\lambda = 0.4475$ and $\lambda = 0.2449$, respectively. There are no theoretical curves with which to compare these results; however, they can be justified heuristically [5] and, as shown in the next section, produce a pattern of radial stress on unloading, consistent with the observed experimental results.

Also plotted in Fig. 7 is a curve of $\theta(1) = q(1)/\rho p(1)$ against the unloading parameter, to determine the value of λ (0.5151) at which outward slip starts. Figure 8 shows a complete unloading regime, giving the relative size and position of the regions of inward slip, outward slip and adhesion during unloading.

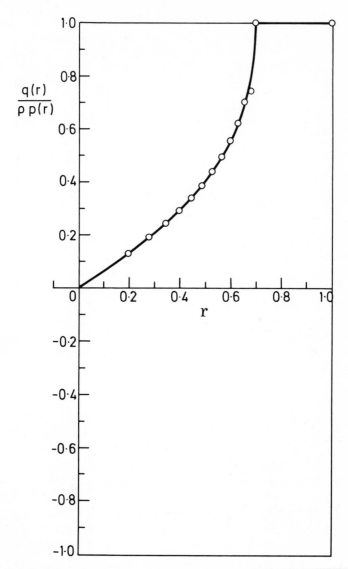

FIG. 2.   Theoretical loading regime, $\theta(r)$ (ref. [6]), with computed numerical values, for $\rho = 0 \cdot 6652$.

J. R. Turner

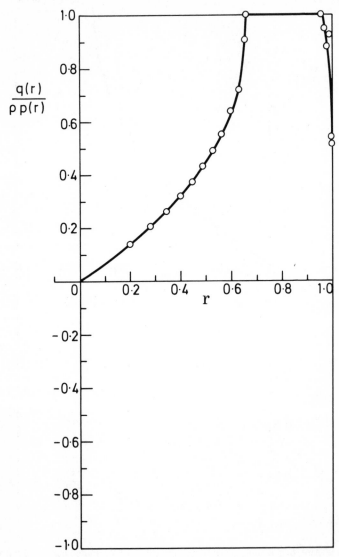

FIG. 3. Computed unloading regime, $\theta(r)$, at $\lambda = 0.8179$, for $\rho = 0.6652$.

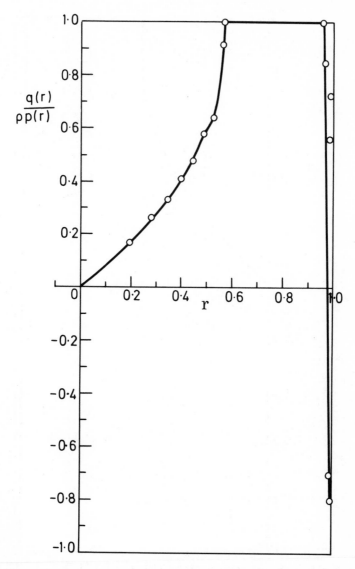

FIG. 4. Computed unloading regime, $\theta(r)$, at $\lambda = 0\cdot5472$, for $\rho = 0\cdot6652$.

J. R. Turner

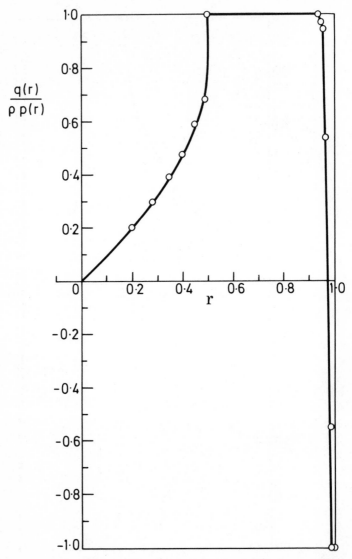

FIG. 5.   Computed unloading regime, $\theta(r)$, at $\lambda = 0.4475$, for $\rho = 0.6652$.

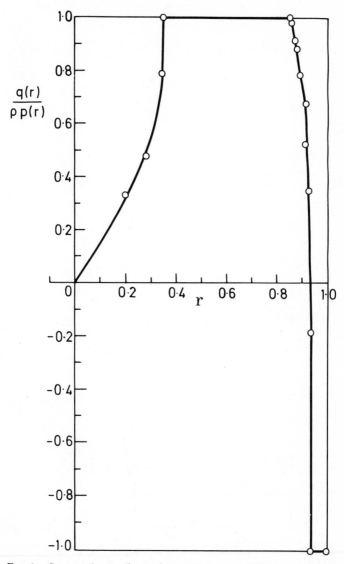

FIG. 6.   Computed unloading regime, $\theta(r)$, at $\lambda = 0.2449$, for $\rho = 0.6652$.

J. R. Turner

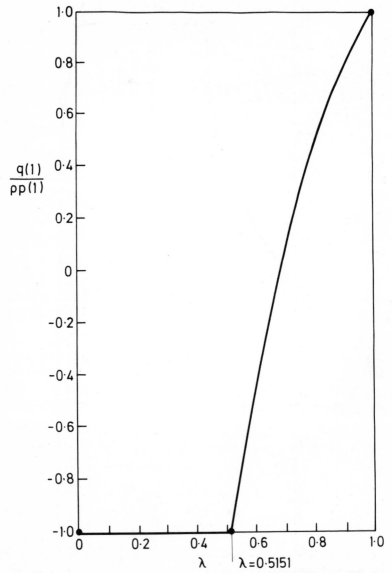

Fig. 7. $q(1)/\rho p(1)$ plotted against the unloading parameter λ, to determine the point of formation of the region of outward slip, for $\rho = 0.6652$.

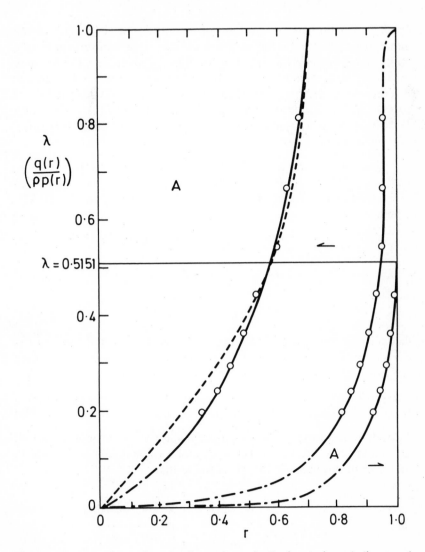

FIG. 8. The unloading regime showing regions of adhesion A, inward slip ⟵ and outward slip ⟶, plotted against the loading parameter λ, for $\rho = 0.6652$. ——— smooth curves through numerical values, — · — · — extrapolated curves, — — — — curve from Fig. 2.

THE SURFACE STRESS DISTRIBUTION

In-plane fracture of the surface of a brittle material occurs when either the radial or the hoop stress reaches a certain critical tensile value. To determine the expected fracture behaviour of a glass plate indented by a steel ball, the surface stresses corresponding to the discrete stress distributions of the last section are obtained below. It can be shown [5] that the radial and hoop stresses in the surface of a half-space, with axisymmetric surface stress distribution $p(r)$, $q(r)$, are, respectively,

$$\sigma_r(r, 0) = -p(r) + 2\sigma_3(r) + 2(1 - \nu)(\gamma\sigma_1(r) - \sigma_2(r))$$

and

$$\sigma_\theta(r, 0) = -2\nu p(r) + 2\nu\sigma_3(r) - 2(1 - \nu)(\gamma\sigma_1(r) - \sigma_2(r)) \tag{33}$$

where

$$\sigma_1(r) = \frac{1}{r}\int_0^\infty \bar{p}(\rho)J_1(\rho r)\,\mathrm{d}\rho$$

$$\sigma_2(r) = \frac{1}{r}\int_0^\infty \bar{q}(\rho)J_1(\rho r)\,\mathrm{d}\rho \tag{34}$$

$$\sigma_3(r) = \int_0^\infty \bar{q}(\rho)J_0(\rho r)\rho\,\mathrm{d}\rho$$

Considering the case $p(r) \equiv 0$ for $r > 1$, and introducing the scalings $q(r) \to \gamma q(r)$, $\sigma_r(r, 0) \to \gamma\sigma_r(r, 0)$ and $\sigma_\theta(r, 0) \to \gamma\sigma_\theta(r, 0)$ as before, then the radial and hoop stresses for $r > 1$ become, as $\nu \to \frac{1}{2}$,

$$\sigma_r(r, 0) = \sigma_1(r) - \sigma_2(r) + 2\sigma_3(r); \qquad \sigma_\theta(r, 0) = -\sigma_1(r) + \sigma_2(r) + \sigma_3(r) \tag{35}$$

Further introducing the finite-element approximation into eqns (34) gives a discrete form for $\sigma_1(r)$, $\sigma_2(r)$ and $\sigma_3(r)$ and, hence, for $\sigma_r(r, 0)$ and $\sigma_\theta(r, 0)$ through eqn (35). It is found that

$$\sigma_1(r) = \mathbf{L}^\mathrm{T}(r)\mathbf{p}; \qquad \mathbf{L}(r) = \frac{1}{r}\int_0^\infty \mathbf{l}(\rho)J_1(\rho r)\,\mathrm{d}\rho$$

$$\sigma_2(r) = \mathbf{N}^\mathrm{T}(r)\mathbf{q}; \qquad \mathbf{N}(r) = \frac{1}{r}\int_0^\infty \mathbf{n}(\rho)J_1(\rho r)\,\mathrm{d}\rho \tag{36}$$

$$\sigma_3(r) = \mathbf{M}^\mathrm{T}(r)\mathbf{q}; \qquad \mathbf{M}(r) = \int_0^\infty \mathbf{n}(\rho)J_0(\rho r)\rho\,\mathrm{d}\rho$$

For the splines used, $\mathbf{L}(r)$, $\mathbf{N}(r)$, $\mathbf{M}(r)$ could be found without further approximation, thus giving exact expressions for the discrete stress distributions. These were calculated for the loading and unloading profiles of the last section, using a FORTRAN program.

It was found for both loading and unloading that $\sigma_\theta(r, 0)$ was almost always compressive, and that fracture was therefore determined by the radial stress. The radial stress distribution for $r > 1$ is plotted for the loading case in Fig. 9, along with its components, the pressure term, $\sigma_1(r)$ (which is the radial stress in the Hertzian case, $q(r) \equiv 0$) and the shear term $(\sigma_2(r) - 2\sigma_3(r))$. The radial stress attains its maximum tensile value of 0.1509 at $r = 1.75$. This should be compared with a maximum value of unity at $r = 1$ in the frictionless case.

The radial stress on unloading also shows a balancing of the pressure and shear terms. Rather than producing a complete set of

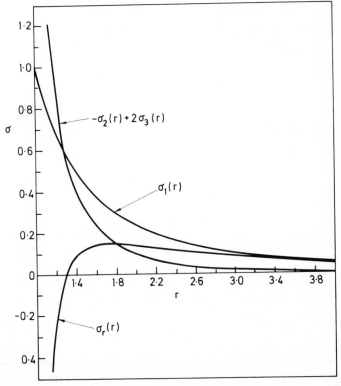

FIG. 9. Surface radial stress $\sigma_r(r)$, outside the contact area $r \geqslant 1$, for the frictional loading case $\rho = 0.6652$, with its pressure, $\sigma_1(r)$, and shear, $-\sigma_2(r) + 2\sigma_3(r)$, components.

curves, the value and position of the maximum radial stress is plotted
in Fig. 10, against both the loading and the unloading parameters, λ.
The maximum radial stress maintains a constant position at $r = 1\cdot75$
during loading, but immediately starts to move towards the contact
area as unloading commences, reaching the outside edge shortly after
the region of outward slip forms. Meanwhile its value increases
linearly during loading. On unloading, it starts to decrease, but at a
slower rate than the increase during loading, until the region of
outward slip forms, at $\lambda = 0\cdot5151$, when it goes through a sharp
increase, reaching a maximum of four times the value on loading (for
this value of ρ) before falling to zero. The explanation for the
behaviour on unloading is that the outer region of adhesion and, later,
the region of outward slip introduce a tensile effect in the half-space.
The tensile effect of the region of adhesion is initially not enough to
balance the total falling stress, giving the initial decrease, but more
than compensates when the annulus of outward slip first forms, giving
the sudden sharp increase. But in the end the falling total stress
dominates and the entire stress state vanishes.

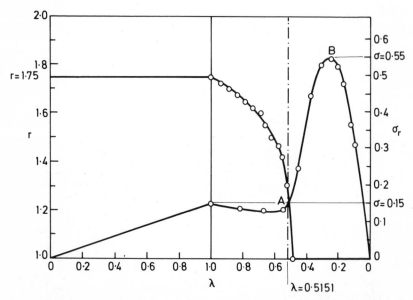

FIG. 10.   Position, r, and value, σ_r, of the maximum tensile radial stress plotted against
the loading and unloading parameters, λ. The Hertz value is $\sigma_r(r) = 1\cdot000$.

The important observation is, however, that the half-space could be loaded to just one-quarter of the fracture load and it would still fracture on unloading for this value of ρ, thus confirming the behaviour observed by Johnson *et al.* [2].

CONCLUSION

A stress analysis problem has been considered in which non-linear effects enter through frictional, and, hence, irreversible, boundary conditions. It has been shown that the solution is given by that stress state which minimises the complementary energy over all the admissible stress states, i.e. those obeying the equilibrium equations and the frictional boundary conditions. Further, using the equilibrium equations, the complementary energy has been expressed as a quadratic functional of the unknown surface stresses alone. In this way the non-linear stress analysis problem was converted to one of quadratic optimisation. The finite-element technique has then been used to find an approximate discrete solution to this latter problem.

The finite-element technique reduced the above optimisation problem to one familiar in Operations Research, a quadratic programming problem, for which there exist standard methods of solution; the quadratic simplex algorithm method has been used here. Further, the surface stress distributions have been calculated for the discrete stress solutions, confirming previously observed experimental behaviour.

ACKNOWLEDGEMENTS

The author wishes to thank Dr D. A. Spence for his continued help while supervising this work, and Ove Arup and Partners for time spent preparing parts of the manuscript while working for them.

REFERENCES

1. SPENCE, D. A., An eigenvalue problem for elastic contact with finite friction, *Proc. Camb. Phil. Soc.*, **73**, 1973, 249.
2. JOHNSON, K. L., O'CONNOR, J. J. and WOODWARD, A. C., The effect of the indentor elasticity on the Hertzian fracture of brittle materials, *Proc. Roy. Soc.*, **A334**, 1973, 95.

3. HERTZ, H., *J. Reine Angew. Math.*, **92**, 1881, 156.
4. MOSSAKOVSKI, V. I., The fundamental general problem of the theory of elasticity for a half space with a circular curve determining boundary conditions, *Prikl. Mat. Mech.*, **18**, 1954.
5. TURNER, J. R. 'A Variational Solution of the Frictional Unloading Problem in Linear Elasticity', D.Phil. Thesis, University of Oxford, 1977.
6. SPENCE, D. A. The Hertz contact problem with finite friction, *J. Elast* **5**, 1975, 297.
7. WAGNER, H. M., *Principles of Operations Research*, Prentice-Hall, 1969.

3

Toughness Changes in Gross Plasticity Cycling

J. C. RADON

Imperial College of Science and Technology

SUMMARY

The J contour integral has been evaluated, both theoretically and experimentally, for the constant K contoured DCB specimen of mild steel BS15 under monotonic loading conditions. It was found that J increases as the crack length increases, reaching a constant value for linear elastic loading. In load cycling (high strain fatigue), the appropriate ΔJ values are much greater than the monotonic values. The above difference was explained by the large increase in total dissipated energy obtained by the cyclic creep phenomenon in load cycling.

Good agreement was found between the J versus crack extension curves in contoured DCB and compact tension specimens. A room temperature J_{IC} value has been obtained in BS15 using the R curve method.

INTRODUCTION

Linear elastic fracture mechanics (LEFM) has been successfully used to analyse the fracture toughness behaviour of high-strength materials where the distribution of the elastic stresses around a crack tip is independent of applied load and specimen geometry, and the extent of plasticity is confined to a relatively small plastic zone at the crack tip. Consequently, in the presence of a crack, failure will occur at nominally elastic stress levels. In the past, much research has been carried out into the theoretical and experimental applications of LEFM to obtain the appropriate fracture toughness value of the material, and recommended testing procedures are now available [1]. The resistance of a material to monotonic or cyclic failures can be described in terms of stress intensity factors which are functions of

41

the applied load, the crack length and the dimensions of the investigated component.

When low- and intermediate-strength alloys, such as mild steel or low-alloy steels, are investigated, it is found that fracture may occur within a region of a relatively large plastic zone. In such cases LEFM will no longer characterise the local stress distribution at the crack tip and the equation used to calculate stress intensity factors, K, based on pure linear elastic behaviour cannot be applied. In such situations the effects of plasticity will be substantial and the factors K will lose their original meaning. A small sharp crack developed in a large plastic zone represents a typical engineering situation for which LEFM is invalid. The analytical and practical solution of the crack growth and, in particular, that of the fatigue crack propagation through the plastically deformed region are therefore important; methods of general yielding fracture mechanics are appropriate in such cases.

The fracture toughness of non-linear elastic materials can be expressed in terms of the J contour integral, the mathematical basis of which is described in Ref. [2]. The J integral has an energy interpretation [3] which leads to an easy experimental determination; the relevant expression is

$$J = -\frac{1}{B}\frac{dU}{da} \qquad (1)$$

where B is the nett fracture width, dU is the potential energy change, and da is an increment in the crack length. In the linear elastic case J is therefore equal to G, the crack extension force, which is related to K by the equation:

$$J = G = K^2/E \qquad (2)$$

In the case of elastic–plastic materials, J cannot be interpreted in terms of the potential energy, but represents the work needed to deflect the specimen arms. This has been discussed by Landes and Begley [4], who have shown that here the J integral loses its physical interpretation in terms of the potential energy, but retains its physical significance as a measure of the intensity of the characteristic crack tip strain field. This interpretation requires further investigation, in particular when applied to cyclic loading.

Equation (1) is valid for non-linear elastic materials and has been applied to determine J values experimentally, using a compliance-type procedure on several specimens of varying crack lengths, but this method [5] may not always be convenient. However, for certain geometries, such as deeply notched compact tension and three-point bend specimens, a simple, approximate method is available [6] which allows J values to be estimated from a single experimental load–displacement curve. In this technique the specimen with a deep crack $(a/w > 0.6)$ is loaded to a particular displacement and J is determined as a function of displacement from the expression

$$J = \frac{2A}{Bb} \tag{3}$$

where A is the area under the load–displacement curve taken at the displacement of interest, B is the thickness, b is the uncracked ligament $(w - a)$, and w is the width of the specimen.

The critical values of J (J_{IC}) are obtained at zero crack extension, but determining the crack initiation is perhaps the most difficult part of the J_{IC} evaluation. To overcome this problem, J is plotted against crack extension [7] in the form of an R curve [8]. J_{IC} is taken at the intersection of the stretch zone line $(J/2\sigma_{flow})$ and the J versus crack extension curve, where σ_{flow} is the mean of the yield stress and the ultimate tensile stress. The stretch zone takes into account the crack front geometry change usually occurring before the actual material separation and is associated with the 'crack opening stretch', COS. Recent experimental work [9] has shown that the COS is equal to J/σ_{ys}, where σ_{ys} is the yield stress of the material. However, the criterion $J/2\sigma_{flow}$ accounts for other cases where the material has a high strain-hardening exponent. A tentative J_{IC} test method based on the above considerations has already been put forward [10] and applied extensively [11] for various materials. It has also been indicated [12] that for elastic–plastic materials J results are not affected when unloading is limited to less than 10% of the maximum load. The deformation theory of plasticity, on which the concept of J integral is based, cannot directly account for the effects occurring during unloading and further work on this aspect is necessary. The J integral concept has been recently applied to load-cycling of a contoured double cantilever beam (DCB) specimen [13]. For each loading cycle a new 'specimen', having a crack length a, was assumed and the deflection caused by the accumulated cyclic deformation recorded. The elastic–plastic work used and ΔJ were then evaluated in the same way as described above.

The present paper describes a further study of the *J* integral application to high strain fatigue and the experimental determination of *J* values under monotonic loading, using a contoured DCB test-piece. Also, the influence of unloading in the *J* integral determination is considered, and J_{IC} values of two different specimen geometries, compact tension and contoured DCB, are briefly discussed.

ANALYSIS

The energy interpretation of *J*, as originally developed for non-linear elastic materials, is illustrated in Fig. 1(a) for the case of monotonic

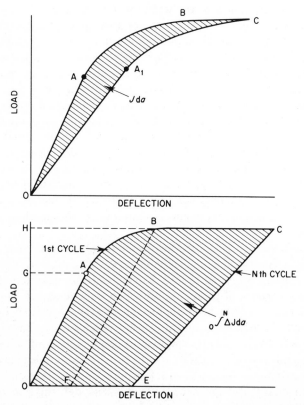

FIG. 1.   The *J* integral concept applied to monotonic and cyclic loading: (a) monotonic loading; (b) cyclic loading.

loading. Line OABC represents the load–deflection curve for a DCB specimen, initial crack length a, in which the loading is linear elastic from O to A; work-hardening occurs from A to B, and along BC the crack extends an amount da at a constant load P, causing an increment of deflection $d\delta$. The line OA_1C is the equivalent load–deflection curve for an initial crack length $(a + da)$. For non-linear elastic materials, the shaded area therefore represents the potential energy change, dU, caused by propagating the crack from a to $(a + da)$. While this simple energy interpretation is not valid for elastic–plastic materials, the J integral characterises the intensity of the crack tip stress–strain field. Furthermore, the applicability of the J integral concept to cyclic loading may raise some doubts because it is valid only within the confines of deformation plasticity theory, which does not account for the unloading. In Ref. [13] this situation was approached by defining an operational value of cyclic ΔJ. It was suggested that the crack growth process could be expressed by the total work contained within the load–deflection (hysteresis) loops and was equal to $\int \Delta J\, da$ (Fig. 1b). This integral, representing the dissipated work during cycling, may be evaluated by subtracting the complementary work of the first loading cycle (area OABHGO) from that of the last cycle (area OFECBHGO). According to the definition of cyclic ΔJ, only the loading parts of each cycle are considered here. Consequently, each cycle, N_i, could be represented by one specimen, S_i, having a specific crack length a_i and a corresponding hysteresis loop, H_i. The effects of intermittent unloading are not considered. These include overlapping of the load–deflection loops, some of which is caused by anelastic deformation, and also the increase of the yield stress due to the cyclic work-hardening.

The area OABFO (Fig. 1b) represents the load–deflection curve obtained in the first loading cycle or the 'first' specimen with a crack length a_0 of an elastic–plastic material. The curve is elastic from O to A and plastic deformation occurs from A to B. This type of loading curve is typical for mild steel. For other metals a non-linear curve starting at O is more likely. BC represents the accumulated deflection caused by cyclic creep and crack growth after N cycles at a constant load range ΔP. The 'last' specimen differs from the 'first' specimen only in respect of crack size, a, and the area OFECBAO simply defines the value of the cyclic J for elastic–plastic material. It should be realised that this area has no interpretation as an energy release in the cyclic process of crack extension or during cyclic creep. This cyclic creep phenomenon has been extensively studied and detailed reviews are available [14]. It consists basically of strain accumulation caused by load-cycling at stress levels above the yield point. In general, the cyclic creep curve (accumulated strain against number of cycles) has a shape analogous to that of the ordinary creep curve. An initial primary stage is followed by a well-defined secondary stage with a constant cyclic creep strain rate, covering 80–90% of the specimen life, and finally a very short tertiary stage near the fracture. Consequently, the cyclic creep deflection represented by FE in Fig. 1(b) is considered here to consist basically of the secondary stage, since experimental evidence indicates that the primary stage is virtually completed during the first loading cycle and is thus represented by OF.

Using eqn (1), the J integral can be evaluated in terms of load (for load-cycling) as follows. The complementary work, U, is obtained from the equation

$$U = -\int \delta \, \mathrm{d}P = U_1 + U_2 \tag{4}$$

In eqn (4), δ, the total deflection, is the sum of the elastic deflection at yield δ_{el} and the plastic deflection δ_{pl}. U_1 is the linear elastic term given by

$$U_1 = \frac{P_{el}\delta_{el}}{2} \tag{5}$$

where P_{el} is the yield load of the specimen. The plastic (non-linear elastic) term of the energy, U_2, is given by

$$U_2 = P_{pl}\delta_{el} + \int_0^{P_{pl}} \delta_{pl} \, \mathrm{d}P \tag{6}$$

where

$$P_{pl} = P - P_{el}$$

It is then assumed that, after P_{el} has been reached, the material follows a stress–strain law:

$$\epsilon_{pl} = A\sigma_{pl}^n \tag{7}$$

where ϵ_{pl} is the plastic strain, $\sigma_{pl} = (\sigma - \sigma_{el})$, n is the strain-hardening exponent and A is a constant. It can be shown, using elementary relations, that

$$\delta_{pl} \propto P_{pl}^n \tag{8}$$

and, hence, by substitution, that

$$U = \frac{P_{el}\delta_{el}}{2} + P_{pl}\left(\delta_{el} + \frac{\delta_{pl}}{(n+1)}\right) \tag{9}$$

Consequently,

$$\begin{aligned} J = &-\frac{1}{B_n}\bigg(P\left(\frac{d(\delta_{el})}{da} + \frac{d(\delta_{pl})}{da}\frac{1}{(n+1)}\right) \\ &- P_{el}\left(\frac{1}{2}\frac{d(\delta_{el})}{da} + \frac{d(\delta_{pl})}{da}\frac{1}{(n+1)}\right) \\ &- \frac{d(P_{el})}{da}\left(\frac{\delta_{el}}{2} + \frac{\delta_{pl}}{(n+1)}\right)\bigg) \end{aligned} \tag{10}$$

where B_n is the nett specimen thickness. Provided that the boundary conditions are prescribed in terms of fixed loads, eqn (10) allows the determination of the J values for any specimen geometry. The values of δ_{el}, δ_{pl}, P_{el} and their derivatives with respect to a may be obtained at the loading points, using the usual stress analysis methods.

In the present case the deflections δ_{el} and δ_{pl} can be calculated with the assumption that the DCB specimen is firmly fixed or 'built-in' at the crack tip. Using linear elastic beam theory [15], the elastic deflection due to the bending moment is obtained by integration from the equation

$$k_{el} = -\frac{d^2(\delta_{el})/dx^2}{(1 + (d(\delta_{el})/dx)^2)^{3/2}} \tag{11}$$

where k_{el} is the elastic curvature. The contoured shape of the DCB specimen may be closely approximated by a straight line (cf. Fig. 5b). The beam height, h, is then a linear function of x, and for $x > 30\,mm$ used here,

$$h = h_0 + h_1(x - 30) \tag{12}$$

For $x < 30\,mm$, $h = h_0$. ($h_1 = 0\cdot172\,mm$ and $h_0 = 12\cdot4\,mm$.) Making the substitution

$$\frac{d(\delta_{el})}{dx} = z_1 \tag{13}$$

in eqn (11), the equation becomes

$$k_{el} = \frac{dz_1/dx}{(1 + z_1^2)^{3/2}} \tag{14}$$

Eqn (14) was integrated between 0 and a, using the modified Euler's method and also the Runge–Kutta method, both methods yielding very similar results. Integration of eqn (13) using Simpson's rule gave the final values of δ_{el}. Plastic deflection, δ_{pl}, was calculated by a similar process, having replaced k_{el} by k_{pl} obtained from non-linear elastic beam theory [15] as

$$k_{pl} = \left(\frac{P_{pl}(2n+1)}{2nB}\right)^n A \frac{x^n}{(h/2)^{(2n+1)}} \qquad (15)$$

The yield load P_{el}, obtained from elementary beam theory, increased with the crack length (Fig. 2) (see also Ref. [13]) and the appropriate values of P_{el} were substituted into eqns (11) and (15) for the integrations.

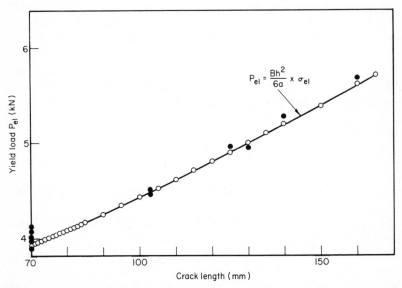

FIG. 2. Theoretical yield load versus crack length. Mild steel BS15. ○, theory; ●, experiment.

Neglecting the singularity, the maximum bending stress in the beam decreased with the crack length (Fig. 3). This stress, σ_{max}, is given by

$$\sigma_{max} = \sigma_{el} + \frac{2P_{pl}(2n+1)}{Bh^2 n} \qquad (16a)$$

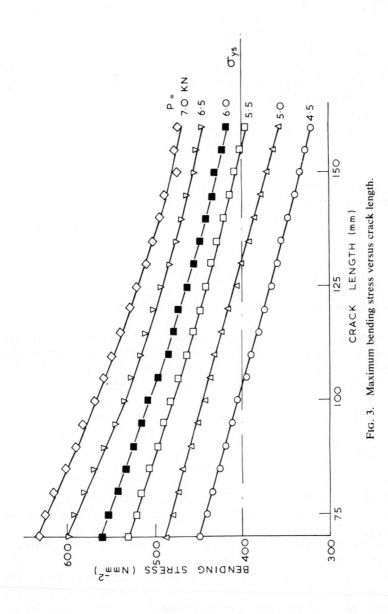

FIG. 3. Maximum bending stress versus crack length.

when $\sigma_{max} > \sigma_{el}$, and

$$\sigma_{max} = \frac{6Pa}{Bh^2} \qquad (16b)$$

when $\sigma_{max} < \sigma_{el}$.

The effect of the shear force on the deflection was also considered. The shear force deflection, δ_s, is obtained by integration from the equations

$$\frac{d\delta_{s,el}}{dx} = \frac{(\tau_{yx})_{y=0}}{G} \qquad (17a)$$

for the linear elastic case, where τ_{yx} is the shear stress, y is the co-ordinate along the beam height and G is the shear modulus, and

$$\frac{d\delta_{s,pl}}{dx} = D(\tau_{yx})_{y=0}^n \qquad (17b)$$

for the plastic case, where $D = 3A(n + 1)/2$. τ_{yx} was obtained from the shear stress distribution in this variable height beam [13, 15].

Subsequently, the rotational effect of the specimen arms at the crack tip was computed, assuming that the arms of the beam could rotate around the crack tip while the far end of the specimen was firmly fixed. It was further assumed that the uncracked ligament would behave entirely within linear elastic conditions; this is a good approximation, since the plastic deformation is confined to the cracked part of the beam and the plastic zone to the crack tip. It can be shown [15] that the deflection due to the rotational effect, δ_R, is:

$$\delta_R = \frac{6Pa^2(w - a)}{EBh_u^3} \qquad (18)$$

where w is the specimen length measured from the loading axis, h_u is the uncracked ligament height and $(w - a)$ is the uncracked ligament length.

A comparison between the bending moment deflections (elastic + plastic), shear force deflections (elastic + plastic) and rotational deflections is presented in Table 1. It may be seen that δ_s is between 2% and 3% of δ_M and δ_R varies between 5% and 8% of δ_M. Figure 4 shows the theoretical load–deflection curves for three crack length values 70 mm, 100 mm and 130 mm.

TABLE 1
Theoretical and experimental deflection values

a (mm)	$\delta_{M,el}$ (mm)	$\delta_{M,pl}$ (mm)	δ_M (mm)	$\delta_{s,el}$ (mm)	$\delta_{s,pl}$ (mm)	δ_R (mm)	δ_{exp} (mm)
70	1·94	5·37	7·31	0·06	0·09	0·41	7·924
80	2·65	5·15	7·80	0·08	0·09	0·47	—
90	2·98	4·3	7·28	0·09	0·1	0·51	—
100	3·87	3·43	7·30	0·09	0·11	0·53	7·745
110	4·23	2·30	6·53	0·1	0·11	0·54	—
120	4·96	1·35	6·31	0·11	0·12	0·56	—
130	5·34	0·99	6·33	0·11	0·13	0·59	—
140	6·13	0·69	6·82	0·12	0·15	0·61	—
150	6·91	0·34	7·25	0·14	0·16	0·63	—
160	7·96	0	7·96	0·15	0·17	0·66	8·83

Monotonic loading $P = 5500$ N.

EXPERIMENTAL

Material and Specimens
The material used was plate mild steel BS15, 11·2 mm thick. The composition and principal mechanical properties are shown in Table 2.

The specimens used were of two types: the contoured double canti-lever beam (DCB) testpiece with a constant stress intensity factor characteristic [16] (Fig. 5) and compact tension specimens of overall dimensions 72 mm × 78 mm, $w = 58$ mm, $a = 28$ mm (Fig. 6). Both types of specimen were cut so that the crack propagated along the rolling direction and the orientation of the crack plane was SL [1b].

TABLE 2
Composition and mechanical properties of mild steel BS15

Element	C	Mn	Si	P	Ni	Cr	Mo	Nb
Weight (%)	0·19	0·59	0·028	0·021	0·02	0·01	0·02	0·005

Young's modulus, $E = 207$ GN m^{-2}
Poisson's ratio, $\nu = 0·3$
Ultimate tensile stress, $\sigma_{uts} = 585$ MN m^{-2}
0·1% Proof stress, $\sigma_{el} = 401$ MN m^{-2}
Minimum elongation, $\epsilon_r = 28\%$

J. C. Radon

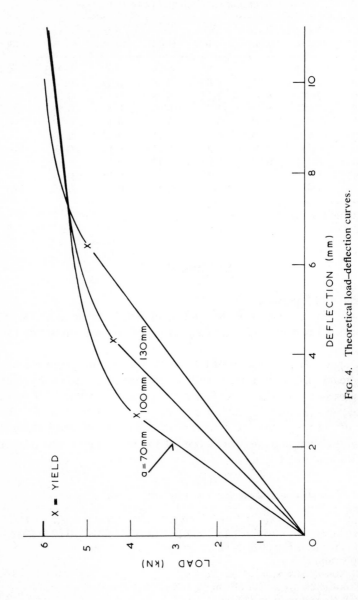

FIG. 4. Theoretical load–deflection curves.

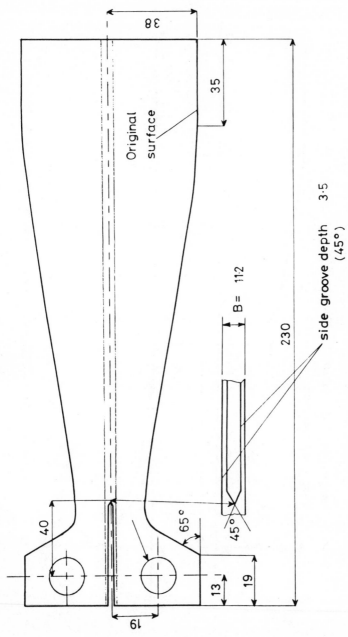

Fig. 5(a). Double cantilever beam specimen (dimensions in mm).

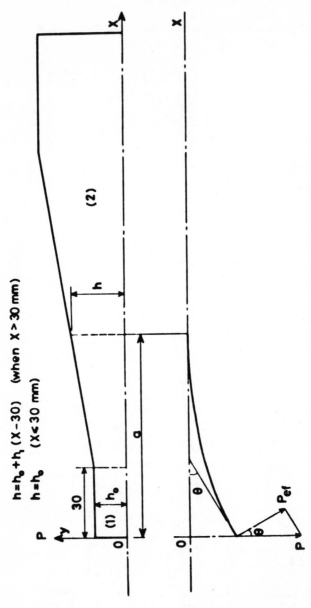

FIG. 5(b).   Specimen geometry used in the theoretical analysis.

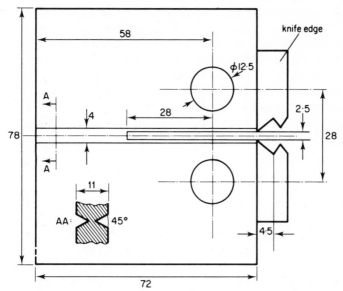

FIG. 6. Compact tension specimen used in the experimental J studies (dimensions in mm).

A side groove 3·5 mm deep was cut on both sides of all specimens to ensure reasonable plane strain conditions. In each contoured DCB specimen a machined slot approximately 65 mm long with a swallowtail front sharpened with a razor blade to form a crack starter was provided.

The fatigue precracking of the DCB specimens was accomplished in a 500 kN Avery hydraulic fatigue machine by load-cycling at a frequency of 10 Hz with a 1·2 kN tensile load. The compact tension specimens were fatigue precracked in a Schenck pulsator fatigue machine at a frequency of 20 Hz with a load of 2 kN. The test was stopped when the ratio $a/w = 0·6$ was reached. All the tests were conducted at 21°C in laboratory air, 50% relative humidity.

J_{IC} Tests

The tests with the contoured DCB testpieces were performed in the 250 kN Avery hydraulic machine. The crack extension at the maximum load reached in the first cycle was recorded with a travelling microscope of ×20 magnification. Five specimens were tested under load cycling conditions at increasing loads of 4·5, 5, 5·5, 6 and 7 kN.

In order to eliminate any influence of previous loading, a second series of tests was performed using decreasing loads. The compact tension specimens were loaded to a range of displacement values as recommended in Ref. [10]. An Instron tensile testing machine (TT-C) of 44 kN load capacity was used. The deflections were measured with a clip gauge positioned on the loadline and the load–deflection curves monitored by an X–Y plotter. Again crack extension was measured with a travelling microscope and confirmed with the 'heat-tinting' technique.

RESULTS AND DISCUSSION

The experimental monotonic load–deflection curves of the DCB specimen at $P = 7$ kN for crack lengths of 70 mm and 160 mm are presented in Fig. 7, together with the corresponding theoretical values. The agreement between the experimental and theoretical results is good. In the same figure the load–deflection loops at a stress ratio of $R = 0$ and $\Delta P = 5 \cdot 5$ kN are shown. As mentioned above, cyclic unloading causes a progressive increase in the permanent deflection of the specimen termed cyclic creep [14]. The total energy (actually the dissipated work during loading) increases and for the same crack length is considerably greater under cyclic conditions than in monotonic loading. This energy is plotted against crack length in Fig. 8. Some experimental results were also obtained and they agree well with the theoretical values. The amount of dissipated energy sufficient to cause linear elastic deformation has also been recorded and, as expected, the elastic–plastic energy is greater than its linear elastic counterpart. The difference between the two energies increases as the load increases and decreases as the crack length increases. This trend may be explained by the increase in yield load with crack length (Fig. 2). The energy curves presented in Fig. 9 show two cases when the load is high enough to rule out the possibility of obtaining linear elastic conditions (settled cyclic state) for the limited geometry investigated here. The slopes of the curves presented in Figs 8 and 9 gave the J values plotted against crack length in Fig. 10. J increases slightly with the crack length and, in the tests where the applied loads were 4·5, 5 and 5·5 kN, it reaches a constant value corresponding to a linear elastic loading (Fig. 4). Under such conditions the value of J is expected to be constant in a constant K testpiece.

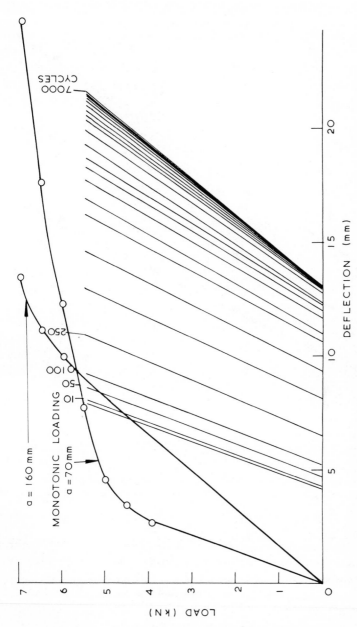

FIG. 7. Experimental load versus deflection loops at $\Delta P = 5.5$ kN (100% unloading) and monotonic loading at $P = 7$ kN: O, theory.

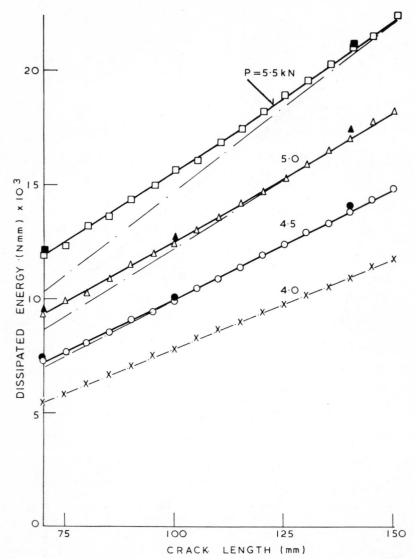

FIG. 8.   Dissipated energy versus crack length: open symbols, theory; closed symbols, experiment; —·—, linear elastic results.

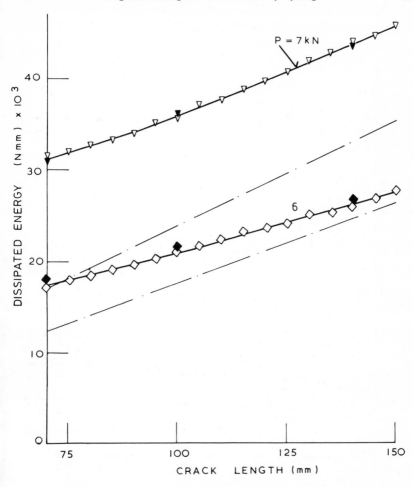

FIG. 9. Dissipated energy versus crack length: open symbols, theory; closed symbols, experiment; —·—, linear elastic results.

The variation of the cyclic J value with crack length is also presented in Fig. 10. Since the amount of total energy available is much greater in cyclic conditions, the cyclic J is higher than the monotonic J value. However, it decreases until linear elastic loading conditions are reached; thereafter, both quantities are the same. It is thought that when evaluating J_{IC} using a single specimen test technique, even the slightest unloading might cause a considerable increase in potential energy and influence the result.

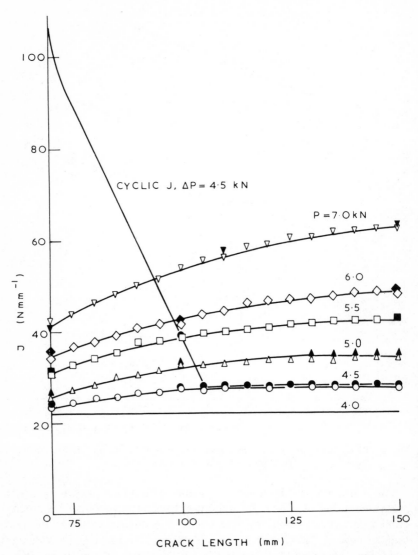

FIG. 10.   J versus crack length: open symbols, theory; closed symbols, experiment.

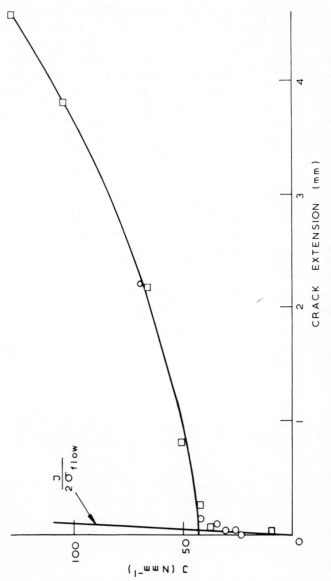

FIG. 11. Mild steel BS15. *J* versus crack extension. ○, double cantilever beam; □, compact tension.

A selection of monotonic J values is plotted against the measured crack extension in Fig. 11 in the form of the usual R curve. The J values corresponding to the crack extensions recorded at the points of maximum load in the first loading cycle of the fatigue crack growth tests performed on DCB specimens with a 70 mm crack starter [13] were taken from Fig. 10. Consequently, they were not affected by any prior unloading. The compact tension results were obtained with the usual technique for J, calculated from eqn (3). For small crack extensions, the agreement between the compact tension and the contoured DCB results is good. It is hoped that a similar agreement might be found for larger crack extensions. The interception of the $J/2\sigma_{flow}$ line with the curve fitted to the data points gives a room temperature J_{IC} value for BS15 of 42 Nmm^{-1}, the value appears to be the same in each of these two specimen geometries. The corresponding value of K_{IC} calculated from eqn (2) is 97 MN m$^{-3/2}$. This is somewhat lower than the linear elastic K_{IC} toughness of 120 MN m$^{-3/2}$ obtained in the parallel program using 3-point bend specimens. In fact the toughness values based on the J_{IC} resistance curve method were always found to be lower than the linear elastic K_{IC}. For some low alloy steels (such as BS4360-50) this difference amounted to approximatley 30% and some factors affecting this low initiation toughness will be discussed in a separate paper.

CONCLUSIONS

(1) The J contour integral has been successfully evaluated both theoretically and experimentally in the constant K contoured DCB specimen of mild steel BS15 in monotonic loading conditions.

(2) J increases as the crack length increases until a constant value is reached for linear elastic loading. In load-cycling (high strain fatigue) the appropriate ΔJ values are greater than the monotonic values. The above difference is caused by the increase in total dissipated energy obtained by the cyclic creep phenomenon in load-cycling. Repeated unloading will then affect any J_{IC} determination.

(3) J_{IC} was obtained for BS15 at room temperature using compact tension and contoured DCB specimens. A satisfactory correlation was obtained for both specimen geometries.

ACKNOWLEDGEMENTS

The material was provided by the Welding Institute, Abingdon, UK. This work is based on tests performed by Dr C. M. Branco and Mr P. M. Castro at Imperial College.

REFERENCES

1. 'Standard Method of Test for Plane Strain Fracture Toughness of Metallic Materials, (a) BSN Draft DD3, 1971; (b) ASTM E399-72, *ASTM Annual Book of Standards*, 1972.
2. RICE, J. R., A path independent integral and the approximate analysis of strain concentrations by notches and cracks, *Trans. ASME, J. Appl. Mech.*, **35**, 1968, 379–386.
3. RICE, J. R., Mathematical analysis in the mechanics of fracture, *Fracture* Vol. 2, (Ed. H. Liebowitz), Academic Press, 1968, Ch. 3.
4. LANDES, J. D. and BEGLEY, J. A., The effect of specimen geometry on J_{IC}, *Fracture Toughness*, ASTM STP 514, 1972, pp. 24–39.
5. BUCCI, R. J., PARIS, P. C., LANDES, J. D. and RICE, J. R., J-integral estimation procedures, *Fracture Toughness*, ASTM STP 514, 1972, pp. 40–58.
6. RICE, J. R., PARIS, P. C. and MERKLE, J. G., Some further results of J-integral analysis and estimates, *Progress in Flaw Growth and Fracture Toughness Testing*, ASTM STP 536, 1973, pp. 231–245.
7. LANDES, J. D. and BEGLEY, J. A., Test results from J-integral studies: an attempt to establish a J_{IC} testing procedure, *Fracture Analysis*, ASTM STP 560, 1974, pp. 170–186.
8. *Fracture Toughness Evaluation by R-Curve Method*, ASTM STP 527, 1973.
9. ROBINSON, J. N. and TETELMAN, A. S., Measurement of K_{IC} on small specimens using critical crack tip opening displacement, *Fracture Toughness and Slow-Stable Cracking*, ASTM STP 559, 1974, pp. 139–157.
10. BEGLEY, J. A. and LANDES, J. D., 'Tentative J_{IC} test method', Presented to the ASTM Committee E-24.01 on the Fracture of Metals, Philadelphia, USA, October 1973.
11. LOGSDON, W. A., Elastic–plastic (J_{IC}) fracture toughness values: their experimental determination and comparison with conventional linear elastic (K_{IC}) fracture toughness values for five materials, *Mechanics of Crack Growth*, ASTM STP 590, 1976, pp. 43–60.
12. CLARKE, G. A., ANDREWS, W. R., PARIS, P. C. and SCHMIDT, D. W., Single specimen tests for J_{IC} determination, *Mechanics of Crack Growth*, ASTM STP 590, 1976, pp. 27–42.
13. BRANCO, C. M., RADON, J. C. and CULVER, L. E., Elastic–plastic fatigue crack growth under load cycling, *J. Strain Anal.*, **12**, 1977, 71.

14. COFFIN, L. F., The influence of mean stress on the mechanical hysteresis loop shift of 1100 aluminium, *J. Bas. Eng., Trans. ASME (D)*, **86**, 1964, 673–680. (See also paper in this Volume by RADON, J. C. and OLDROYD, P. W. J., 'Plasticity of steels in reversed cycling'.)
15. TIMOSHENKO, S., *Strength of Materials*, Part II, Advanced Publications, van Nostrand Reinhold, 1970.
16. BRANCO, C. M., RADON, J. C. and CULVER, L. E., Growth of fatigue cracks in steels, *J. Mat. Sci.*, **11**, 1976, 149–155.

4

Variable-Thickness, Imperfect, Circular Plates at Large Deflections

G. J. TURVEY

University of Lancaster

SUMMARY

The suggestion is advanced that the design of laterally loaded, variable-thickness, circular plates may be rationalised by utilising 'designed-in' imperfections in conjunction with large deflection theory. Using a modified dynamic relaxation procedure to analyse the axisymmetric, large-deflection equations, it is shown that considerable reductions in both additional deflections and bending stresses arise when the imperfections are of the order of the plate thickness. Non-dimensional design data are presented for plates with linear-thickness tapers and single-wave-type initial imperfections, which are concave to the loading.

NOTATION

A_r, A_t, A_1	extensional stiffnesses
$A_{r_i}, A_{t_i}, A_{1_i}$	extensional stiffnesses at node i
a	plate diameter
$b^j (j = u, w)$	Gerschgörin bounds
$b_i^j (j = u, w)$	Gerschgörin bounds at node i
D_r, D_t, D_1	flexural stiffnesses
$D_{r_i}, D_{t_i}, D_{1_i}$	flexural stiffnesses at node i
E	Young's modulus
e_r, e_t	radial and tangential strains
h_0	plate thickness at $r = 0$
h_r	plate thickness at an arbitrary radius r
k_r, k_t	radial and tangential curvatures
$k_j (j = u, w)$	damping factors

G. J. Turvey

$\bar{k}_j(j = u, w)$	modified damping factors
M_r, M_t	radial and tangential stress couples
N_r, N_t	radial and tangential stress resultants
q	lateral load
$\bar{q}(= q r_0^4 E^{-1} h_0^{-4})$	dimensionless lateral load
r	arbitrary radius
$r_0(= a/2)$	reference radius
u, u^a, u^b	in-plane displacements
w, w^a, w^b	total deflections
w^0	initial deflection
w_1	additional deflection
$\bar{w}_1(= w_1 h_0^{-1})$	dimensionless additional deflection
w_0	initial deflection at the centre of the plate
$\bar{w}_0(= w_0 h_0^{-1})$	dimensionless initial deflection at the plate centre
α	thickness taper ratio
ν	Poisson's ratio
δr	mesh interval
δt	time increment
$\rho_j(j = u, w)$	fictitious densities
$\sigma_{r,t}^b$	radial and tangential bending stresses
$\bar{\sigma}_{r,t}^b(= \sigma_{r,t}^b r_0^2 E^{-1} h_0^{-2})$	dimensionless radial and tangential bending stresses
$\sigma_{r,t}^m$	radial and tangential membrane stresses
$\bar{\sigma}_{r,t}^m(= \sigma_{r,t}^m r_0^2 E^{-1} h_0^{-2})$	dimensionless radial and tangential membrane stresses
$(\ )^{\cdot}$	differentiation with respect to r
$(\ )_{,t}$	differentiation with respect to time t

INTRODUCTION

The products of the aeronautical, marine and mechanical engineering industries often employ variable-thickness, circular plates in a primary structural role. Typical applications of their use are to be found, for example, in steel pressure vessels and liquid storage tanks. The design of these plates is usually based on small-deflection theory and the effects of initial imperfections are ignored. This approach to design may be regarded as simple, but conservative. A more rational procedure would be to allow the plate to function in the large-deflection regime and make use of 'designed-in' imperfections to reduce stress levels for a range of design loads. Unfortunately, there are insufficient design data available to enable the designer to adopt such a procedure for variable-thickness, circular plates, with the necessary degree of confidence (the lack of data is less acute for rectangular plates [1]). At the present time, this rational design procedure would have to be based to a great extent on the work of Nylander [2], Murthy and Sherbourne [3] and Turvey [4], all of which is only partially relevant. Nylander's large-deflection results apply to uniformly loaded, clamped and simply supported circular plates of uniform thickness with affine imperfections. However, these results should be used with caution, since they are based on simple, two-parameter approximations, which may be inaccurate for stresses [4], and, in addition, since real imperfections are not affine. By way of complementing this information, Murthy and Sherbourne and Turvey provide extensive large-deflection data for uniformly loaded, variable-thickness circular and annular plates, but both of these papers ignore initial imperfections. Thus, the recognition of a lack of relevant design information has provided the stimulus to undertake the study of the large-deflection behaviour of imperfect, variable-thickness, circular plates, which is described in the later sections of the paper.

The analysis is based on the axisymmetric forms of the large-deflection, circular-plate equations. These equations are solved by the finite-difference, iterative procedure, dynamic relaxation (DR) [5]. The present application of the DR method differs from many previous applications, in that the majority of the parameters which control the stability and, by implication, the convergence of the computations are rationally determined in accordance with the procedures suggested by Cassell and Hobbs [6]. The advantage of this approach, as has been pointed out in Ref. [4], is that solution convergence is achieved with far fewer iterations.

As mentioned above, the prime objective of the work is to analyse the effect of the magnitude of the initial out-of-plane imperfection (in the direction of the load) and the effect of variable thickness on the non-linear elastic response of circular plates. However, in order to minimise the computational effort, it has been necessary to severely limit the range of parameters considered. Thus, only uniform loads and linear thickness variations are included. Furthermore, the plate boundary is assumed to be either simply supported or clamped and the initial imperfection (single-wave) is always in the positive coordinate direction, i.e. snap-buckling behaviour is excluded, though the computer program could be readily extended to handle this phenomenon.

PLATE GEOMETRY

The positive coordinate directions are shown in Fig. 1(a), which represents a vertical section through an initially flat plate. On the same figure the two dashed lines, AOB and AO'B, represent, respectively, the mid-planes of the initially flat and curved (imperfect) plates considered herein and the length of the line OO' is the magnitude of the initial deflection, w_0, at the plate centre. This type of rotationally symmetric, single-wave, initial deflection applies to all the results presented in Figs 3–9 and is given by the equation

$$w^0 = w_0 \cos (\pi r / a) \tag{1}$$

where $0 \le r \le a/2$.

The majority of the computed results apply to three initial imperfection ratios: $w_0/h_0 = 0$, 1 and 2·5.

The practical difficulties associated with the fabrication of variable-thickness plates suggest that it is sufficient to include only linear variations in the present study. These are defined in terms of a parameter α, the taper ratio. Hence, the plate thickness at an arbitrary radius, r, is given by

$$h_r = h_0(1 - 2\alpha r a^{-1}) \tag{2}$$

where $0 \le r \le a/2$. Thus, positive values of α apply to plates which decrease in thickness from the centre outwards and vice versa, as

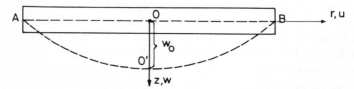

FIG. 1a.   Coordinate system for flat and imperfect circular plates.

shown in Fig. 1(b). Here, too, computations are undertaken for three α values $(-\tfrac{1}{2}, 0$ and $+\tfrac{1}{2})$ since these represent the range of taper ratios likely to occur in practice.

FIG. 1b. Axisymmetric linear thickness variations.

ANALYSIS

Axisymmetric, Imperfect, Circular Plate Equations

The compatibility, constitutive and equilibrium equations for the large-deflection analysis of imperfect, circular plates may be deduced from Refs [1–4] as:

Compatibility equations

$$
\left.
\begin{aligned}
e_r &= u^{\cdot} + \tfrac{1}{2}w_1^{\cdot 2} + w^{0 \cdot} w_1^{\cdot} \\
e_t &= r^{-1}u \\
k_r &= -w_1^{\cdot\cdot} \\
k_t &= -r^{-1}w_1^{\cdot}
\end{aligned}
\right\}
\tag{3}
$$

where $w^0 =$ initial deflection and $w_1 =$ additional deflection.

Constitutive equations

$$
\left.
\begin{aligned}
N_r &= A_r e_r + A_1 e_t \\
N_t &= A_1 e_r + A_t e_t \\
M_r &= D_r k_r + D_1 k_t \\
M_t &= D_1 k_r + D_t k_t
\end{aligned}
\right\}
\tag{4}
$$

where, for isotropic materials,

$$
\text{and} \quad
\left.
\begin{aligned}
A_r &= A_t = Eh(1 - \nu^2)^{-1}, A_1 = \nu A_r \\
D_r &= D_t = Eh^3(1 - \nu^2)^{-1}/12, D_1 = \nu D_r
\end{aligned}
\right\}
\tag{5}
$$

Equilibrium equations

$$
\left.
\begin{aligned}
N_r^{\cdot} + r^{-1}(N_r - N_t) &= 0 \\
M_r^{\cdot\cdot} + r^{-1}(2M_r^{\cdot} - M_t^{\cdot}) + N_r(w^{\cdot\cdot} + r^{-1}w^{\cdot}) + N_r^{\cdot}w^{\cdot} + q &= 0
\end{aligned}
\right\}
\tag{6}
$$

where $w = w^0 + w_1$.

Boundary Conditions
Plate centre $(r = 0)$

The restriction to axisymmetric behaviour implies the following conditions at the centre of the plate:

$$u = 0 \tag{7}$$

and w, M_r, M_t, N_r and N_t are all symmetric about $r = 0$.

Plate edge $(r = a/2)$

Two types of flexural boundary condition are assumed at the plate edge, viz simply supported and clamped. In each case full in-plane edge restraint is assumed. Hence, the two sets of edge boundary conditions are, respectively,

$$u = w_1 = M_r = 0 \tag{8}$$

and

$$u = w_1 = w_1^{\cdot} = 0 \tag{9}$$

The DR Method

As the details of the DR method of analysis are fully explained elsewhere [5], only a brief outline is necessary here.

In order to apply the method, inertia and damping terms must be introduced on the right-hand sides of eqns (6), as follows:

$$N_r^{\cdot} + r^{-1}(N_r - N_t) = \rho_u u_{,tt} + k_u u_{,t}$$
$$M_r^{\cdot\cdot} + r^{-1}(2M_r^{\cdot} - M_t^{\cdot}) + N_r(w^{\cdot\cdot} + r^{-1}w^{\cdot}) + N_r^{\cdot}w^{\cdot} + q \tag{10}$$
$$= \rho_w w_{,tt} + k_w w_{,t}$$

These equations are then re-arranged into a form typical of initial value problems:

$$\left.
\begin{aligned}
u_{,t}^a &= [(1 - \bar{k}_u)u_{,t}^b + \delta t \rho_u^{-1}\{N_r^{\cdot} + r^{-1}(N_r - N_t)\}]/(1 + \bar{k}_u) \\
w_{,t}^a &= [(1 - \bar{k}_w)w_{,t}^b + \delta t \rho_w^{-1}\{M_r^{\cdot\cdot} + r^{-1}(2M_r^{\cdot} - M_t^{\cdot}) \\
&\quad + N_r(w^{\cdot\cdot} + r^{-1}w^{\cdot}) + N_r^{\cdot}w^{\cdot} + q\}]/(1 + \bar{k}_w)
\end{aligned}
\right\} \tag{11}$$

where $\bar{k}_j = \frac{1}{2}k_j \delta t \rho_j^{-1}$ and $j = u, w$. They may be used in the DR method without further modification, as has been the case in the majority of previous applications of the method. However, by setting $\delta t = 1$ and evaluating ρ_j by means of the equation

$$\rho_j = \frac{1}{4}b^j \qquad (j = u, w) \tag{12}$$

where b^j are the Gerschgörin bounds of the stiffness matrix, the stability and convergence characteristics of the calculations are greatly improved, as has been pointed out in Ref. [4]. Details of the calculation of the b^j terms are given in the Appendix.

The modifications to the equilibrium equations, which allow their transformation to eqns (11), introduce an extra variable into the problem, namely time (parametric). Hence, a set of displacement–time compatibility equations has to be introduced:

$$\left.\begin{array}{l} u^a = u^b + u^a_{,t} \\ w^a = w^b + w^a_{,t} \end{array}\right\} \tag{13}$$

(N.B. Since $\delta t = 1$ it is no longer necessary to include it as a multiplying factor to the velocity components in eqn 13.)

Equations (11), (13), (3) and (4), when expressed in interlacing, finite-difference form (see Fig. 2), represent the set of equations for the DR iterative procedure, and a proper choice of damping factors, $\bar{k}_j$ ($j = u, w$), ensures convergence to the required static solution.

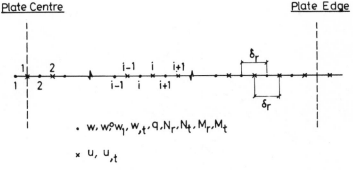

FIG. 2.   Interlacing, finite-difference mesh.

RESULTS

The DR computer program, based on the foregoing analysis, was partially verified by using it to reproduce a selection of the results of Refs [3, 4], which apply to initially flat, circular plates. Further checking was not possible, since comparable initially curved, circular plate results do not appear to be available. These preliminary computer runs also facilitated the choice of a suitable mesh size for the main computations. Thus, all the subsequent computations were based on a $10\frac{1}{2}$ interval mesh.

The results presented in Figs 3–6 apply to simply supported circular plates. Figure 3 shows curves of additional plate centre deflection versus lateral pressure for three values of initial deflection amplitude and taper ratio. The main features of this figure are that the magnitude of the additional deflection reduces significantly as the taper ratio decreases and the initial imperfection amplitude increases. Figure 4(a) depicts the variation in the bending stress at the centre of the plate as a function of the lateral pressure. It is observed that this stress is relatively insensitive to variations in the taper ratio, but is greatly reduced as the imperfection amplitude increases. By contrast, the plate centre membrane stress decreases only marginally as the imperfection amplitude increases (see Fig. 4b), though this again is insensitive to taper ratio changes.

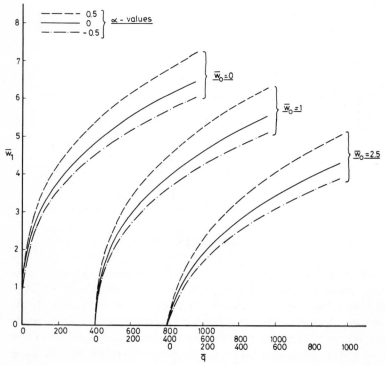

FIG. 3. Additional plate centre deflection versus lateral pressure for a simply supported plate ($\nu = 0\cdot3$).

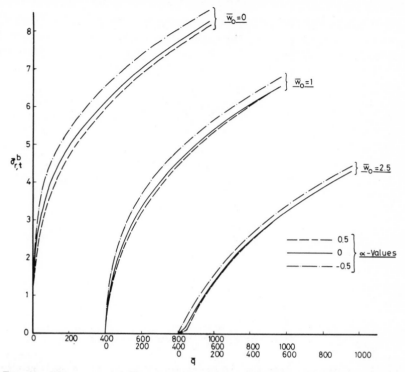

FIG. 4a. Plate centre bending stress versus lateral pressure for a simply supported plate ($\nu = 0.3$).

In Fig. 5 the additional plate centre deflections, expressed as a percentage of their corresponding flat plate values, are plotted against lateral pressure for a range of initial imperfection amplitudes. These curves show that significant reductions in additional deflection occur even with moderate imperfection amplitudes. For example, when $\bar{w}_0 = 1$ and $\alpha = 0$, the reduction in additional deflection varies between $\approx 35\%$ and $\approx 15\%$ over the lateral pressure range considered. These values are reduced or increased slightly for positive and negative taper ratios, respectively.

A similar presentation of the data for bending and membrane stresses at the plate centre is given in Figs 6(a) and 6(b), respectively. Figure 6(a) indicates that over the same lateral pressure range the percentage reduction in the bending stress is rather greater than that associated with the additional deflection. In this case the figures

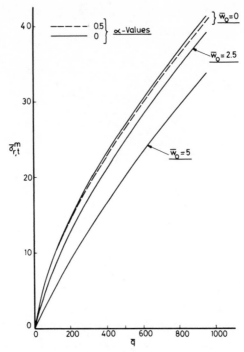

FIG. 4b.   Plate centre membrane stress versus lateral pressure for a simply supported plate ($\nu = 0.3$).

corresponding to those quoted above are $\approx 45\%$ and $\approx 20\%$, respectively, and, moreover, these values are independent of taper ratio. However, by comparison, Fig. 6(b) shows rather small reductions in the central membrane stress.

All the clamped-edge plate results are presented in Figs 7–9. Figures 7, 8(a) and 8(b) show the additional plate centre deflection, bending stress and membrane stress versus lateral pressure for three values of taper ratio and initial deflection amplitude. These curves are broadly similar to their simply supported plate counterparts (Figs 3, 4a and 4b). However, the stresses appear to be rather more dependent on taper ratio for clamped than for simply supported plates. Curves corresponding to Figs 5, 6(a) and 6(b) are not presented for the clamped plate, since they would be essentially identical with these figures, i.e. clamping the plate edge does not affect the percentage

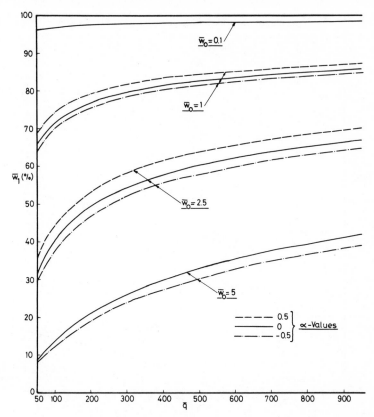

FIG. 5. Additional plate centre deflection (as a percentage of the corresponding flat plate value) versus lateral pressure for a simply supported plate ($\nu = 0{\cdot}3$).

reduction in the deflection and stress levels arising from the initial curvature of the plate.

The plate edge bending and membrane stresses are plotted as functions of the lateral pressure in Figs 8(c) and 8(d), respectively. Figure 8(c) shows that the bending stresses reduce slightly as the imperfection amplitude increases, but that they are predominantly affected by the taper ratio. The plate edge membrane stresses are, however, almost independent of the initial imperfection amplitude and, therefore only the flat plate stresses are shown. These stresses also are very dependent on the plate taper ratio.

76 *G. J. Turvey*

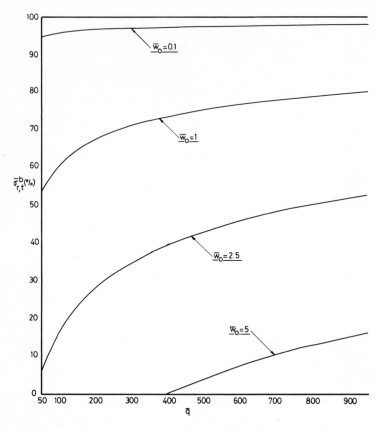

FIG. 6a. Plate centre bending stress (as a percentage of the corresponding flat plate value) versus lateral pressure for a simply supported plate ($\nu = 0\cdot3$).

The last figure, Fig. 9, presents the reduction in edge bending stress as a percentage of the corresponding flat plate stress for a range of lateral pressures and initial imperfection amplitudes. From this figure it appears that, for a given imperfection amplitude, the greatest stress reductions occur when the taper ratio is $-\frac{1}{2}$.

CONCLUSIONS

An analysis of uniformly loaded, variable-thickness, imperfect, circular plates, based on a partially rationalised DR analysis, has been

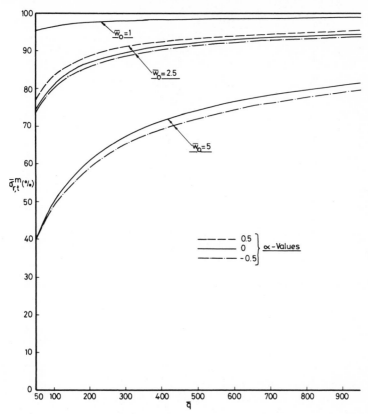

FIG. 6b. Plate centre membrane stress (as a percentage of the corresponding flat plate value) versus lateral pressure for a simply supported plate ($\nu = 0.3$).

described. Results have been presented in a dimensionless form for clamped and simply supported plates, which possess a linear thickness taper and/or a single-wave type of imperfection in the direction of the lateral load. These results show that, for a given lateral pressure, both the additional deflection and the bending stress at the centre of the imperfect plate are less than their corresponding flat plate counterparts. Furthermore, these reductions are of considerable significance for imperfections of the order of the plate thickness. The plate taper ratio only affects the percentage reduction in deflection in a minor way, i.e. it is slightly smaller for positive ratios and vice versa. On the other hand, the bending stresses at the edge of the plate

78

G. J. Turvey

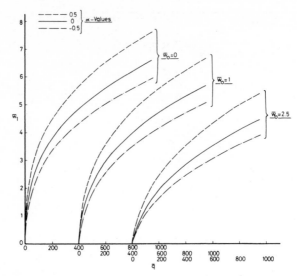

FIG. 7. Additional plate centre deflection versus lateral pressure for a clamped plate
($\nu = 0\cdot3$).

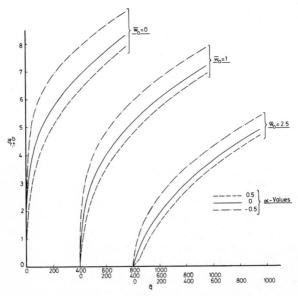

FIG. 8a. Plate centre bending stress versus lateral pressure for a clamped plate
($\nu = 0\cdot3$).

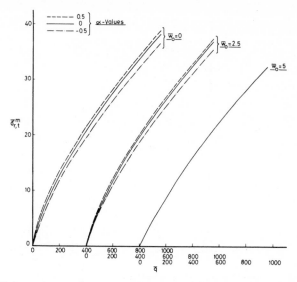

FIG. 8b. Plate centre membrane stress versus lateral pressure for a clamped plate ($\nu = 0\cdot3$).

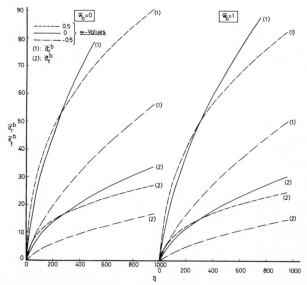

FIG. 8c. Plate edge bending stress versus lateral pressure for a clamped plate ($\nu = 0\cdot3$).

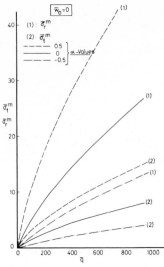

FIG. 8d. Plate edge membrane stress versus lateral pressure for a clamped plate
($\nu = 0\cdot3$).

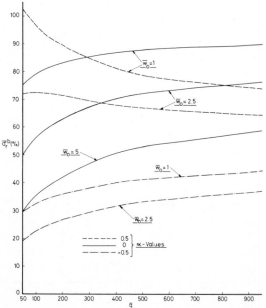

FIG. 9. Plate edge bending stress (as a percentage of the corresponding flat plate
value) versus lateral pressure for a clamped plate ($\nu = 0\cdot3$).

decrease with increasing initial imperfection amplitude, but are very dependent on the plate taper ratio. The membrane stresses are, however, largely independent of the imperfection amplitude, though the edge membrane stresses are very dependent on the taper ratio. Finally, in the quest to reduce stress and deflection levels, it is hoped that the quantification of the above stress and deflection characteristics may enable 'designed-in' imperfections to be used with more confidence.

ACKNOWLEDGEMENTS

The use of the computing facilities of the Department of Engineering and the assistance of Mr Ö. Ercan Ataer with the preparation of the figures are greatly appreciated.

REFERENCES

1. RUSHTON, K. R., Large deflexion of plates with initial curvature, *Int. J. Mech. Sci.* **12**, 1970, 1037–1051.
2. NYLANDER, H., Initially deflected thin plate with initial deflection affine to additional deflection, *Internat. Assoc. Bridge and Struct. Engrg.*, **11**, 1951, 347–374.
3. MURTHY, S. D. N. and SHERBOURNE, A. N., Nonlinear bending of elastic plates of variable profile, *A.S.C.E. Proc., J. Eng. Mech. Div.*, **100**, 1974, 251–265.
4. TURVEY, G. J., Large deflection of tapered annular plates by dynamic relaxation, *A.S.C.E. Proc., J. Eng. Mech. Div.*, (to be published).
5. DAY, A. S., An introduction to dynamic relaxation, *The Engineer*, **219**, 1965, 218–221.
6. CASSELL, A. C. and HOBBS, R. E., Numerical stability of dynamic relaxation analysis of non-linear structures, *Int. J. Num. Meth. Eng.*, **10**, 1976, 1407–1410.

APPENDIX: CALCULATION OF THE GERSCHGÖRIN BOUNDS, b^j $(j = u, w)$

The calculation of the Gerschgörin bounds follows the procedure given in Ref. [6]. Here, only the final forms of the finite-difference expressions for a typical node, i, are given.

Calculation of b_i^u

The finite-difference expression for this bound is,

$$b_i^u = \delta r^{-1}(\bar{N}_{r_{i+1}} + \bar{N}_{r_i}) + (2r_i + \delta r)^{-1}(\bar{N}_{r_{i+1}} + \bar{N}_{r_i} + \bar{N}_{t_{i+1}} + \bar{N}_{t_i}) \quad \text{(A.1)}$$

where

$$\bar{N}_{r_i} = \bar{N}_{r_i}^c + \bar{N}_{r_i}^v$$

$$\bar{N}_{t_i} = \bar{N}_{t_i}^c + \bar{N}_{t_i}^v$$

$$\bar{N}_{r_i}^c = A_{r_i}(2r^{-1} + \tfrac{1}{2}\delta r^{-2}|w_{i+1}^0 - w_{i-1}^0|) + \tfrac{1}{2}A_{1_i}\{(r_i + \tfrac{1}{2}\delta r)^{-1}$$
$$+ (r_i - \tfrac{1}{2}\delta r)^{-1}\}$$

$$\bar{N}_{r_i}^v = \tfrac{1}{2}A_{r_i}\delta r^{-2}|w_{1_{i+1}} - w_{1_{i-1}}|$$

$$\bar{N}_{t_i}^c = A_{1_i}(2\delta r^{-1} + \tfrac{1}{2}\delta r^{-2}|w_{i+1}^0 - w_{i-1}^0|) + \tfrac{1}{2}A_{t_i}\{(r_i + \tfrac{1}{2}\delta r)^{-1}$$
$$+ (r_i - \tfrac{1}{2}\delta r)^{-1}\}$$

$$\bar{N}_{t_i}^v = \tfrac{1}{2}A_{1_i}\delta r^{-2}|w_{1_{i+1}} - w 1_{i-1}|$$

and the superscripts c and v denote, respectively, those quantities which are constant and variable throughout the iterative procedure.

Calculation of b_i^w

The finite-difference expression for this bound is

$$b_i^w = \delta r^{-2}(\bar{M}_{r_{i+1}} + 2\bar{M}_{r_i} + \bar{M}_{r_{i-1}}) + \tfrac{1}{2}(r_i\delta r)^{-1}\{2(\bar{M}_{r_{i+1}} + \bar{M}_{r_{i-1}}) + \bar{M}_{t_{i+1}} + \bar{M}_{t_{i-1}}\}$$
$$+ N_{r_i}\{4\delta r^{-2} + (r_i\delta r)^{-1}\} + \bar{N}_{r_i}\{\delta r^{-2}|w_{i+1} - 2w_i + w_{i-1}|$$
$$+ \tfrac{1}{2}(r_i\delta r)^{-1}|w_{i+1} - w_{i-1}|\} + \tfrac{1}{2}\delta r^{-2}|N_{r_{i+1}} - N_{r_{i-1}}|$$
$$+ \tfrac{1}{4}\delta r^{-2}|w_{i+1} - w_{i-1}|(\bar{N}_{r_{i+1}} + \bar{N}_{r_{i-1}}) \quad \text{(A.2)}$$

where

$$\bar{M}_{r_i} = 4\delta r^{-2}D_{r_i} + (r_i\delta r)^{-1}D_{1_i} = \bar{M}_{r_i}^c$$

$$\bar{M}_{t_i} = 4\delta r^{-2}D_{1_i} + (r_i\delta r)^{-1}D_{t_i} = \bar{M}_{t_i}^c$$

N.B. Modified bounds should strictly be used at the plate boundaries, but in practice the above expressions may be used at these locations as well.

5

Non-Linear Phenomena in Polyhedral Sandwich Shells

G. M. Parton and I. Smith

University of Durham

SUMMARY

A general introduction to the polyhedral shell family of structures is followed by consideration of three principal forms of non-linear behaviour displayed by these structures:

(1) *The time-dependence of deflection is discussed and its relevance to the deflection and collapse limit states is shown.*
(2) *Non-linear behaviour exhibited by the joints is related to the boundary conditions used in the numerical model.*
(3) *The effect of the aspect ratio of the domes on stiffness and on stress levels is demonstrated.*

Each of these forms of behaviour is associated with the limit state of either deflection and stiffness or ultimate load and fracture mode.

INTRODUCTION

Most of the research in the field of structural engineering in the University of Durham has been concerned with various forms of lightweight structures and structural elements, such as cable structures, sandwich plates and shells. One of the most extensive projects has been the development of a family of polyhedral sandwich shells suitable for the roofing and enclosure of small and medium-sized spaces between 8 and 20 m in span or diameter. The intention is to form structures which will approximate in shape to a spherical dome membrane or a cylindrical vault, but will be composed of a regular series of identical flat plates. Such a geometry would be of great structural efficiency and at the same time would be specially suitable for large-scale industrial production followed by quick and easy site erection.

The use of sandwich plate for the facets of the polyhedra again tends to promote an efficient use of material in the structure and at the same time gives the environmental advantages of heat and sound insulation and a convenient tracking of services as an integral part of

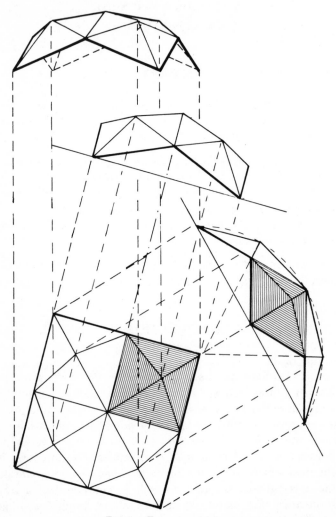

FIG. 1. Four-segment dome.

the structure. The surface condensation which is often such an undesirable feature of shell structures is eliminated.

Figure 1 shows the simplest of the dome geometries, which has four identical pyramids (one is shaded), each of which is made from four identical sandwich triangles. Figure 2 shows the sixfold symmetry of one of the larger structures, which has, altogether, thirty-six identical triangular plates.

The related family of geometries in the project are vault configurations. The simplest is shown in Fig. 3, where 8 m spans have been built using only one row of pyramids, with half-pyramids filling in the voids. Figure 4 shows a more complex vault.

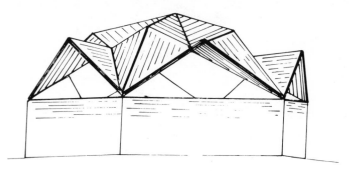

FIG. 2.   Six-segment dome with extensions.

MATERIALS

Several forms of non-linear behaviour are exhibited by the polyhedra, but the most notable is associated with the materials used in the sandwich plates. Here two groups of materials can be recognised, viz. those for the faces and the cores.

The faces, for which a wide variety of possibilities exist, include aluminium with many different surface finishes. The others used here are all organic materials, including timber products such as hardboard and plywood, laminar thermosetting polymers and GRP. All the latter group display a variety of non-linear mechanical properties. Other face possibilities are the cement-based products, which display almost as wide a range of non-linear characteristics.

The cores tend to be of three kinds. Those used in the models in Durham have been some kind of foam polymer; polyurethane (given

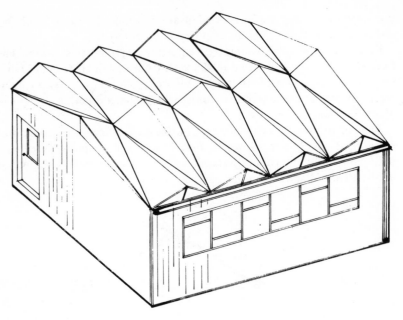

FIG. 3. Simple polyhedral vault.

by ICI) and PVC foam (from BTR) have formed the cores of most of the model plates. Cellular aluminium and timber constructions have been common in the aircraft and motor industries, though the former is probably unsuitable for the thicker plates required in building work. Finally, cement- and silicate-based cores are being examined here and elsewhere, using fillers such as polymer beads, wood flour and many other light filler materials. Most of these will have their own strange mechanical properties.

The most notable characteristic common to all sandwich plates, which affects their behaviour under load, is the low shear stiffness caused by the low modulus of the core. In the more familiar structural elements almost all the transverse deformations are due to flexure. The occurrence of large shear displacements produces unusual deflected forms and also causes interesting and significant characteristic behaviour at the angles or joints between the plates.

As a further complication some of these face and core materials will be liable to show considerable time-dependent displacements.

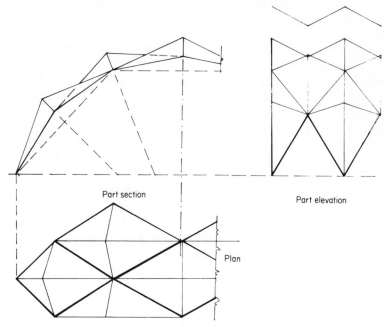

Part section

Part elevation

Plan

FIG. 4. Three-segment polyhedral vault.

NON-LINEAR BEHAVIOUR

The three different forms of non-linear behaviour displayed by the polyhedra will be described briefly.

Time-Dependent Deformations

The total displacement of any typical point in a structure may be considered as the sum of two component displacements:

(1) The instant component, occurring concurrently with the application of the load or load increment. (This may or may not be Hookean, but usually will approximate to it during the low-stress phase of loading.)

(2) The time-dependent component. Any single increment of this can be expected to approximate to some exponential function of time and stress, in which the constants will increase as the stress increases, giving a proportionally larger time-dependent displacement component and a larger effective time-span.

Useful information on time-dependent deformations is not easy to obtain when the subject of study is such a large and complex structure. Because of the effects of past loading history on any current behaviour, it would be necessary, ideally, to construct a new model for each stress level; this has not been achieved in these investigations, but it has been possible to gather some data which give a background to proposed future work. Most of the data have been obtained during the final stages of testing of a pair of the larger four-segment domes. In each, successive load increments were applied at the same time each morning for many days and the displacement was monitored during each day.

The results are presented in a number of different ways so that the efficacy of different numerical techniques can be seen. All relevant details and dimensions may be found in Ref. [1].

Figure 5 shows graphs for three lower load increments of direct deflection versus time and versus (time)$^{1/2}$. Figure 6 shows the similar graph for a middle-range load increment. In each case the

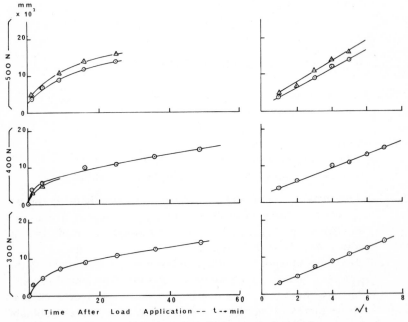

FIG. 5. Time-dependent displacements: low loads.

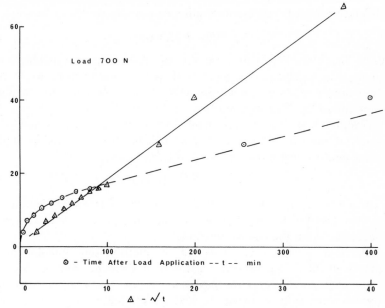

FIG. 6. Time-dependent displacements: 700 N loads.

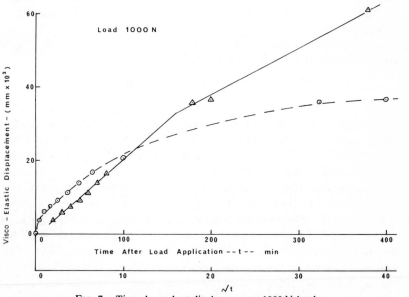

FIG. 7. Time-dependent displacements: 1000 N loads.

displacement Δ is represented quite well by the simple expression

$$\Delta = Wb + 1.85 \times 10^{-2} t^{1/2} \qquad (1)$$

for a load duration of up to about 2 h, after which the displacement increase is significant. In eqn (1) W is the load in newtons; b is a constant or modulus. It will be seen that the second term is not a function of the load magnitude.

Figure 7 is typical of larger loads and shows a distinct change in the slope of the (time)$^{1/2}$ graph. For such loads the relationship of displacement to time must be stated in two parts:

$$\Delta = Wb + 1.85 \times 10^{-2} t^{1/2} - !1.35 \times 10^{-2} (t - 50)^{1/2} \qquad (2)$$

where the step function (denoted by !) operates at about 150 min.

The change from the time-dependence of eqn (1) to that of eqn (2) occurs at about the 800 N/panel load.

Another approach is suggested by plotting the time-dependent displacement Δ_t against the logarithm of the time. Figures 8 and 9 show such graphs for a succession of increments in the higher loads, the final one (Fig. 9) being the load which caused the four-segment dome to fail by developing sufficient deflection to cause a joint fracture. (It is worth mentioning that even then the structure did not collapse; the geometry was modified so as to carry the load in a different way.)

Superimposed upon the deflection versus log time curves in Fig. 9 are lines derived from the function

$$\Delta_n = 0.05 \log t + !0.427 \log (t/140) + !2.863 \log (t/1100) \qquad (3)$$

Again, it will be seen that these expressions are not functions of the load. However, they appear to model the dome's behaviour under these conditions. Moreover, the values of deflection obtained from eqns (2) and (3) are nowhere more than about 20% apart below the second 'critical time' of 4000 min.

To appreciate the significance of the time-dependent deflections, it is necessary to look at the other 'instantaneous' (i.e. elastic) component of deflection for the range of load increments. Figure 10 (taken from Ref. [1]) shows the result of plotting these alone, without the time-dependent components, for the four-segment dome. The structure is seen to become progressively stiffer as the load increases, the modulus of deformation (i.e. b in eqns 1 and 2) decreasing approximately linearly from about 1.5×10^{-3} mm/N to about $1.0 \times$

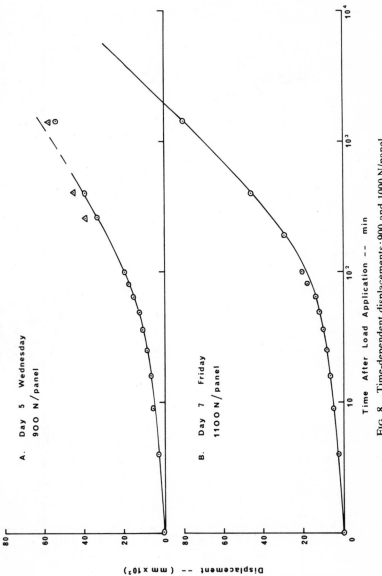

FIG. 8. Time-dependent displacements: 900 and 1000 N/panel.

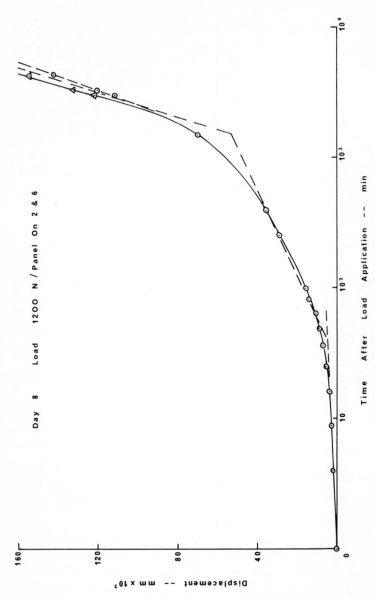

FIG. 9.   Time-dependent displacements: 1200 N/panel.

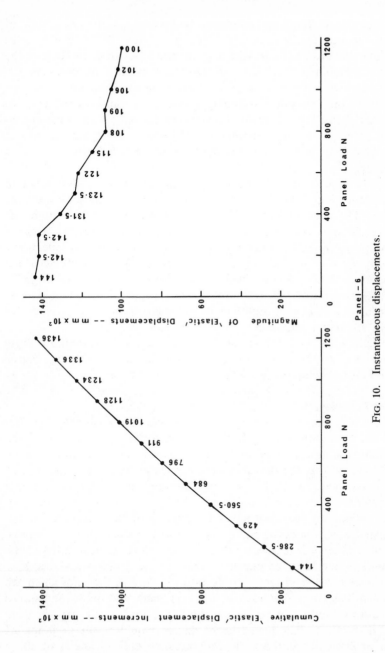

FIG. 10. Instantaneous displacements.

10^{-3} mm/N as the load increases from zero to the maximum of 1200 N/panel. This is to be expected, since single, flat sandwich plates exhibit similar behaviour, which seems to be due to an increase in the direct membrane resistance of the faces as they deform. All the structures tested so far have displayed similar behaviour. For this structure the membrane-stiffening of the dome causes a reduction in 'instant deflection' of about the same magnitude as the time-dependent component. The structure would appear to respond linearly to the load if deflection measurements were all taken after one day's loading.

There appear to be three useful conclusions to be drawn. First, time-dependent deflections are of little significance in the lower load range. The dead weight of the structure will always be within this range. Second, the deflection can be represented quite well if either eqn (2) or eqn (3) is built into the numerical model of deflections. Third, there is a load (in this case about 1200 N/panel) at which time-dependent deflections will become critical and effectively impose the design limit on the structure, if it is accepted that sustained loads will have a maximum duration of the order of 1–2 days. (In service these are snow, wind and occasional human access.)

Much work still has to be done before the behaviour of these domes can be thoroughly understood and predicted with any degree of precision; some of this is already in preparation.

Joint Behaviour

Figure 11 shows some typical joint details for the models. The first two are 'unstiffened' joints in the sense that the cores are butt-jointed without the insertion of any stiffening element. The third is fully stiffened, so that the face loads are carried over, and the core is also stiffened. The fourth has a core-stiffening effect but no moment transfer.

In attempting to reach an understanding of the behaviour of the polyhedra, it is necessary to know how the unstiffened joints behave when a moment is applied to them. Figure 12 is an idealisation of the change in geometry produced by this. The first diagram shows how the inner faces tend to penetrate into the core at the joint when the moment is tending to decrease the bevel angle, and the second shows the different effect of increasing the bevel. One would expect, from the first effect, that lower bevel angles would result in joints of greater flexibility as the angle decreases, so that for angles of about

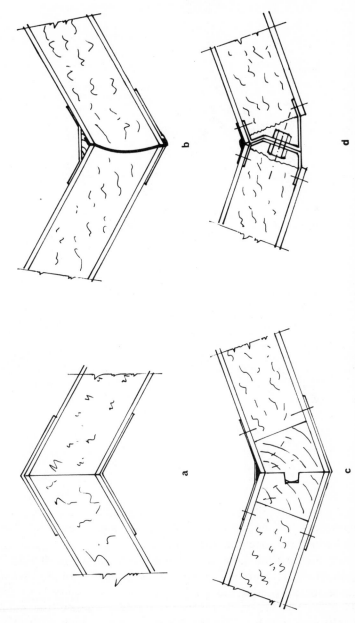

FIG. 11. Joint details. (a) Plane butt joint; G.R.P. cover strips. (b) Aluminium or G.R.P. face bedding strip, with G.R.P. cover strip. (c) Keyed wooden edge strip; metal cover plates set in mastic cement. (d) Cold formed steel edge channels; bolt inserted through slot in soffit.

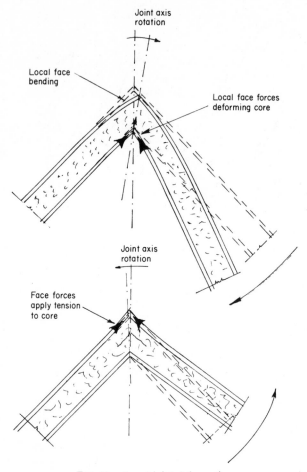

FIG. 12.   Local joint deformations.

80° or less the joint stiffness has become negligible when compared with the stiffness of the flat plate. There is not such a marked effect when the moment is 'opening' the joint.

   The way in which the behaviour of the joints may be dealt with in forming the numerical model depends on the geometry. In the case of the four-sector polyhedra the joints within the pyramids have un-stiffened cores and are at relatively low angles, whereas the joints between the pyramids have shear stiffeners and very high angles (i.e.

they are almost flat). Thus, the numerical models [1–4] are made to treat the former as hinges and the latter as moment-bearing; this appears to produce a good approximation to the correct deflections, though the stress configuration is not quite so good. Of course, it would be possible to improve this by refining the numerical model so as to have a thin row of flexible elements along the joint.

In general, it is safe to conclude that for low-angle, unstiffened joints there will be: (1) little moment transfer, (2) no shear transfer, (3) a large concentration of membrane stresses along the joint.

The Effect of Aspect Ratio

The effect of aspect ratio (i.e. height at centre/radius of circumscribing circle of the base) is important in several ways. Most obviously, a dome of high aspect ratio will require more material than a dome of a lower one and will be more expensive in consequence. Again, a high dome will be more expensive to heat than a low one, both because of the increased area for heat transfer and because of the tendency for the warm air to accumulate at the top of the structure. There are three other points of significance to the structural analyst.

(1) As the aspect ratio is reduced, the membrane stresses in the shell and the horizontal support reactions increase also. The latter effect may, in practice, be of greater significance than the former in determining the maximum permissible prototype size.

(2) Again, as the aspect ratio decreases, the dome's overall stiffness will decrease. This can be the limiting criterion of prototype size. It can either mean that working, pseudo-elastic displacements are unacceptable or that a lower limit of load is established at which time-dependent displacements become dangerous. It is considered that the former can be dealt with adequately by the existing numerical models, but the latter could not be dealt with confidently without testing.

(3) Because of the considerable increase in membrane stress and strain as the aspect ratio is lowered, it is predictable that the mode of failure of the dome may be different. For example, it would be possible for the dome to fail catastrophically by 'flipping through' at lower aspect ratios even though tests had shown that this did not happen with higher domes.

An extensive group of tests is reported in Ref. [2] which investigated these points in a set of four-segment domes in which only the aspect ratio was changed. Probably the most important information gained was that even the lowest dome showed no tendency to flip-through even after large and obvious local failures had developed in the ridge joints. (Indeed, after the initial change of geometry due to the failure, the dome always appeared to be stiffened by the failure.)

The graphs in Fig. 13 [5] show the relationship between the aspect ratio and (i) the load to cause material failure and (ii) the dome flexibility in mm/N per panel.

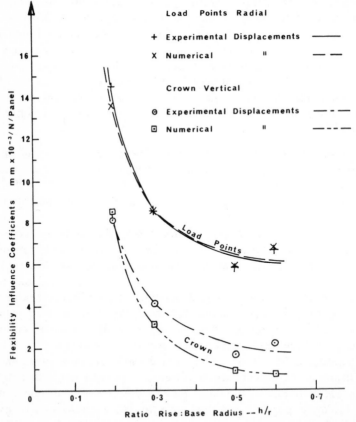

FIG. 13. Experimental and numerical flexibility. Influence coefficients vs. ratio rise : base radius. (From ref. [5], FIG. 7.19.1.)

It is probably no less significant that the visual appeal of the two higher domes was much better than that of the others. One has to consider the architect and, perhaps, even the client in these matters.

REFERENCES

1. PARTON, G. M., 'The structural behaviour of polyhedral sandwich shells', Ph.D. Thesis, Durham, 1974.
2. BETTESS, P., 'Analysis of polyhedral domes structures composed of flat plates of sandwich material', Ph.D. Thesis, Durham, 1971.
3. MANOS, G., 'Analysis of polyhedral domes sandwich structures', Ph.D. Thesis, Durham, 1975.
4. PARTON, G. M., 'Polyhedral sandwich domes: construction and analysis', IASS World Congress on Space Enclosures, Concordia University, Montreal, 1976.
5. SMITH, I., 'Failure of polyhedral sandwich domes', M.Sc. Thesis, Durham, 1977.

6

Applications of BERSAFE in Non-Linear Stress Analysis Problems

H. A. MONEY AND A. G. JAMES

Central Electricity Generating Board

SUMMARY

This paper details two studies which are in progress at the Midlands Region Scientific Services Centre of the Central Electricity Generating Board. The first considers a two-dimensional (axisymmetric) problem using well-documented routines from the BERSAFE finite-element system to quantify strains in a commonly occurring component. The second considers a three-dimensional problem using the latest developments in the BERSAFE system, to extend the applicability of fracture mechanics techniques. The two cases are presented together to demonstrate the versatility of BERSAFE.

The two-dimensional analysis examines the accumulation of permanent strain in the thread of a bolt. Part of the strain is determined by a plastic analysis and part by a creep analysis. The problem is complicated by the rapid accumulation of strain during the early part of the load history, which causes difficulties in the collection and application of materials data.

The three-dimensional study investigates the spread of plasticity under monotonically increasing tensile load in an edge-cracked bar. Three-dimensional analysis is necessary to examine the practically important intermediate region between the limiting states of plane stress and plane strain. Considerable computing effort and cost are involved in obtaining an adequate mesh refinement at the tip of the crack. In this paper it is shown that by considering stresses at a number of reference points within each element, a reasonable representation of three-dimensional crack-tip plasticity can be obtained without the computing cost becoming prohibitive.

THE TWO-DIMENSIONAL ANALYSIS

Steam Turbine Bolts

Most high-pressure steam turbine casings consist of two large flanged castings which are bolted together. The bolts may be up to 1·5 m long and 100 mm in diameter. A large number of such bolts are used on a turbine, which means a considerable financial outlay if they have to be replaced. Figure 1 shows the bolts on the inner casing of a high-pressure turbine.

Various undercuts are used to improve the performance of bolts and different methods of increasing the flexibility of the nut have been tried. Elegant refinements have little value on turbine bolts, because the hostile environment in which they are used encourages scaling and thus precludes the use of tight tolerances.

The bolts are therefore of a straightforward waisted design with a medium-fit fine thread at each end (Fig. 2). In service they are tightened using a hydraulic or thermal tensioning device to provide a known shank strain, free of any torsional component. When the turbine is run up to speed, the temperature of the flange increases. In

FIG. 1.   Photograph of a high-pressure turbine showing inner casing bolts.

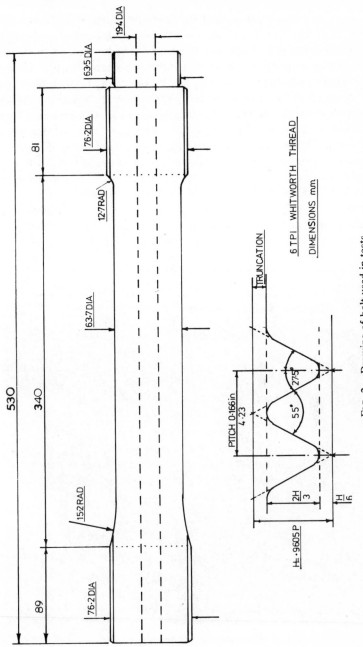

FIG. 2. Drawing of bolt used in tests.

some cases steam is used to heat the bolt and flange simultaneously; in other cases the heat is radiated to the bolt and conducted through the nuts, resulting in a temporary increase in the load on the bolt due to differential expansion, until it reaches the temperature of the flange. By the time the turbine reaches its operating temperature, the 'yield' point of the bolt material has dropped by 25% and the highly stressed zones of the thread have suffered irrecoverable straining by a mechanism consisting of plasticity and primary creep in unknown proportions. At the stabilised operating temperature secondary creep will occur. There may be many thermal cycles in the 30 000 h between statutory overhauls. During overhaul the bolts are released and then retightened back to the original load for further service.

Irrecoverable strain accumulates in the threads at a variable rate depending on the number of temperature cycles and retightenings,

FIG. 3.　Photograph of fracture face.

and on machining tolerances, etc. An occasional failure occurs. A code of practice has been devised covering design features, service life and the effect of retightening in order to reduce the failures to an inconsequential level. Because there were no data on strain accumulation at the stress concentrations in the thread, the penalty for retightening Durehete D1055 turbine bolts was set at the conservative figure of 15 000 h equivalent life usage.

On some machines it has been found necessary to strip and reassemble the turbine after only a few hours' service, and this quickly uses up the bolts' calculated design life. A research programme into actual life usage on retightening promised to be cost-effective. Experimental work was therefore undertaken at the Midlands Region Scientific Services Department of the CEGB, principally to establish material behaviour in typical bolts under representative conditions. This work continues and will be published by Mellor [1], whose contribution to this paper is acknowledged. It was soon realised that it was virtually impossible to measure the strain at the critical location. Failure nearly always occurs at the first engaged thread root, showing a nearly flat, apparently brittle fracture face (Fig. 3). Micrographs of the material at the thread roots show creep cavities, the frequency of which only hints at the strain applied to the material. Work was therefore started to see whether it was possible to predict the strain accumulation at the first thread root during start-up following tightening, using a finite-element model.

Stress Analysis

The first problem was to design a finite-element model of the nut and bolt assembly in which the load distribution along the thread was correct. This was done by inserting one-dimensional finite elements between the thread faces of the nut and bolt. These transmit load across the interface, but do not contribute to the stiffness or displacement of the two parts [2, 3]. When a satisfactory elastic behaviour had been achieved, the model had to be truncated in order to fit within the tighter restrictions of the non-linear version of the computer program. This was done by removing the low-stressed areas and fixing the boundaries thus created with the elastic displacement appropriate to the initial tightening load. As the version of the program used in this analysis does not update the displacements during iterations, this truncation is acceptable.

From the elastic analysis it was clear that the root of the first thread would become plastic during the initial tightening of the bolt. A simple tensile specimen produced a stress–strain curve (Fig. 4) which was split into increments for the plastic version of BERSAFE [4]. Using these data, the room temperature plastic strains were calculated.

During heating the limit of proportionality drops and the stress–strain curve changes shape. At high temperatures creep effects also become significant. The effort needed to determine the rates of these two processes at different temperatures did not seem to be justified. Instead a plastic analysis was performed using a stress–strain curve appropriate to the operating temperature, 565°C (Fig. 4), and the resulting stress field, plotted in Fig. 5, used as the starting point for a creep analysis. The plastic strain was added to the creep strain to give the total irrecoverable strain at any time.

A two-part analysis has the advantage that the creep data can be obtained practically, since the maximum stress on any testpiece is

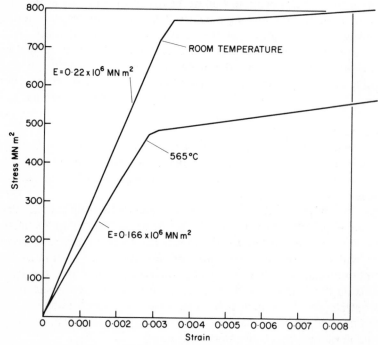

FIG. 4. Stress–strain curves used in the plastic step.

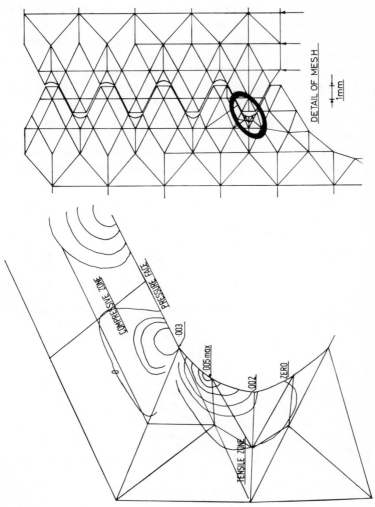

FIG. 5. Plastic strains at first thread root (0·15% strain in shank, 565°C).

not much above yield and only one temperature need be considered. The primary creep strain that takes place at a lower rate during the heating phase is ignored, and, hence, the assessment of the damage occurring during the first few hours could be incomplete.

The bolt problem is one of stress relaxation. The truncation of the finite-element model for the creep analysis is more suspect than in the plastic case because the stress in the whole body changes with time. On the other hand, the single inversion of the stiffness matrix in the version of BERSAFE used for this problem effectively freezes the displacements. On balance, it seemed a reasonable starting assumption.

It was necessary to obtain creep data for very short times in order to calculate the required strains. Special tests were conducted using simple specimens made from the same batch of material as used for full-scale bolting tests. Effort was concentrated on recording strain

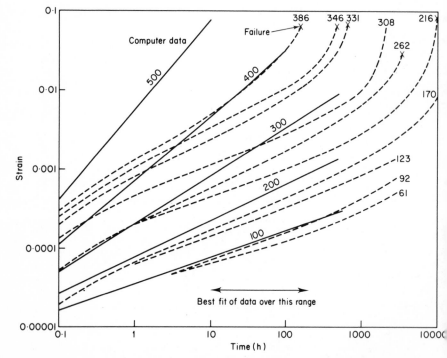

FIG. 6(a).   Creep data Durehete D1055. Comparison of computer data and measured data. (565°C. Stress levels in MN m^{-2} are indicated against the curves.)

during the first few minutes on load. Figure 6(a) shows the raw data. From a cross-plot of these data (Fig. 6b), it can be seen that there is considerable scatter in the strain values. In extrapolating the data towards zero time some of the curves cross. This sort of inconsistency causes problems in the interpolating routines used to find materials properties within the program. Attempts to fit an equation to the data demanded considerable adjustment of the curves. The final

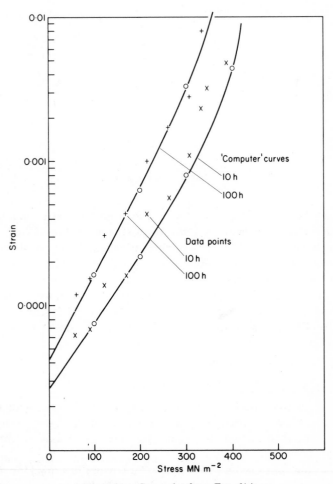

FIG. 6(b). Cross-plot from FIG. 6(a).

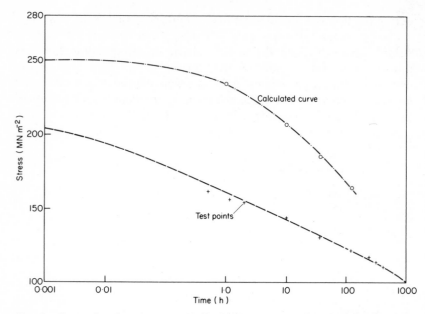

FIG. 7. Comparison between test and calculated stress relaxation. Stress in bolt shank (proportional to load).

data was the best that could be achieved from the limited number of tests, but was nevertheless a fabrication at times below 10 h.

An analysis was performed using these data. Figure 7 shows the reduction of stress in the shank of the bolt determined by theory and by test. A contour plot of equivalent strain is given in Fig. 8, which shows the extremely small size of the high-strain zone at the first thread root (also depicted in Fig. 5). During the 120 h since loading, the ratio of the stress at the first thread root to the stress in the shank changed from 2·25 to 1·53.

Tests of Full-Scale Bolts

Two tests were performed for comparison with the calculations. The first consisted of tightening a bolt onto a block of steel, raising it to 565°C and then immediately cooling it. This cycle took 8 h and the bolt lost 20% of its initial tension.

The second test used a 2 MN hydraulic tension test machine with a servo-control to maintain a constant strain. The bolt was assembled in

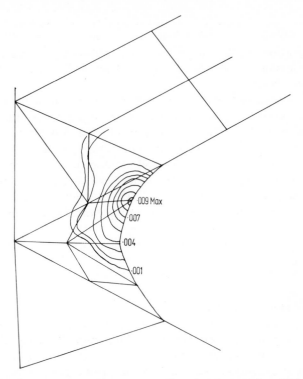

FIG. 8.   Sketch of creep strain contours at first thread root after 120 h (including plastic strains). Shank creep strain 0·000 57.

this machine and brought up to temperature before the load was applied. The test bolt should have been capable of carrying 725 kN at 0·15% strain, but when this strain was reached, the load was only 615 kN and it dropped by a further 20% within 15 min. This low value of load may have been due to the thread bedding-down and adding to the extension of the bolt for no increase in load, but creep relaxation data from British Steel show considerable variation in initial stress for the same initial strain. The test involving the simple heating cycle was more representative of the service situation. Again some of the load loss on this test may have been due to bedding-down in the threads, but it was clear that the calculation seriously underestimated the stress relaxation during the first few hours of service.

A number of longer-term tests were conducted in which bolts were tensioned onto blocks of steel and allowed to relax for 1000 h. In these tests an average of 60% of the load was lost in the first loading and the magnitude of the initial strain made very little difference to the proportion of the load that was lost. Subsequent loadings showed that hardening was significant for two load cycles, but after that had very little effect.

The conclusion that has to be drawn from this work is that the analysis demands more data than have so far been obtained. The experiments to determine the input data for the computer program are almost as costly and time-consuming as the full-scale tests; they can, however, accommodate the fits and limits of assembly and correctly model the heating cycle. On the other hand, the calculation is much quicker and cheaper, and gives the strain distribution in the thread. It will therefore be pursued into longer times where scatter in the data is reduced.

THE THREE-DIMENSIONAL ANALYSIS

Plasticity at a Crack-Tip

High stresses in the vicinity of the tip of a crack may cause a small enclave of yielded material at that point and the resulting distribution of stresses is of fundamental importance in determining the likelihood of crack propagation. The size of the plastic zone is an important parameter in fracture mechanics; its presence will modify the value of stress intensity associated with the crack and it may also be related to parameters such as crack-opening displacement [5]. Consideration of the stresses at the crack-tip [6] gives rise to the expressions

$$2r_y = \frac{1}{\pi}(K/\sigma_y)^2 \quad \text{(plane stress)} \tag{1}$$

and

$$2r_y = \frac{1}{3\pi}(K/\sigma_y)^2 \text{ (plane strain)}$$

where r_y is the plastic zone size, K is the stress intensity factor associated with the crack and σ_y the uniaxial tensile yield strength. Dugdale's method [7] or use of the Westergaard stress functions [5, 8] leads to the expression

$$2r_y = \frac{\pi}{8}(K/\sigma_y)^2 \tag{2}$$

The foregoing equations refer only to the limiting cases of plane stress and plane strain. Dislocation methods have been used to model the elastic–plastic deformation at the crack-tip in plane stress but the most important studies of crack-tip plasticity have been carried out using numerical methods, particularly finite-element analysis. For example, Rice and Rosengren [9], using elements designed to model the $1/r^{1/2}$ crack-tip singularity, showed a plastic zone extending a distance $0.175\,(K/\sigma_y)^2$ along a line at 70° to the x_2 axis, for a power law hardening material. Hellen [10] modelled the development of plastic zones in plane strain; more recently Carlsson and Larsson [11] modelled the development of plasticity in plane strain for a number of practical testpiece shapes and demonstrated the dependence of plastic zone size on $(K/\sigma_y)^2$ within the loading ranges $K = (A_1 - A_2)\sigma_y a^{1/2}$, where a is crack length and A_1 and A_2 depend on testpiece geometry.

Plane stress and strain will only occur in very thick or very thin sections; in most practical cases the plastic zone at the crack-tip will be formed at some intermediate stress state depending on testpiece thickness. Section thickness has a marked effect on crack propagation; in fracture toughness tests, for example, results for thin testpieces may be an order of magnitude higher than for thick pieces. More information is needed on the development of plasticity in the intermediate range between plane stress and strain, but investigations in this range where σ_{33} varies from zero (plane stress) to $(\sigma_{11} + \sigma_{22})/2$ (plane strain) involve three-dimensional solutions. In principle, the finite-element method seems to offer the most promising possibility of three-dimensional elastic and elastic–plastic stress analyses, and at Midlands Region Scientific Services Department of the CEGB a project is under way to consider the feasibility of such a study. Some of the initial computations carried out during this investigation have been used to give the extent of the plastic zones at different levels of monotonically increasing load at the centre and near the edge of models of a rectangular edge-cracked bar loaded in simple tension. By varying the dimension of the model in the σ_{33} direction three intermediate stress states between plane stress and plane strain were considered.

Finite-Element Analysis

It is common practice in the examination of crack-tip problems to include in the finite-element mesh a large number of elements in the crack-tip region to represent adequately the highly localised stress field in that area. The computing effort and cost involved in a three-dimensional finite-element analysis is necessarily high and to refine the mesh to the degree used in most two-dimensional elastic investigations for a three-dimensional elastic–plastic computation would be prohibitively expensive, even with the large computing capability available to the authors. The purpose of these initial computations was therefore to investigate the accuracy which could be obtained using a lesser degree of refinement in three dimensions. In order, to obtain stress values at points sufficiently localised to the crack-tip region, use was made of the stresses computed at a number of reference integrating points within each element in the course of determining nodal stresses.

Method

The solutions considered are for a linear elastic ideally plastic material obeying a von Mises yield criterion. Stresses were analysed using the program BERSAFE phase III [4]. The program incorporates isoparametric elements using up to fifth-order integration and gives stresses at gauss points within these elements. In the two-dimensional analysis second-order integration is used giving 4 gauss points per element; in the three-dimensional analysis third-order integration gives 27 reference points per element. The gauss point stresses are more accurately determined than the nodal stresses. The program performs an elastic analysis and subsequently scales the stresses so that the greatest gauss point stress value becomes equal to the yield stress. The load is then stepped up to its maximum value in a series of predetermined increments, the stresses being calculated for each increment.

The three-dimensional program was used to calculate the variation of σ_{11}, σ_{22} and σ_{33} (see Fig. 9) in the yielded region ahead of the crack-tip in a model of an edge-cracked tensile testpiece at a number of stages of monotonically increasing load. For symmetry reasons, only one half of the testpiece was considered. The specimens were characterised by the parameters $l/w = w/a = 2.5$, $\nu = 0.3$ and $E/\sigma_y = 274$, and three values of t were taken $0.125w$, $0.5w$ and w. Figure 9 shows the actual finite-element model, which comprised 28 three-dimensional isoparametric elements with quadratic sides. Of these, 20 were quadrilateral elements and 8, in a small region around the crack-tip, were triangular. The linear dimension of the crack-tip elements was within 4% of the crack length and the gauss point arrangement was such that the stresses were calculated at 36 points

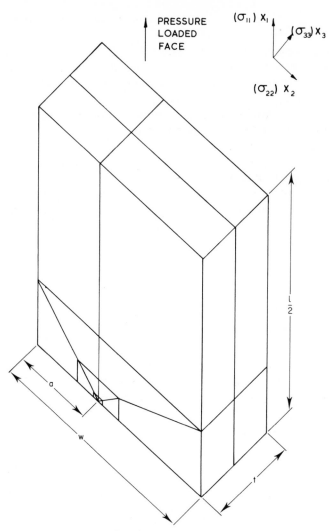

FIG. 9. Finite-element model.

across the crack-tip within a radius equal to 0·5% of the crack length, at 72 points within 2% and at 108 points within 3·5%. The computations were made on an IBM 370 computer. Ten load increments, taking an average 18 s CPU time each, gave a total computation time of around 3 min.

The three-dimensional finite-element mesh of quadrilateral and triangular elements was generated from a two-dimensional hand-drawn face using the program BERTHA [12]. The parameter controlling the x_3 direction dimension was altered to vary the thickness of the model. By prescribing the direction of displacements of nodes in the top surface or by allowing the top face to rotate, it was possible to consider cases where bending is either allowed or restrained. The model was loaded by applying a negative pressure load to the end of the model.

Development of Plasticity

The development of yield zones for three intermediate stress states between plane stress and plane strain has been studied; the model thicknesses were $0.125w$, $0.5w$ and w. Figures 10 and 11 show the

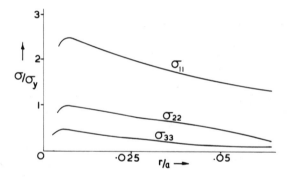

FIG. 10.   Stress versus radial distance from crack-tip; $t = w/8$.

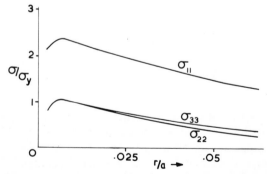

FIG. 11.   Stress versus radial distance from crack-tip, $t = w$.

variations of the three principal stresses within the yield zone for model thickness of $0.125w$, and w. The stresses are plotted against increasing radial distance from the crack-tip and are those calculated at points along a line making a small angle (approximately 6°) to the direction of crack propagation, i.e. the x_2 axis. It is seen that σ_{33} is approximately equal to σ_{22} in the thick model and $\sigma_{22}/2$ in the thin model. Only small changes occurred for a number of loading cases in the range $K = 0.19\sigma_y a^{1/2}$ to $K = 0.6\sigma_y a^{1/2}$, the ASTM limit for small-scale yielding.

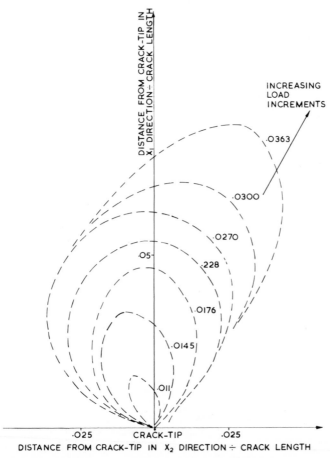

FIG. 12.   Plastic zone extent for seven load increments ($t = w$).

To model the development of plasticity with increasing load, a boundary was drawn at each increment considered between gauss points at which the von Mises equivalent stress has exceeded the yield stress and non-yielded points. Figure 12 shows such a plot with smoothed boundaries for the thickest piece ($t = w$) as the load is increased to $K = 0{\cdot}35\sigma_y a^{1/2}$. It is interesting to compare this picture with the plane strain results of Hellen [10] and Carlsson and Larsson [11]. The same picture emerges of plasticity spreading initially im-

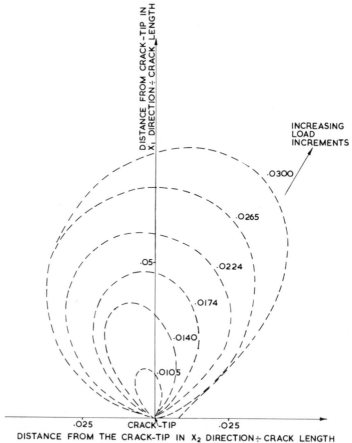

FIG. 13.   Plastic zone extent for six load increments ($t = w/2$).

mediately above the crack-tip and then developing at an angle to the crack-tip extension line $x_1 = 0$. The angle in this case is about 68°. For the model of intermediate thickness $(t = 0 \cdot 5w)$ the plastic zones with increasing load are shown in Fig. 13. In this case the development is similar to the previous case except that the final direction of plasticity development seems to have moved in a clockwise direction such that the angle of development is now about 63°. Finally, in the thinnest section (Fig. 14) the plasticity is seen to spread out on a much broader front in the x_1x_2 plane, being centred on an angle of about 48° to the $x_1 = 0$ axis.

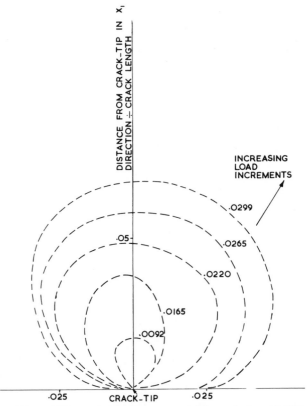

FIG. 14.   Plastic zone extent for five load increments $(t = w/8)$.

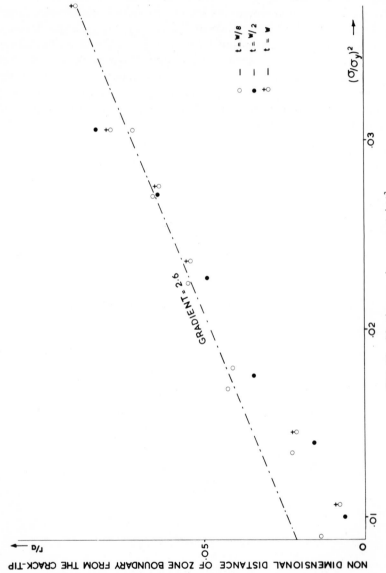

FIG. 15.   Plastic zone size versus $(\sigma/\sigma_y)^2$.

RELATION BETWEEN ZONE SIZE AND APPLIED LOAD

Analytical relationships (e.g. eqns 1 and 2) suggest a square-law relationship between zone size and applied load. In Fig. 15 the extent of the plastic zones measured from Figs 12, 13 and 14 for the angles of propagation recorded in the last section is plotted against $(\sigma/\sigma_y)^2$ for all testpiece thicknesses. A best-fit line through the origin has a gradient of 2·6.

Further computations were carried out on the thinnest model to extend the load up to $K = 0\cdot6\sigma_y a^{1/2}$, the ASTM maximum for small-scale yielding; yield zones for this level of loading are plotted in Fig. 16. The zone now extends outside the smaller elements at the tip of the crack and above $K = 0\cdot4\sigma_y a^{1/2}$ the frequency of occurrence of gauss points is not high enough to get an accurate definition of the extent of yielding. However, up to $K = 0\cdot4\sigma_y a^{1/2}$ zone radius versus $(\sigma/\sigma_y)^2$ is plotted in Fig. 17; a square-law relationship still describes the development of plasticity fairly closely, the gradient of the line being within 2% of that in Fig. 15.

Discussion

Figures 15 and 17 suggest a relationship of the form

$$\frac{r_y}{a} = 2\cdot6(\sigma/\sigma_y)^2 \tag{3}$$

for the size of plastic zone. For uniaxial tensile stressing of an edge-cracked sheet subject to no bending constraint, Bowie and Neal [13] and Brown and Srawley [14] suggest that K is given by

$$K = 2\cdot0\sigma(\pi a)^{1/2} \tag{4}$$

for a testpiece width to crack length ratio (w/a) of 2·5.

It follows from eqns (3) and (4) that

$$r_y = 0\cdot2[K/\sigma_y]^2 \tag{5}$$

The coefficient in eqn (5) is well within the range of those reported for the limiting cases of plane stress and plane strain where values between 0·08 and 0·4 have been found.

Finally, it was found that the technique is accurate enough to distinguish between the plasticity developing in planes near to the centre $(x_3 = t/2)$ and those near the edge $(x_3 = 0$ and $t)$ of the plate.

It is concluded, therefore, that within the limits of computing

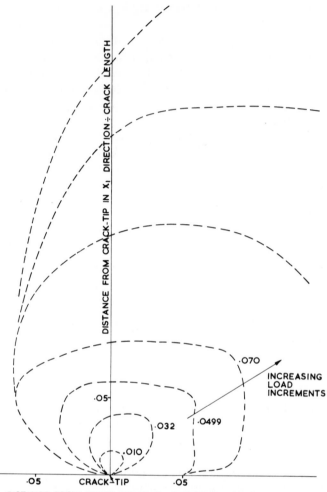

FIG. 16.   Plastic zone development up to $K = 0 \cdot 6\sigma_y(a)^{1/2}$, $t = w/8$.

resources available for this work a fair description of crack-tip plasticity was obtained. The program used is currently limited to 1000 gauss points, but this could easily be increased to give greater accuracy of representation of the zones without increasing the computing time to a prohibitive level. Also, a modification to include

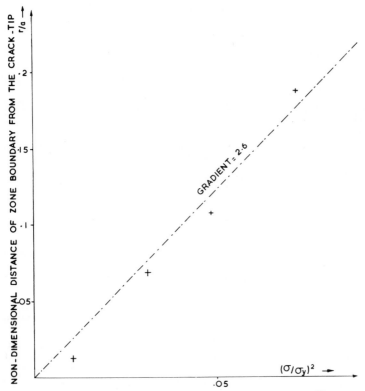

FIG. 17. Plastic zone size versus $(\sigma/\sigma_y)^2$. Loads up to $K = 0 \cdot 4\sigma_y(a)^{1/2}$, $t = w/8$.

elements containing a $1/r^{1/2}$ singularity which are already available for elastic analysis should increase accuracy.

ACKNOWLEDGEMENT

The authors wish to thank the Central Electricity Generating Board for the facilities to complete this work and permission to publish the paper.

REFERENCES

1. MELLOR, H. G., Private communication.
2. MONEY, H. A., 'Plastic and Elastic Analysis of Turbine Bolts', C.E.G.B. Report SSD/MID/N9/76, 1976.
3. MONEY, H. A., 'Stress Problems in Turbine Casing Bolts', M.Phil. Thesis, University of Nottingham, 1975.
4. HELLEN, T. K. and HARPER, P. G., 'A Users Guide to BERSAFE phase III level 1 for Plasticity and Creep Analysis', CEGB Report RD/B/N3597, 1976.
5. BURDEKIN, F. M. and STONE, D. B. W., The crack-opening displacement approach to fracture mechanics in yielding materials, *J. Strain Anal.*, **1**, 1966, 145–153.
6. KNOTT, J. F., *Fundamentals of Fracture Mechanics*, Butterworths, 1973.
7. DUGDALE, D. S., Yielding of steel sheets containing slits, *J. Mech. Phys. Solids*, **8**, 1960, 100–104.
8. WESTERGAARD, M. M., Bearing pressures and cracks, *J. Appl Mech.*, **6**, 1939, A49–A53.
9. RICE, J. and ROSENGREN, G. F., Plane strain near a crack tip in a hardening material, *J. Mech. Phys. Solids*, **16**, 1968, 1–12.
10. HELLEN, T. K., Finite Element Methods in Fracture Mechanics', Ph.D. Thesis, University of London, 1976.
11. CARLSSON, S. G. and LARSSON, A. J., Influence of non-singular stress terms and specimen geometry on small-scale yielding at crack-tips in elastic–plastic materials, *J. Mech. Phys. Solids*, **21**, 1973, 263–277.
12. MOYSER, G., 'BERTHA—A Program to Extend Certain Two-dimensional Finite Element Meshes to Three Dimensions', CEGB Report RD/B/N3661, 1976.
13. BOWIE, O. L. and NEAL, D. M., Single edge cracks in rectangular tensile sheet, *J. Appl. Mech.*, **32**, 1965, 708–709.
14. BROWN, W. F. and SRAWLEY, J. E., Plane Strain Crack Toughness Testing, ASTM, STP 410, 1966.

7

Finite-Element Analysis for Plasticity and Creep: An Introduction

C. Patterson
University of Sheffield

SUMMARY

A discussion is presented of the initial stress, initial strain and tangent stiffness approaches to the finite-element solution of stress analysis problems where non-linear constitutive laws prevail, with particular reference to convergence.

The cases of incremental plasticity as an initial stress problem and creep as an initial strain problem are examined in detail.

INTRODUCTION

Both plasticity and creep problems appear to the finite-element analyst as special cases of the more general problem of stress analysis under prevailing non-linear constitutive relations. In this paper a brief examination is made of some of the techniques which have been used to solve such problems. This is followed by more detailed discussions of incremental plasticity analysis by the Initial Stress Method and creep analysis by the Initial Strain Method. A useful bibliography may be found in the textbook by Zienkiewicz [1].

Non-linearities occur in stress problems due to non-linear constitutive behaviour and also when displacements can no longer be considered small; in the latter case non-linear strain–displacement relations are required and the equations of static equilibrium can assume non-linear form. It is assumed here that these conditions do not arise, although it should be mentioned that once the non-linear problem has been formulated, the available solution techniques are still broadly those discussed here.

There is a basic common strategy underlying most solution techniques. It is that the original problem, which may be highly non-linear, is reduced to a sequence of weakly non-linear problems, usually by considering that the load is incremented from zero to its true value in suitably small steps. The resulting non-linear problems are then solved sequentially using a convergent iterative process, with the solution to an appropriate linear problem supplying a starting trial solution. This approach appears reasonable on physical grounds for quasistatic processes.

The basic complexity occurring in a non-linear finite-element analysis is that the system of equations for the degrees of freedoms of the problem is no longer linear. Indeed, a large system of non-linear simultaneous algebraic equations must be solved. It is instructive, by way of introduction, to examine the much simpler problem of solving a single non-linear algebraic equation in one variable, for most of the important systematic features are already evident. Two familiar ways of solving the non-linear equation $y(x) = 0$ are the Newton and Newton–Raphson methods [2]. In the former some trial solution x_n is found which is sufficiently close to the exact solution but such that $y(x_n) \neq 0$. An improved trial solution is

$$x_{n+1} = x_n + \Delta x_n$$

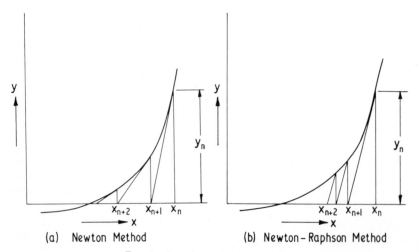

(a)   Newton Method   (b) Newton–Raphson Method

FIG. 1.   Iterative solution methods.

with

$$\Delta x_n = -\frac{y(x_n)}{\left(\dfrac{dy}{dx}\right)_n}$$

Proceeding iteratively, the exact solution can be found to any desired accuracy. The Newton–Raphson method is similar, but a constant value of the slope, $(dy/dx)_0$ say, is employed throughout. Both processes are illustrated in Fig. 1. Features of general significance are (1) that for either method to work the initial trial solution must be sufficiently close to the exact solution to give a convergent iterative procedure; (2) that the Newton process converges more rapidly than the Newton–Raphson process but at the cost of repeatedly evaluating a slope; and (3) that many similar procedures are, in principle, possible which could accelerate the convergence.

NON-LINEAR STRESS PROBLEMS

The linear stress analysis problem requires the simultaneous satisfaction of

(1) the strain–displacement relations,
(2) the constitutive (Hookean) relations and
(3) the equations of static equilibrium

together with the displacement and traction boundary conditions. In the finite-element method this is usually achieved by using a finite-element family of fields to give a set of trial fields for the displacements and using the relations (1) and (2) to define corresponding strain and stress fields. The potential energy of the problem is defined over the trial fields and achieves a stationary minimum value if the residual vector $\mathbf{R}(\boldsymbol{\delta})$ vanishes, i.e. if

$$\mathbf{R}(\boldsymbol{\delta}) = \mathbf{K}\boldsymbol{\delta} - \boldsymbol{\phi} = 0 \qquad (1)$$

Here $\boldsymbol{\delta}$ is the parameter vector of the trial fields, $\mathbf{K}$ is the stiffness matrix of the system and $\boldsymbol{\phi}$ is the generalised load vector. The solution to eqns (1), under well-defined restrictive conditions, is a valid approximate solution to the exact problem and converges in a well-defined manner to the exact solution as the mesh is refined [3, 4].

Where initial stresses σ_0 and initial strains ϵ_0 are admitted, the constitutive relations (i.e. the generalised Hooke's law) become

$$(\sigma - \sigma_0) = \mathbf{D}(\epsilon - \epsilon_0) \tag{2}$$

and in eqns (1) the generalised force assumes the form

$$\phi = \phi_0 + \phi_{\epsilon_0} + \phi_{\sigma_0} \tag{3}$$

in which the right-hand terms derive from the applied loads, initial strains and initial stresses, respectively.

The non-linear problems under consideration are such that the linear constitutive relations, eqn (2), are replaced by non-linear relations which may be given explicitly as

$$F(\sigma, \epsilon) = 0 \tag{4}$$

or in terms of relations between stress–strain increments

$$F'(\Delta\sigma, \Delta\epsilon) = 0 \tag{5}$$

The basic solution strategy is to formulate a linear finite-element description of the problem using the constitutive relations of eqns (2) and then, by adjustment of any (or all) of σ_0, ϵ_0 and $\mathbf{D}$, to seek a solution to the linear finite-element problem (eqn 1) which simultaneously satisfies the constitutive relations of eqn (4) or eqn (5) as appropriate. The non-linear constitutive relations are thereby satisfied and an approximate solution found. Clearly, an iterative solution procedure is required in adjusting the parameters. Moreover, if incremental relations (eqn 5) hold true, an incremental approach must also be adopted.

Variation of $\mathbf{D}$ is called the Variable (or Secant) Stiffness Method. In principle this method is quite valid, but since it requires that the system stiffness matrix be re-evaluated at each step and subsequently inverted, it is very expensive on computer time and is not favoured.

In contrast, variation of σ_0 or ϵ_0 only changes the load vector ϕ so that the system stiffness matrix remains unchanged and need only be set up and inverted once. The first alternative is called the Initial Stress Method and requires for its operation that the strains (or their increments) be explicit functions of the stresses (or their increments). The second is called the Initial Strain Method; it needs constitutive relations which give stress quantities as explicit functions of strain quantities.

The steps in a typical iterative solution for the Initial Stress Method are:

(1) Solve the initial elastic problem using externally applied loads.
(2) Determine the change in stress values ($\Delta\sigma_0$) required to satisfy the constitutive relations using the computed strains.
(3) Use $\Delta\sigma_0$ as an initial stress and compute the associated load vector $\Delta\phi$. From this vector compute the elastic displacement increments which will hold this force in equilibrium, using the relationship

$$\Delta\delta = \mathbf{K}^{-1}\Delta\phi \qquad (6)$$

(4) Update the displacements and strains and repeat iteratively from step (2) above until the displacement increments are sufficiently small. The final field quantities are the solution to the problem.

If the problem to be solved is highly non-linear, convergence of this iterative procedure may be either very slow or even non-existent. In such a case the load should be increased incrementally from zero to full load and the displacements, etc., determined iteratively for each load value.

The same basic procedure applies in the Initial Strain Method, except that strains are adjusted to comply with computed stresses.

If the constitutive law is such that both the stresses and strains may be determined explicitly, then either method may, in principle, be used. However, in a given context one of the methods is usually preferable on the grounds of speed of convergence. Thus, in softening materials the corrective initial strains become progressively larger as loading increases while the corresponding initial stresses remain small; the Initial Stress Method is therefore indicated.

In both methods the same stiffness matrix is used throughout. This corresponds directly to the Newton–Raphson process with one variable. As such it converges less rapidly than the related Newton approach but is usually much preferred on the grounds of cost; the iterative procedure is relatively cheap, whereas setting up and inverting a new stiffness matrix is costly. In problems where effective stiffnesses change significantly in the course of a calculation it may be profitable to update the stiffness matrix periodically while employing initial stress/strain between updates. This gives a process which is a fusion of the Newton and Newton–Raphson processes, since a given slope is used for several iterations and then periodically updated. In order to employ this stratagem at a given stage, constitutive relations are required in the form

$$d\boldsymbol{\sigma} = \mathbf{D}_T(\boldsymbol{\epsilon})\,d\boldsymbol{\epsilon} \tag{7}$$

The use of $\mathbf{D}_T$ in the setting up of the stiffness matrix gives a stiffness matrix, $\mathbf{K}_T$, which is appropriate to small changes about the given operating point. $\mathbf{D}_T$ is obtained from the derivatives of the appropriate constitutive relationship at the particular strains. $\mathbf{K}_T$ is called the tangent stiffness matrix.

Because of the basic economy of effort involved in the initial stress/strain approaches, these methods are preferred in weakly or moderately non-linear problems. However, they are inefficient in terms of speed of convergence, and sooner or later in grossly non-linear problems it pays to update the stiffness matrix periodically with a new tangent stiffness matrix. In a specific problem a balance between the initial stress/strain approach, the use of a tangent stiffness matrix and the use of acceleration in the iteration loop is used to give optimal results.

As in the case of linear problems, the initial choice of an adequately refined mesh is of considerable importance. If at any stage in the iterative process the mesh used is markedly incomplete, then errors occur in solving the appropriate intermediate linear problem due to incompleteness. As these errors tend to accumulate in the iterative process, there is a strong requirement for an adequately complete mesh. The conditions for and the nature of convergence in non-linear problems are largely unknown. However, it is unlikely that the quality of convergence will be better than that of linear methods i.e. energy convergence [3, 4]. In addition, the solution to a non-linear problem need not be unique, so that the physical reasonableness of a computed solution should always be established.

Just as in the linear case, know-how is of considerable importance in practical problem-solving. Furthermore, when tackling new problems, convergence studies, in which both the element meshes and increment sizes are reduced to give convergence, should always be carried out. Finally, where feasible, alternative methods should be used in order to 'validate' the converged solution.

INCREMENTAL PLASTICITY: INITIAL STRESS

Under increasing load an elastic–plastic material behaves elastically until an appropriate scalar yield function attains a critical value [5], i.e.

$$F(\boldsymbol{\sigma}, k) = 0 \tag{8}$$

Here k is a work-hardening parameter which accounts for the change in yield condition with the amount of plastic work done. Thereafter, both elastic and plastic straining occurs so as to maintain stresses on the yield surface i.e. to maintain satisfaction of eqn (8).

A commonly used yield function is that due to von Mises:

$$F(\boldsymbol{\sigma}, k) = \sigma^* - \bar{\sigma}(k) \tag{9}$$

in which $\bar{\sigma}(k)$ is the uniaxial yield stress and σ^* the von Mises equivalent stress, given by

$$(\sigma^*)^2 = \tfrac{3}{2}(\sigma_x^2 + \sigma_y^2 + \sigma_z^2 + 2(\tau_{xy}^2 + \tau_{yz}^2 + \tau_{zx}^2) - 3p^2) \tag{10}$$

where

$$p = \tfrac{1}{2}(\sigma_x + \sigma_y + \sigma_z)$$

If the load is increased infinitesimally after yield, the infinitesimal total strain consists of two parts, the elastic and plastic strains,

$$d\boldsymbol{\epsilon} = d\boldsymbol{\epsilon}_e + d\boldsymbol{\epsilon}_p \tag{11}$$

The elastic strain is still related to the stress increment through the usual linear constitutive relations and the plastic strain is given by an appropriate flow rule so that stresses remain on the yield surface. A commonly used flow rule is that of Prandtl–Reuss in which the plastic strain increment, $d\boldsymbol{\epsilon}_p$, is proportional to the deviator stress, $\boldsymbol{\sigma}'$, i.e.

$$d\boldsymbol{\epsilon}_p = \lambda \boldsymbol{\sigma}' \tag{12}$$

The deviator stress is given by

$$\boldsymbol{\sigma}' = \boldsymbol{\sigma} - p\mathbf{e} \tag{13}$$
$$\mathbf{e} = (1, 1, 1, 0, 0, 0)$$

When engineering strains are used, the Prandtl–Reuss flow rule, in association with the von Mises yield function, assumes the form

$$d\boldsymbol{\epsilon}_p = \lambda \, \frac{\partial F}{\partial \boldsymbol{\sigma}} \tag{14}$$

Thus, the general elasto-plastic strain increment can be expressed as

$$d\boldsymbol{\epsilon} = \mathbf{D}^{-1} d\boldsymbol{\sigma} + \lambda \frac{\partial F}{\partial \boldsymbol{\sigma}} \tag{15}$$

in which $\mathbf{D}$ is the usual Hookean matrix relating elastic stresses and strains and λ is a parameter, at present undetermined.

For variations on the yield surface

$$\frac{\partial F^{\mathrm{T}}}{\partial \boldsymbol{\sigma}} d\boldsymbol{\sigma} + \frac{\partial F}{\partial k} dk = 0 \tag{16}$$

Furthermore, on defining A so that

$$A\lambda = -\frac{\partial F}{\partial k} dk \tag{17}$$

it follows that

$$\frac{\partial F^{\mathrm{T}}}{\partial \boldsymbol{\sigma}} d\boldsymbol{\sigma} - A\lambda = 0 \tag{18}$$

Equations (15) and (18) can be combined to give

$$\begin{bmatrix} d\boldsymbol{\epsilon} \\ 0 \end{bmatrix} = \begin{bmatrix} \mathbf{D}^{-1} & \dfrac{\partial F}{\partial \boldsymbol{\sigma}} \\ \dfrac{\partial F^{\mathrm{T}}}{\partial \boldsymbol{\sigma}} & -A \end{bmatrix} \begin{bmatrix} d\boldsymbol{\sigma} \\ \lambda \end{bmatrix} \tag{19}$$

from which λ can be eliminated, with the result

$$d\boldsymbol{\sigma} = \mathbf{D}^* \, d\boldsymbol{\epsilon} \tag{20}$$

in which

$$\mathbf{D}^* = \mathbf{D} - \frac{\mathbf{D} \dfrac{\partial F}{\partial \boldsymbol{\sigma}} \dfrac{\partial F^{\mathrm{T}}}{\partial \boldsymbol{\sigma}} \mathbf{D}}{\left(A + \dfrac{\partial F^{\mathrm{T}}}{\partial \boldsymbol{\sigma}} \mathbf{D} \dfrac{\partial F}{\partial \boldsymbol{\sigma}} \right)} \tag{21}$$

$\mathbf{D}^*$ is the elastic–plastic matrix and directly relates the elastic–plastic strain increment to the stress increment. It was first used by Yamada, Yoshimura and Sakurai [6] and Zienkiewicz [7]. Up to this point $\mathbf{D}^*$ is explicitly determined save for the parameter A, which can be shown to be the slope of the plot of uniaxial stress at yield versus plastic strain. For non-work-hardening materials this is simply zero.

The typical elastic–plastic analysis proceeds as follows:

(1)   Apply a load increment and determine, elastically, the resul-

tant stress field; scale this so that the largest von Mises stress has just reached the yield surface and scale the load accordingly. In short, reach initial yield.

(2) Apply a load increment and determine the incremental elastic stresses and strains.

(3) Evaluate the von Mises stress at gauss points and, where yield has occurred, determine by linear interpolation that portion of the strain increment which is purely elastic (i.e. is required to initiate yield) and that which is elastic–plastic. Similarly, determine the stress increment, $\Delta\sigma_p$, associated with plastic flow.

(4) For the elastic–plastic strain use eqn (20) to determine the corresponding elastic–plastic stress increment $\Delta\sigma$. Under infinitesimal changes the stress increment, $\Delta\sigma_c$, given by

$$\Delta\sigma_c = \Delta\sigma - \Delta\sigma_p \qquad (22)$$

is a corrective stress which when added to the original total stress will bring it to the yield surface. For finite changes, however, the elastic–plastic constitutive law (eqn 20) is only approximate and the corrected stress may·not be exactly on the yield surface. In such cases the corrective stress should be scaled to bring the resultant to yield. The scaled corrective stress is the required 'initial' stress.

(5) Determine the residual forces deriving from the initial stress and, hence, using the elastic relationships, the corrective displacements and stresses. Update the stress field by addition of these stresses.

(6) Repeat iteratively from step (3) above until the residual forces are suitably small. The stress field associated with the given load is then determined.

(7) Return to step (2) above until the load is fully applied.

This simple iterative process, without acceleration, should converge rapidly (i.e. in not more than 15 cycles) for sufficiently small load increments.

CREEP: INITIAL STRAIN

The phenomenon of creep [8, 9] arises in materials in which, for constant stresses, the strains are time-dependent. In engineering metals creep strain rates are usually very low but are quite significant at elevated temperatures and stress levels. In some cases a full stress–strain history is required for a creep analysis, while in others state–dependent creep laws give adequate information. Only the latter will be considered here.

A creeping material, in an infinitesimal time interval, experiences an infinitesimal creep strain increment which appears to the material as an 'initial' strain, such as a thermal strain. The consequent loading leads to an infinitesimal stress readjustment. As time elapses, these processes evolve in dynamic equilibrium. Clearly, the Initial Strain Method is well suited to the analysis of creep.

There are two basic requirements of a creep analysis. First, for any given stress distribution the creep strain rate must be determinate. An estimate of the creep strain increment over a time interval Δt may then be made. Second, an iterative procedure is required which gives a self-consistent creep strain increment and consequent stress increment for the time interval. Since the creep strain rates depend non-linearly on stress, this problem is non-linear.

In a multiaxial stress field the instantaneous creep strain rate, $\dot{\boldsymbol{\epsilon}}_c$, is determined in the following manner. It is assumed that the creep strain rate is proportional to the deviator stress, i.e. that

$$\dot{\boldsymbol{\epsilon}}_c = \gamma \dot{\boldsymbol{\sigma}} \tag{23}$$

and that the parameter γ may be determined from uniaxial data using the von Mises equivalent stress σ^* and strain ϵ^*. In the uniaxial case, where only σ_x is finite,

$$\sigma^* = \sigma_x \quad \text{and} \quad \dot{\epsilon}^* = \dot{\epsilon}_x$$

and eqn (23) reads

$$\dot{\epsilon}^* = \gamma \sigma'_x = \frac{2}{3} \gamma \sigma^*$$

so that

$$\gamma = \frac{3}{2} \frac{\dot{\epsilon}^*}{\sigma^*} \tag{24}$$

Clearly, γ may be determined, using eqn (24), from experimental uniaxial data. Alternatively, a creep law may be appropriate. Once γ is determined, the creep strain increment in the time interval Δt is

$$\boldsymbol{\epsilon}_c = \dot{\boldsymbol{\epsilon}}_c \Delta t = \gamma \Delta t \boldsymbol{\sigma}' \tag{25}$$

When the steady state creep law [8, 9],

$$\dot{\epsilon}_c^* = C(\sigma^*)^n \tag{26}$$

applies, then

$$\gamma \Delta t = \frac{3}{2} C(\sigma^*)^{n-1} \Delta t \tag{27}$$

Again, under the time-hardening law

$$\epsilon_c^* = At^n(e^{\sigma^*/\sigma_0} - 1) \tag{28}$$

and

$$\gamma \Delta t = \frac{3}{2\sigma^*} \{\epsilon_c^*(t + \Delta t) - \epsilon_c^*(t)\}$$

Consequently, the relationship

$$\gamma \Delta t = \frac{3A}{2\sigma^*} \{(t + \Delta t)^n - t^n\}(e^{\sigma^*/\sigma_0} - 1) \tag{29}$$

may be used in numerical analyses.

An appropriate initial strain iterative procedure is:

(1) Apply the given mechanical loads and thermal loads if present, and solve the resulting elastic problem to get the starting stress distribution σ.

(2) Assume constant stresses over the time interval Δt and determine the creep strain rate, the creep strain increment (initial strain) and, thence, the creep load (see eqn 3).

(3) Obtain, elastically, the displacement increment $\Delta \delta$ caused by this load and, thence, the consequent stress increment $\Delta \sigma$.

(4) Take $(\sigma + \frac{1}{2}\Delta \sigma)$ as a refined estimate of the average stress over the time interval.

(5) Iterate from step (2) above until the change in $\Delta \sigma$ is acceptably small. Usually two iterations are sufficient.

(6) Increment the starting stresses and displacements by $\Delta \sigma$ and $\Delta \delta$, respectively, to get the starting values for the next time increment.

(7) Advance to the next time increment and repeat from step (2) above as desired.

This iterative procedure works quite well in practice provided that Δt is kept sufficiently small. Otherwise the process can become numerically unstable. For large Δt the breakdown of convergence is immediate, but for intermediate values apparent stability can be observed for substantial elapsed times to be followed by a more or less sudden breakdown of convergence. Qualitatively, the reason for this instability is that the creep strain rates, and therefore the creep strain increments and creep loads, are functionally dependent upon the stress field. But, in turn, the creep loads determine the stress field increments, so that there is a form of closed-loop behaviour. A perturbation in the current stress field implies a related perturbation in the stress field increment and, consequently, in the updated stress field. Because round-off error is inevitable, such perturbations must occur, so that, for numerical stability, the loop gain must be held at less than unit magnitude. This implies that in any situation there is a maximum permitted Δt which, incidentally, is dependent on the stress levels in the problem and may not be exceeded even in a steady state creep situation.

REFERENCES

1. ZIENKIEWICZ, O. C., *The Finite Element Method in Engineering Science*, McGraw-Hill, 1971.
2. HARTREE, D. R., *Numerical Analysis*, Oxford University Press, 1958.
3. OLIVEIRA, E. R. DE A., Theoretical foundations of the finite element method, *Int. J. Solids Struct.*, 4, 1968, 929–951.
4. PATTERSON, C., Sufficient conditions for convergence in the finite element method for any solution of finite energy, *The Mathematics of Finite Elements and Applications* (Ed. J. R. Whiteman), Academic Press, 1973, pp. 213–224.
5. JOHNSON, W. and MELLOR, P. B., *Engineering Plasticity*, Van Nostrand Reinhold, 1973.
6. YAMADA, Y., YOSHIMURA, N. and SAKURAI, T., Plastic stress–strain matrix and its application for the solution of elastic–plastic problems by the finite element method, *Int. J. Mech. Sci.*, 10, 1968, 343–354.
7. ZIENKIEWICZ, O. C., VALLIAPPAN, S. and KING, I. P., Elasto-plastic solution of engineering problems. Initial stress finite element approach, *Int. J. Num. Meth. Eng.*, 1, 1969, 75–100.
8. FINNIE, I. and HELLER, W. R., *Creep in Engineering Materials*, McGraw-Hill, 1959.
9. PENNY, R. K. and MARRIOTT, D. L., *Design for Creep*, McGraw-Hill, 1971.

8

The Creep Collapse of Pressurised Non-Uniformly Heated Tubes

D. J. F. EWING

Central Electricity Research Laboratories

SUMMARY

When a slightly irregular circular tube is pressurised at high tempera-ture, it eventually fails through creep collapse. The out-of-roundness generates destabilising bending moments which grow under creep. Following Hoff, Jahsman and Nachbar [1] and Bargmann [2], the phenomenon can be analysed by a 'sandwich approximation' (the tube wall being replaced by a 'sandwich' having the same elastic stiffness and formed by two membranes). This paper analyses two kinds of problem: (1) a long tube heated uniformly along its axis but non-uniformly around its circumference and (2) a tube of finite length and at uniform or non-uniform temperature. An analytical solution is found for a closed tube under uniform temperature. Problem (1) is two-dimensional and its equations can be set up directly from equili-brium considerations. The equations for problem (2) are found most easily from a two-part variational principle, using the 'reduced deflection method' to calculate the elastic strain energy; the equations of problem (1) are recovered as a special case.

Comparisons with an independent finite-element calculation of problem (1) by Harper [3] are made, and the limitations of the 'sandwich approximation' are discussed.

NOTATION

a	tube radius
$B = \exp(\gamma - \beta/T)/t_R$	creep temperature function (eqn 7) (units: h^{-1})
$C(\dot{w})$	creep power rate (eqn 48)
$c = a^3 q/D$	dimensionless pressure ratio

137

$D = Eh^3/12(1 - \nu^2)$	elastic bending stiffness
d	parameter defined by eqn (64)
E	Young's modulus
$\mathbf{e}_1, \mathbf{e}_2, \mathbf{e}_3$	orthogonal triad of unit vectors in circumferential, axial and inward radial directions (Fig. 1)
$F(Z)$	function of Z defined by eqn (26)
h	shell thickness
$h_s = h/(3)^{1/2}$	sandwich thickness in Hoff *et al.*'s 'sandwich approximation'
$J(X)$	function of X defined by eqn (34)
k	axial wavenumber (multiple of π/L)
L	length of tube
M	circumferential bending moment per unit axial length (Fig. 1)
m	circumferential wavenumber (integer)
N	circumferential membrane thrust per unit axial length (Fig. 1)
n	secondary creep law index
P, Q	constants in eqn (29)
p, p_R	compressive stress, reference stress ($15\cdot44$ MN m^{-2}, i.e. 1 tonf in^{-2})
q	applied external pressure
R	radius of curvature of deformed shell
r	stiffness reduction factor (eqn 71)
S	shear force per unit axial length (Fig. 1)
T	absolute temperature
t, t_R	time in hours, reference time (1 h)
t^*, t^{**}, t^{***}	time constants defined by eqns (25), (33), (70)
$U_{\text{eff}}(\dot{\mathbf{w}})$	effective elastic power-rate, including 'geometric-stiffness' terms (eqn 53)
$u, v, w,$	circumferential, axial and inward radial displacements of tube mid-surface from perfect tube geometry
V	amplitude in eqn $Y = V \cos m\theta \sin kz$
W	amplitude in eqn $w = W \cos m\theta \sin kz$
$\mathbf{w} = u\mathbf{e}_1 + v\mathbf{e}_2 + w\mathbf{e}_3$	displacement vector: $\dot{\mathbf{w}} = \partial\mathbf{w}/\partial t$
X	amplitude of Z in eqn $Z = X \cos m\theta \sin kz$
Y	dimensionless displacement ($w = Yh_s/2$)
Z	dimensionless bending moment ($M = -qaZh_s/2$)
$\bar{Z}$	value of Z averaged round the circumference

α_1, α_2	roots of $\alpha^2 - P\alpha + Q = 0$
β, γ	creep temperature-variation constants (eqn 7)
ΔD	change in diameter
ϵ	tensile strain (rate $\dot{\epsilon}$, elastic part $\dot{\epsilon}_e$, creep part $\dot{\epsilon}_c$)
κ	curvature change (rate $\dot{\kappa}$, elastic part $\dot{\kappa}_e$, creep part $\dot{\kappa}_c$)
ν	Poisson's ratio (0·3)
σ	tensile stress

Subscripts

c	creep
C,D,F	collapse, doubling, Fourier averaging
e	elastic
oo	original (before pressure is applied)
o	starting (after pressure is applied)
R	reference
s	sandwich

Derivatives

For a general quantity f, $\dot{f}$ denotes $\partial f / \partial t$, f' denotes $\partial f / \partial \theta$ and (except for t^* and t^{**}) f^* denotes $a \partial f / \partial z$. $\int_A$ means integration from $z = 0$ to L, and from $\theta = 0$ to 2π. δ denotes arbitrary variation.

INTRODUCTION

The motivation for the present creep study arose in January 1975, when it was required to calculate the allowable period of overheating of a steel reheater penetration sheath tube at a power station, should the supply of gas coolant temporarily fail. The temperature was estimated to vary from 520°C at the top to 280°C at the base. The classical Hoff–Bargmann analysis [1, 2] based on an all-round temperature of 520°C, gave an uncomfortably short lifetime of only 34 h (making worst-case assumptions about the initial out-of-round-ness and creep law acting). A more realistic calculation of the lifetime, allowing for the strengthening effect of the cold material, was therefore required.

This paper analyses two variants of the problem: (1) a long tube under non-uniform circumferential temperature and (2) a tube of finite length and/or non-uniform temperature. These problems are two- and three-dimensional, respectively. A variational method is used to set up the equations for the two cases, and in case (1) the equations can be checked by setting them up from equilibrium considerations directly.

The calculations are based on the 'sandwich' representation of shells in creep introduced by Hoff, Jahsman and Nachbar [1]. The rationale of this representation is as follows. The stress and strain distributions across a bending shell are unmanageably complicated because of creep effects. Accordingly, the actual shell is replaced by a sandwich of two membranes in each of which the stress is uniform with thickness. The dimensions of the sandwich are chosen to equalise the elastic stretching and bending stiffness of the tube wall.

The long tube calculations were later reworked by Harper [3], using the finite-element method. His results suggest that the sandwich approximation overestimates lifetimes; possible reasons are discussed.

THE LONG AXIALLY UNIFORM TUBE

This case has the advantage that the equilibrium equations and compatibility equations can be set up directly (thus providing a check on the variational method derived later on), because the problem is two-dimensional.

Problem Formulation

A cross-section showing the problem geometry is shown in Fig. 1. A tube of nominal mid-surface radius $a = 216$ mm (8·5 in) and thickness $h = 22·2$ mm (0·875 in) is acted on by a pressure $q = 4·20$ MN m^{-2} (0·2723 tonf in^{-2}, i.e. 610 lbf in^{-2}). (The problem was originally specified in tonf-inch-hour units, and the calculations were, in fact, carried out in these units.) The shell is under a varying temperature $T(\theta)$, where θ is the angle measured from the apex of the shell as in Fig. 1(a). The only effect of the axial constraint is to suppress axial movement. Any initial geometric out-of-roundness leads to destabilising circumferential bending moments in the tube wall. If the temperature is high enough, these lead eventually to creep collapse. The problem is to calculate the tube's useful working life for the worst initial out-of-roundness likely to be met in practice.

Let w denote inward deflection of the mid-surface from its nominal

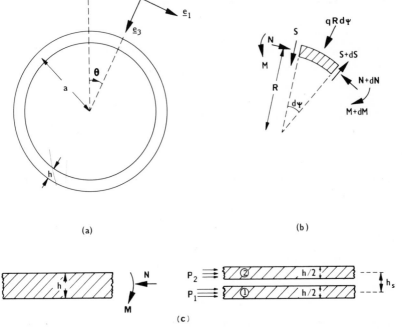

Fig. 1. (a) Cross-section of tube showing coordinate system used; Z and e_2 (not shown) come out of the paper. (b) Forces acting on a deformed element. (c) Sandwich representation of tube wall stresses.

(perfectly circular) shape, given initially before the pressure q is applied as

$$w = W_{oo} \cos m\theta \qquad (1)$$

The design code to which the tube was built gave W_{oo} as, at worst, 3·87 mm (0·1525 in). A straightforward elastic calculation [4, 5] shows that the out-of-roundness after the pressure, q, is applied is $W_o \cos m\theta$, where

$$W_o = (m^2 - 1)W_{oo}/(m^2 - 1 - c) \qquad (2)$$

in which

$$c = a^3 q/D \qquad (3)$$

and

$$D = Eh^3/12(1 - \nu^2) \qquad (4)$$

D being the tube's elastic bending stiffness. Note that $m = 2$, $c = 3$ corresponds to $q = 3D/a^3$, i.e. to the classical buckling load of linearised elastic theory [4, 5]. Here, taking $E = 193$ GN m^{-2} (12 500 tonf in^{-2}) $q = 0 \cdot 2723$ tonf in^{-2} and $\nu = 0 \cdot 3$, c is equal to $0 \cdot 2181$, the value used in the calculations.

The most dangerous wavenumber is $m = 2$, for which $W_o = 4 \cdot 18$ mm ($0 \cdot 1645$ in).

The creep law assumed under uniaxial tension σ is

$$\dot{\epsilon}_c = B(\sigma/\sigma_R)^n \tag{5}$$

in which σ_R is a reference stress, taken as $15 \cdot 44$ MN m^{-2} (1 tonf in^{-2}), and B is a function of temperature. This non-linear law creates a complicated stress and strain distribution across the tube wall, which Hoff *et al.* [1] approximated by an equivalent structural sandwich, i.e. two membranes of thickness $h/2$ in which the stresses are assumed uniform. The separation h_s of the membranes is chosen to be $h/(3)^{1/2}$. This ensures that the sandwich has the same elastic bending stiffness as the original wall. Its stretching stiffness has already been equalised by taking the membrane thickness as $h/2$.

It is convenient to work in terms of *compressive* stresses p_θ (circumferential) and p_z (axial). The through-thickness stress p_n and the shear stress p_{nz} are assumed small compared with p_θ. In order-of-magnitude terms, $p_n \sim q$, while $p_\theta \sim qa/h$. p_n and p_{nz} can therefore be neglected when considering stress–strain laws, as is normal in shell theory. Since there is supposed to be no axial strain, p_z must be half p_θ (since otherwise an axial deviatoric stress would arise which would cause axial creep). Equation (5), therefore, generalises [1] to

$$\dot{\epsilon}_{c\theta} = -\left(\frac{3}{4}\right)^{(n+1)/2} B|p_\theta/p_R|^{n-1}(p_\theta/p_R) \tag{6}$$

assuming that the von Mises law for creep under triaxial compression applies. p_R is $15 \cdot 44$ MN m^{-2} (1 tonf in^{-2}). Axial stresses and strains play no further part in the calculation and, accordingly, the subscript θ will be dropped in the rest of this section.

The 'worst-case' creep law assumed had the creep index $n = 5$ and

$$B(T) = \exp(\gamma - \beta/T)/t_R \tag{7}$$

where the reference time t_R was 1 h; γ was assumed to be $23 \cdot 05$ and β was assumed to be 30 940K. These were each one standard deviation away from their experimental means; there is an estimated 92%

chance that the real creep rates are slower than those used in the example, and the expected creep rates are about three times as slow. At the reference temperature $T_R = 793K$ (520°C), $B = B_R = 1/8\,592\,000\,h^{-1}$.

The Compatibility Equation

The moment–curvature relationship is, as usual,

$$\dot{M} = D\dot{\kappa}_e = D(\dot{\kappa} - \dot{\kappa}_c) \tag{8}$$

where subscripts e, c denote elastic and creep parts. From Fig. 1(c),

$$M = hh_s(p_1 - p_2)/2 \tag{9}$$

(M is reckoned positive if it tends to increase κ, i.e. to compress the inner fibre.) From Fig. 1(c) and eqns (6), (8),

$$\dot{\kappa} = (\dot{\epsilon}_2 - \dot{\epsilon}_1)/h_s = \left(\frac{3}{4}\right)^{(n+1)/2} B(p_1^n - p_2^n)/h_s p_R^n + \dot{M}/D \tag{10}$$

when (as here) n is an odd integer. (From now on this will be assumed in the algebra; for general n replace

$$(p/p_R)^n \text{ by } |p/p_R|^{n-1} p/p_R \tag{11}$$

whenever it occurs.)

The Equilibrium Equation

In the notation of Fig. 1(b) the equations of equilibrium of an element are:

$$N + dS/d\psi = q(R + h/2) \simeq qR \tag{12}$$

$$S = dN/d\psi \tag{13}$$

$$dM/d\psi = R\,dN/d\psi \tag{14}$$

R is the current radius of curvature of the centre-line in the deformed configuration, which subtends an angle $d\psi$ to its centre of curvature. It can be written

$$R = a - w - w'' \tag{15}$$

where primes (″) denote $d/d\theta$. The error is of order w^2/a provided that the centre-line strain is of order h^2/a^2 or smaller, i.e. provided that the deformation is basically one of pure bending (as would be expected physically). It can be verified later that the solution finally derived meets this condition. To the same order of accuracy (i.e. neglecting terms of order w/a compared with those retained) $d/d\theta$ and $d/d\psi$ are identical. So, combining the above,

$$N + N'' = qR = q(a - w - w'') \tag{16}$$

$$N = q(a - w - A_1 \cos \theta - A_2 \sin \theta) \tag{17}$$

$$M = \bar{M} = -qa(w + A_1 \cos \theta + A_2 \sin \theta) \tag{18}$$

Equation (17) follows by integrating eqn (16), and eqn (18) then follows by substituting for N and integrating eqn (14). A_1, A_2 and $\bar{M}$ are constants of integration. A_1 and A_2 represent so far arbitrary rigid-body displacements. From eqn (18), $\bar{M}$ is the average value of M, i.e.

$$\bar{M} = \frac{1}{2\pi} \int_0^{2\pi} M \, d\theta \tag{19}$$

since $\bar{w} = 0$ by centre-line incompressibility.

Combined Equations

A dimensionless displacement Y and a dimensionless bending moment Z are defined by the equations

$$Y = 2w/h_s \tag{20}$$

$$Z = -2M/qah_s \tag{21}$$

Since the hoop thrust N equals $(p_1 + p_2)h/2$ (see Fig. 1c), it follows from eqn (9) that the membrane stresses p_1 and p_2 are

$$p_1 = N/h + 2M/hh_s = \frac{qa}{h}(1 - Z) \tag{22a}$$

$$p_2 = N/h - 2M/hh_s = \frac{qa}{h}(1 + Z) \tag{22b}$$

(Here, and later, the working is to first order in h/a.) So, from eqns (10) and (15), since $\kappa = 1/R$ and $\dot{M}/D = -ch_s\dot{Z}/2a^2$, it follows that

$$\dot{w} + \dot{w}'' = -d/dt(1/\kappa) = a^2\dot{\kappa}$$
$$= \left(\frac{3}{4}\right)^{(n+1)/2} a^2 B(qa/hp_R)^n\{(1 - Z)^n - (1 + Z)^n\} - ch_s\dot{Z}/2 \tag{23}$$

or, on rearranging,

$$t^* \, d/dt(Y + Y'' + cZ) = -(B/B_R)F(Z) \tag{24}$$

where

$$t^* = \left(\frac{4}{3}\right)^{(n+1)/2} (hp_R/aq)^n (h^2/12a^2nB_R) \tag{25a}$$

and

$$= 27\cdot 7 \text{ h in the numerical example} \tag{25b}$$

$$F(Z) = \{|1 + Z|^{n-1}(1 + Z) - |1 - Z|^{n-1}(1 - Z)\}/2n \tag{26a}$$

$$= Z + 2Z^3 + Z^5/5 \ (n = 5) \tag{26b}$$

(A more exact calculation gives $t^* = 27\cdot 8$ h, but it did not seem worth reworking the calculations to account for this.) So far the equilibrium equations have not been used; however, from eqns (17) and (18),

$$Y + Y'' = Z + Z'' - \bar{Z}$$

so that the final equation to be solved, on eliminating Y, is

$$t^* \, \mathrm{d}/\mathrm{d}t(Z + Z'' + cZ - \bar{Z}) = -(B/B_R)F(Z) \tag{27}$$

Finally, it is necessary to calculate w from Z or from M. This requires the determination of the arbitrary constants A_1, A_2 in eqn (18). Strictly speaking, this cannot be done, since no end conditions are available. For definiteness, the centroid of the tube was assumed to remain stationary; this required the integrals of $w \sin \theta$ and $w \cos \theta$ to both vanish. This provides a means of determining A_1 and A_2 which was used in the finite-difference programs to solve eqn (27) in the general (non-symmetric) case.

An alternative way of presenting the solution in a form directly useful to the designer is in terms of a diameter change. This is independent of A_1 and A_2. However, it is only useful for m even (the diameter remains constant for m odd).

Analytical Solutions for Uniform All-round Temperature

For uniform all-round temperature $T = T_R$ and $B = B_R$ and it can be shown that $\bar{Z} = 0$, $Y = Z$. The constants A_1, A_2 are zero by symmetry. The uniform-temperature version of eqn (27) was obtained by Bargmann [2], who improved Hoff *et al.*'s original analysis [1] by including the elastic thrust terms. Hoff *et al.*'s equation omitted the cZ term. An approximate analytical solution to the problem when the initial out-of-roundness is proportional to $\cos m\theta$ can be derived by assuming that

$$w = W \cos m\theta \tag{28}$$

throughout the deformation, substituting into eqn (27), and applying a suitable averaging procedure. This gives an ordinary differential equation for W which can be integrated analytically (for n an odd integer) to give W as a function of time. In practice, the case $m = 2$ is of principal interest since this represents the fastest-growing deformation.

Two averaging procedures are available: direct averaging (the 'collocation method') and Fourier averaging. The collocation method is used in Refs. [1] and [2]; it involves integrating eqn (27) from $\theta = 0$ to $\theta = \pi/m$ while putting $Y = Z = X \cos m\theta$. $X = 2W/h_s$ is the dimensionless amplitude with initial value X_0 calculated from W_0 (eqn 2). With $n = 5$ (the case of immediate practical interest) there results the equation

$$(m^2 - 1 - c)t^* \, dX/dt = X + PX^3 + QX^5 \qquad (29)$$

where $P = 4/3$, $Q = 8/75$. With Fourier averaging, both sides of eqn (27) are first multiplied by $\cos m\theta$ before the integration, which is now from 0 to 2π. This process is equivalent to extracting the $\cos m\theta$ Fourier coefficient of Y on the assumption that all other harmonics are negligible. Equation (29) again results, but now with $P = 3/2$, $Q = 1/8$.

To integrate eqn (29), the right-hand side is factorised as

$$X + PX^3 + QX^5 = X(1 + \alpha_1 X^2)(1 + \alpha_2 X^2) \qquad (30)$$

where

$$\alpha_1 = \frac{1}{2} \{P + (P^2 - 4Q)^{1/2}\} \text{ and } \alpha_2 = P - \alpha_1 \qquad (31)$$

For collocation averaging, $\alpha_1 = 1 \cdot 2479$, $\alpha_2 = 0 \cdot 0855$; for Fourier averaging $\alpha_1 = 1 \cdot 4114$, $\alpha_2 = 0 \cdot 0886$. The solution of eqn (29) with $X = X_0$ at time $t = 0$ is

$$t/t^{**} = J(X_0) - J(X) \qquad (32)$$

where

$$t^{**} = (m^2 - 1 - c)t^*, \qquad (33)$$

and

$$J(X) = \int_X^\infty dx/(x + Px^3 + Qx^5) \qquad (34a)$$

$$= [\alpha_1 \ln(1 + 1/\alpha_1 X^2) - \alpha_2 \ln(1 + 1/\alpha_2 X^2)]/2(\alpha_1 - \alpha_2) \qquad (34b)$$

Thus, the time needed to reach a given amplitude factor X is given directly by eqn (32). In particular, the time for complete collapse is

$$t_C = t^{**}J(X_o),$$ (35)

The most dangerous mode is for $m = 2$. In the numerical examples $X_o = 0.6511$.

Conversely, the deformation after a given time may be required. This is obtained by solving eqn (32) for X in terms of T. The solution is easily found by iteration, using Newton's rule. If X is the current guess, the updated guess, X_{new}, is

$$X_{new} = X + (J(X) - J(X_o) + t/t^{**})(X + PX^3 + QX^5)$$ (36)

Convergence is normally very rapid, as is usual with Newton's method.

The method extends in principle to other odd integers n, but becomes laborious for large n.

Numerical Solution for Non-Uniform Circumferential Temperature

For a non-uniform circumferential temperature distribution, a solution can be calculated by finite differences. This also allows one to compare the merits of 'collocation' and 'Fourier' averaging for the approximate analytical solution in the uniform-temperature case. A Fortran program provisionally called CREEP3 was accordingly written, based on the following analysis.

Suppose the solution is calculated up to and including time t: call this known solution $\hat{Z}$. Let Z denote the wanted solution at the next time step $(t + \Delta t)$. Proceeding as in the well-known Crank–Nicolson method for the diffusion equation, eqn (27) is first approximated as

$$(t^*/\Delta t)[Z'' + Z(1 + c) - \bar{Z} - \hat{Z}'' - \hat{Z}(1 + c) + \bar{\hat{Z}}]$$
$$= -(B/2B_R)(F(Z) + F(\hat{Z}))$$ (37)

bars denoting average with respect to θ. This equation is now solved by discretising Z with respect to θ into $(N + 1)$ mesh-point values Z_0, $Z_1, \ldots, Z_N$ at angular intervals $\Delta\theta = \pi/N$ (this is for the symmetric case, $Z(\theta) \equiv Z(-\theta)$; the unsymmetric case is considered later). The system to be solved is then

$$Z_{j+1} - G_j(Z_j) + Z_{j-1} - H(Z) = C_j \quad (j = 1,2,\ldots,n-1)$$ (38)

where

$$H(Z) = (\Delta\theta^2/2N)(Z_0 + 2Z_1 + \cdots + 2Z_{N-1} + Z_N)$$ (39)

is the finite-difference approximation to $\bar{Z}\Delta\theta^2$, and where

$$C_j = \hat{Z}_{j+1} - \hat{G}_j(\hat{Z}_j) + \hat{Z}_{j-1} - H(\hat{Z}) \tag{40}$$

$$G_j(Z_j) = (2 - (1 + c)\Delta\theta^2)Z_j - (B_j\Delta t\Delta\theta^2/2B_R t^*)F(Z_j) \tag{41}$$

$$\hat{G}_j(\hat{Z}_j) = (2 - (1 + c)\Delta\theta^2)\hat{Z}_j + (B_j\Delta t\Delta\theta^2/2B_R t^*)F(\hat{Z}_j) \tag{42}$$

B_j is the value of B at the node point θ_j. θ_0 is the apex ($\theta = 0$).

These are $(N - 1)$ equations in $(N + 1)$ unknowns $Z_0, \ldots, Z_N$. The two remaining equations needed come from the symmetric boundary conditions and correspond to setting $j = 0$, $j = N$ in eqn (38), where $Z_{-1} = Z_1$ and $Z_{N+1} = Z_{N-1}$.

This system of non-linear equations was solved iteratively by Newton's method, using as the initial guess the old solution $\hat{Z}$. The iteration was stopped when the root mean square error in the solution (as calculated from the residuals (RHS − LHS) in eqn 38) is negligible (less than one part in 10^{10}). Convergence was normally obtained in two or three loops, except towards the end of the tube's lifetime, when bending moments were getting large. It was arranged for the program to stop when either the RMS value of Z exceeded 7 or convergence of the Newton process was not achieved in ten iterations.

Results
Uniform all-round temperature case ($T = T_R$)

The numerical values used have already been given. The main ones, repeated here for convenience, are

$a = 216$ mm		tube radius
$h = 2\cdot25$ mm		tube thickness
$c = 0\cdot2181$		dimensionless pressure parameter
$t^* = 27\cdot7$ h		time constant
$W_{oo} = 3\cdot874$ mm		initial out-of-roundness before the pressure is applied
$m = 2$		wavenumber
$T_R = 793$K		temperature at $\theta = 0$

When the temperature is uniform (i.e. $T = T_R$), the numerical and analytical solutions can be compared. The inward displacement W at the pole ($\theta = 0$) was calculated (1) allowing for the elastic effects ($c = 0\cdot2181$) and (2) ignoring the elastic effects ($c = 0$). Values obtained by the finite-difference program CREEP3 (with a time step of $\frac{1}{4}$ h and a space step of 2·5 deg) were also calculated for comparison. The results are given in Table 1.

The values of t_C are calculated to the nearest quarter of an hour by CREEP3 and correspond to the time at which the program predicted that excessive deformation has been reached, and stopped running. With $c = 0.2181$ it takes 32·6 h according to the collocation method, and 30·3 h according to the Fourier method, to reach the deformation of 16·7 mm attained after 30 h according to CREEP3.

It will be clear from the foregoing that whereas the times required to reach a given deformation calculated by the Fourier method and by CREEP3 are quite close (within 1% up to 30 h in the above example), the collocation method appears to overestimate such times by about 8–9%. CREEP3 tends to overestimate displacements which occur in a given time (as was seen by comparing runs with successively finer meshes), while the Fourier method slightly underestimates them. This is because the neglected sixth, tenth . . . order harmonics excited by the $m = 2$ mode are destabilising. Thus, the Fourier method gives both more conservative and more accurate answers than the collocation method, so is to be preferred.

Results in the Fourier and collocation columns were evaluated directly by HP-65 electronic calculator, using eqns (32)–(36).

Non-uniform temperature

The temperature profile was assumed to vary linearly with height ($\theta = 0$ being the highest and hottest place) in the form

$$T = \frac{1}{2}[(1 + \cos \theta)T_R + (1 - \cos \theta)T_{min}] \tag{43}$$

The majority of the initial out-of-roundness profiles investigated were

$$w = W_{oo} \cos 2\theta \tag{44a}$$
$$w = W_{oo} \sin 2\theta \tag{44b}$$

for $W_{oo} = 3.874$ mm (0·1525 in). The antisymmetric (sine) profile was investigated for completeness; it was feared that the symmetric (cosine) profile would not necessarily provide the worst case. There is a partial analogy with the elastic buckling of a pin-ended shallow arch (the hotter material at the crown is restrained by cooler material at the edges) and it is well known that such an arch buckles elastically into an antisymmetric S wave. This fear proved to be unfounded; the profiles are very similar, with the symmetric one marginally worse (see Fig. 2).

TABLE 1

Values of W calculated by different methods (uniform all-round temperature case)

t(h)	$c = 0$ W(mm)			$c = 0.2181$ W(mm)		
	Collocation	Fourier	Numerical	Collocation	Fourier	Numerical
0−	3·87	3·87	3·87	3·87	3·87	3·87
0+	3·87	3·87	3·87	4·18	4·18	4·18
5	4·25	4·27	4·27	4·65	4·68	4·68
10	4·70	4·75	4·76	5·23	5·31	5·32
20	5·95	6·13	6·18	7·04	7·37	7·44
30	8·28	9·05	9·21	12·0	15·8	16·7
35	10·8	13·2	13·8	—	—	—
40	19·7	—	—			
t_{C}(h)	41·2	38·2	38·0[a]	34·4	31·8	31·5[a]

[a]Program stopped.

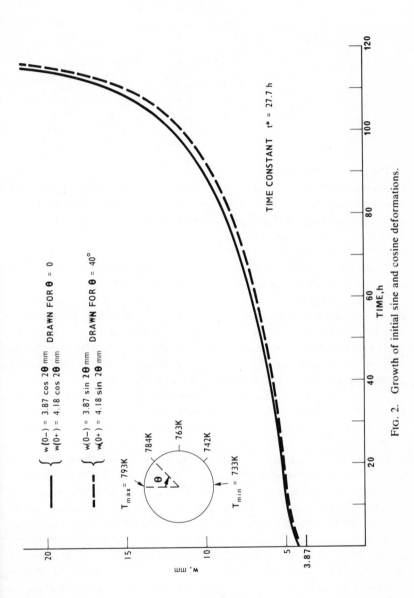

FIG. 2. Growth of initial sine and cosine deformations.

TABLE 2
Variation of collapse time with T_{min}

T_{max} (K)	793	793	793	793	793
T_{min} (K)	793	733	673	613	553
$\Delta\theta$ (deg)	2·5	5	5	5	5
Δt (h)	0·25	2·5	2·5	6·25	12·5
t_D (h)	21	73	191	280	414
t_C (h)	31·5	122·5	335	575	800
t_C/t_{CF}	0·99	3·85	10·5	18·1	25·2
$B_R/\bar{B}$	1	2·76	4·18	5·22	6·07

Table 2 shows the effect of varying T_{min} while holding T_{max} (i.e. T_R) constant (at 793K) for the cosine profile (eqn (44a)). Δt, $\Delta\theta$ are the time and angular steps used in the calculations. Also shown are the ratios t_C/t_{CF} (where t_{CF} is the critical time as calculated by the Fourier-averaging method for uniform temperature T; eqn 32b) and $B_R/\bar{B}$. $\bar{B}$ is the average value of B around the circumference and B_R is the value attained at the crown $\theta = 0$, the hottest point. t_D is the time taken to double the original inward deflection at the crown ($\theta = 0$).

The last entry of Table 2 ($T_{max} = 793K$, $T_{min} = 553K$) corresponds to the most unfavourable case in the original reheater sheath tube problem. The intermediate cases are included to show in detail how the trend develops. The lifetime of the tube is over 400 h before the original worst case out-of-roundness can double, and about 800 h until complete collapse; so a temporary loss of coolant is not a serious problem in this particular context. Figures 2–4 show typical results for an intermediate case. It is assumed that the centroid of the tube does not move (in order to eliminate the otherwise indeterminate rigid body motion).

To allow comparisons with independent calculations, Table 3 presents further results with $T_{min} = 553K$. ΔD_c is the creep change in diameter along the symmetry axis. The initial distortion, including elastic effects, is $\Delta D = 0·329$ in. (ΔD and ΔD_c are quoted in inches because these were the units used in the calculations.) At early times Z decreases at the crown (the hottest point), indicating creep relaxation. After 200 h the deformation gradually builds up, doubling at approximately 400 h with final collapse at about 800 h. The calculations in Table 3 were carried out with an angular interval of 10 deg and a time interval of 1 h (to examine the early history).

It is useful to have a conservative 'rule of thumb' for estimating t_C

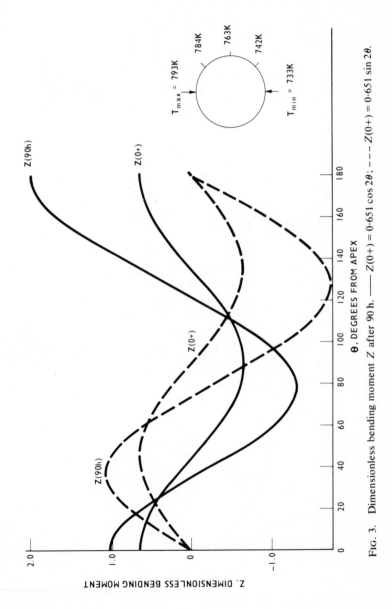

FIG. 3. Dimensionless bending moment Z after 90 h. ——— $Z(0+) = 0.651 \cos 2\theta$; – – – $Z(0+) = 0.651 \sin 2\theta$.

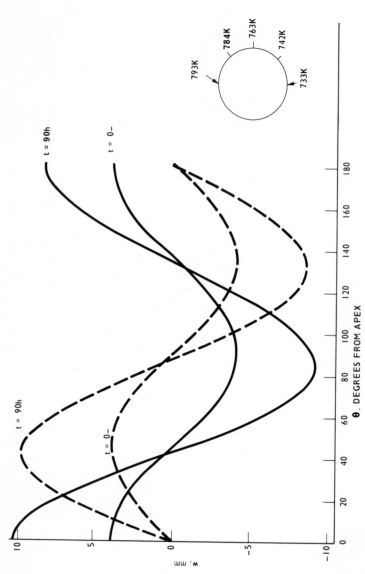

FIG. 4.   Deformations after 90 h. —— $w(0-) = 3\cdot87\cos 2\theta$ mm; --- $3\cdot87\sin 2\theta$ mm.

TABLE 3
Additional results for $T_{max} = 793K$, $T_{min} = 553K$

t (h)	0	20	40	60	80	100	200
Z (crown)	0·651	0·259	0·210	0·207	0·214	0·222	0·272
Z (base)	0·651	0·822	0·877	0·917	0·956	0·996	1·216
$\bar{Z}$	0	−·141	−·152	−·145	−·135	−·124	−·061
ΔD_c (in)	0	0·015	0·022	0·028	0·035	0·041	0·172

independently of the computer. From Table 2, t_C is almost always larger than $t_{CF}B_R/\bar{B}$. (t_{CF} is calculated analytically from eqn 29 with $P = 3/2$, $Q = 1/8$, and $\bar{B}$ is the value of B averaged around the circumference.) So $t_{CF}B_R/\bar{B}$ provides a conservative estimate of the lifetime (except for uniform temperature when it is slightly high). The pessimism of this 'rule of thumb' worsens as T_{min} decreases; e.g. when $T_{min} = 553K$, the rule gives $31·8 \times 6·07 = 193$ h as the collapse lifetime as compared with over 800 h by computer.

In conclusion, two general points may be made about the results. First, they are very sensitive to temperature; an all-round temperature rise of 14°C will roughly double B (from eqn 7) and so halve the lifetimes. Second, it is the bending stresses which are responsible for the high creep rates. The average membrane pressure is qa/h (i.e. $40·9$ MN m^{-2} numerically). Taking the worst case, of an all-round temperature T_{max} equal to 793K, this uniform compression produces a uniform creep strain per 1000 h of:

$$(40·9/15·44)^5 \times (1/8\ 592\ 000) \times 1000 = 1·5\% \tag{45}$$

and yet the lifetime of the whole tube is only 31 h because of the growth of the assumed initial out-of-roundness.

VARIATIONAL FORMULATION

When axial variations have to be allowed for, the method of setting up equilibrium equations directly, as used so far, is not practicable, because the problem becomes three-dimensional and it is difficult to decide which terms are small. Instead a two-part variational principle will be used.

Equilibrium Equations

As in the axially uniform long-tube case, geometry changes will be allowed for, but only to the same order as in the linearised theory of elastic buckling. This can conveniently be done using the 'reduced deflection method' [6] to calculate the effective elastic strain energy power rate $U_{\text{eff}}(\dot{\mathbf{w}})$, allowing for the destiffening effect of the circumferential hoop thrust and for the rotation of the line of action of the pressure as the tube surface deforms. ($U_{\text{eff}}(\dot{\mathbf{w}})$ is a power rate, with units energy time^{-2}, because $\dot{\mathbf{w}}$ is a velocity.)

Let $\mathbf{u}$ denote displacement from the reference (perfectly circular) shape. Split $\mathbf{u}$ up into two parts, a 'primary' part $\mathbf{u}_0$ representing the deformation of the perfectly circular shell (which is a pure compression) and a remainder, the 'reduced deflection', $\mathbf{w}$. $\mathbf{u}_0$ is of negligible magnitude, but it corresponds to a hoop thrust $N_\theta = qa$ (end effects being neglected). $\mathbf{w}$ is written

$$\mathbf{w} = u\mathbf{e}_1 + v\mathbf{e}_2 + w\mathbf{e}_3 \qquad (46)$$

where $\mathbf{e}_1$, $\mathbf{e}_2$ and $\mathbf{e}_3$ are unit vectors in the coordinate directions θ, z and inward normal (Fig. 1). The additional stresses and strains set up by $\mathbf{w}$ will be called *auxiliary* ones.

The variational statement of the equilibrium equations in terms of $\dot{\mathbf{w}}$ is

$$\delta U_{\text{eff}}(\dot{\mathbf{w}}) = \delta C(\dot{\mathbf{w}}) + \int_A \dot{q}\delta\dot{w} \, dA \qquad (47)$$

for all $\delta\dot{\mathbf{w}}$ meeting the boundary conditions for $\mathbf{w}$. In eqn (47) the pressure is constant, so $\dot{q}$ and the last integral vanish; C is the creep power rate. In standard matrix notation

$$C(\dot{\mathbf{w}}) = \int_V \boldsymbol{\epsilon}_c^T \mathbf{D}\boldsymbol{\epsilon}(\dot{\mathbf{w}}) \, dV \qquad (48)$$

where $\boldsymbol{\epsilon}(\dot{\mathbf{w}})$ is the generalised strain rate calculated from $\dot{\mathbf{w}}$. V (eqn 48) and A (eqn 47) are the reference volume and area of the 'perfect' tube. U_{eff} is defined by

$$U_{\text{eff}}(\dot{\mathbf{w}}) = U(\dot{\mathbf{w}}) + P_0(\dot{\mathbf{w}}) - W''(\dot{\mathbf{w}}) \qquad (49)$$

and is a homogeneous quadratic function of $\dot{\mathbf{w}}$. $P_0(\dot{\mathbf{w}})$ is the 'pre-stress power rate' or 'initial-stress power rate' in the notation of Refs [6] and [7], respectively, and $W''(\dot{\mathbf{w}})$ is the load-change power rate due to rotation of the line of the acting pressure, i.e. is the coefficient of t^2 in the expansion of $W(\mathbf{u}_0 + \dot{\mathbf{w}}t)$ as a power series in t. $W(\mathbf{u})$ is the work done by the pressure expressed as a function of $\mathbf{u}$. The combination $P_0 - W''$ is often called the 'geometric stiffness' term in a finite-element context [7]. Equation (47) is the generalisation to include the linearised-buckling effect of the standard creep virtual work principle given by Zienkiewicz and Cormeau [8].

Compatibility Equations

The variational statement for the compatibility equation is, in matrix notation,

$$\int_V (\dot{\boldsymbol{\sigma}} - \mathbf{D}\dot{\boldsymbol{\epsilon}}_e)^T \mathbf{D}^{-1} \boldsymbol{\delta}\dot{\boldsymbol{\sigma}} \, dV \equiv 0 \quad \text{for all } \boldsymbol{\delta}\dot{\boldsymbol{\sigma}} \tag{50}$$

where $\mathbf{D}$ is the material elastic rigidity matrix (or generalised elastic stiffness). This corresponds to the standard compatibility law $\dot{\boldsymbol{\sigma}} = \mathbf{D}\dot{\boldsymbol{\epsilon}}_e$. It is important to include the $\mathbf{D}^{-1}$ term since otherwise non-homogeneous (or even dimensionally inconsistent) averaging may result. Dimensional inconsistency occurs if $\boldsymbol{\sigma}$ is a generalised stress with components of different dimensions (e.g. bending moments and membrane thrusts) and the simple scalar produce $\boldsymbol{\sigma} \cdot \boldsymbol{\delta}\boldsymbol{\sigma}$ is formed. The first term in eqn (50) corresponds to the variation in the elastic complementary energy power rate (units, energy s^{-2}).

Equation (50) neglects geometry changes. This is consistent with the normal approximations of linearised-buckling theory.

Equations (47) and (50) provide a two-part variational principle applicable to the creep collapse of a general pressurised shell. Independent approximations can be made to the stress and strain distributions. The principle is of the same order of accuracy as that of Sanders, McComb and Schlechte [9], but with the advantage that each part has a direct physical interpretation.

Specialisation to the Cylinder

The following additional physical assumptions and approximations are now made.

(1) The most important auxiliary bending moment is the circumferential one, and the most important auxiliary membrane stress resultant is the axial one. (Some auxiliary membrane stress is inevitable; it is impossible to deform a cylindrical shell fixed at both ends or even one end without some mid-surface stretching.)

(2) As a corollary of (1) above, only the circumferential auxiliary

bending moment M and the axial auxiliary membrane stress are allowed for when forming the auxiliary bending and membrane energies. Consistent with this, the auxiliary circumferential and shear membrane strains $\epsilon_\theta, \epsilon_z$ are assumed to vanish. This implies that

$$w = u', \qquad v' = -u^* \tag{51}$$

where the asterisk denotes $a\partial/\partial z$.

(3)   As a further corollary of (1) above, all creep strains are neglected except for the circumferential auxiliary bending strain κ. This is evaluated using the sandwich approximation.

From approximation (3) above, the basic bending moment versus creep curvature change relationship is unaltered, so leading back to eqn (24), since, as before,

$$\dot{M} = D\dot{\kappa}_e, \quad \dot{\kappa}_e = \dot{\kappa} - \dot{\kappa}_c, \quad \dot{\kappa}_c = (-h_s/2a^2)BF(Z)/t^*B_R \tag{52}$$

Equation (50) has not really been needed in this derivation, but becomes useful later in averaging processes.

From eqns (47) and (49) it can be shown that

$$U_{\text{eff}}(\dot{\mathbf{w}}) = \frac{D}{2a^3} \int_A \{(\dot{w} + \dot{w}'')^2 + 12a^2 v^{*2}/h^2\}\, d\theta\, dz$$
$$+ \frac{1}{2}q \int_A (\dot{u}'^2 - \dot{u}''^2)\, d\theta\, dz \tag{53}$$

under the physical assumptions made. The integration is over the tube's area A, from $z = 0$ to L and from $\theta = 0$ to 2π. The first integral corresponds to the ordinary strain energy power rate of linear elastic theory. The second integral (which is negative) shows the effect of the 'primary' hoop thrust (qa) in destiffening the structure, and also includes the effect of the rotation of the line of action of q. A third integral representing the destiffening effect of the axial primary thrust $(N_z = \frac{1}{2}N_\theta)$ has not been included, because it turns out to be of smaller order of magnitude. As mentioned earlier, N_z arises because axial creep is suppressed.

The creep power rate reduces simply to

$$C(\dot{\mathbf{w}}) = \int_A \kappa_c D\dot{\kappa}\, dA$$
$$= \frac{h_s D}{2a^3} \int_A (\dot{w} + \dot{w}'')(BF(Z)/t^*B_R)\, dA \tag{54}$$

Since it is linear in the velocity field, $\delta C(\dot{u}) = C(\delta\dot{u})$.

On writing $\dot{w} = f''$, $\dot{u} = f'$, $\dot{v} = -f^*$ (using eqn 51 and substituting into eqn 54), the following equation for f is obtained by the usual calculus-of-variations argument, after suitable integration by parts:

$$(d^2 + d^4)^2 f + c(d^4 + d^6)f + (12a^6/h^2)\frac{\partial^4 f}{\partial z^4} = -(d^2 + d^4)h_s BF/2t^* B_R$$

$$(55)$$

(where d denotes $\partial/\partial\theta$). Or, using eqn (24), the following equation is finally obtained:

$$\frac{d}{dt}\{(\partial^4/\partial\theta^4 + \partial^6/\partial\theta^6)(Y - Z) + (12a^6/ch^2)\partial^4 Y/\partial Z^4\} = 0 \qquad (56)$$

A particular case of eqn (56) is when there is no axial variation. It can be integrated four times with respect to θ, giving rise to a cubic in θ on the right-hand side (containing the four arbitrary constants of integration), but because the displacement must be single-valued, this cubic collapses to a single (time-varying) constant, i.e. eqn (27) results. Thus, the variational method gives the same final equation as the direct method.

The Finite-Length Tube with Uniform Circumferential Temperature

An important special case is that of uniform circumferential temperature, so that B varies in the z direction only. If the initial distortion is proportional to $\cos m\theta$, then the subsequent distortion can be taken to be so, too, to sufficient accuracy. Substituting $Y = Y_1 \cos m\theta$, $Z = Z_1 \cos m\theta$ into eqn (56), there results

$$\frac{d}{dt}\{m^4(m^2 - 1)(Y_1 - Z_1) - (12a^2/ch^2)a^4 d^4 Y_1/dz^4\} = 0 \qquad (57)$$

while the averaged equations (eqns 24) again give

$$t^* \, d/dt(m^2 Y_1 - Y_1 - cZ_1) = B/B_R \int_0^{2\pi} F(Z_1 \cos m\theta) \cos m\theta \, d\theta/\pi$$

$$(58)$$

$$= (B/B_R)(Z_1 + 3Z_1^3/2 + Z_1^5/8) \text{ when } n = 5 \qquad (59)$$

This averaging procedure is equivalent to use of eqn (50) directly, and so is the 'Fourier' method of averaging mentioned previously. Equation (50) simplifies to

$$\int_0^L \int_0^{2\pi} (\dot{M} - D\dot{\kappa}_e)\delta\dot{M} \, dz \, d\theta \tag{60}$$

and $Z = Z_1 \cos\theta$ is substituted in both the $\dot{M}$ and $\delta\dot{M}$ parts.

These equations can be solved by finite differences using the methods used for the case of non-uniform circumferential temperature. As the aspect ratio $L/2a$ decreases, the wavenumber m of the most dangerous mode increases.

Analytical Solution for a Closed Tube

For a closed tube at uniform temperature it is possible to construct an approximate solution analytically. The pressure on the closed ends produces a uniform axial membrane thrust of $qa/2$ per unit circumferential length, and, as before, this is half the hoop thrust N_θ. Axial creep remains suppressed, and the foregoing analysis applies. The effect of the circular ends is to make $u = w = 0$. They do not constrain v, because axial movement can be accommodated by pure bending of the ends. The axial bending moment vanishes at the ends, so d^2w/dz^2 also vanishes. The end effects and the elastic destiffening of the end thrust will again be neglected.

Now consider the approximate solution

$$Y = V \cos m\theta \sin kz, \quad Z = X \cos m\theta \sin kz \tag{61}$$

which satisfies the required boundary conditions for u, w. k is of form $l\pi/L$, where $l = 1,2,3,\ldots$. The case $k = 0$ corresponds to the long-tube case analysed earlier. Initially, before the pressure is applied, $X = 0$ and $V = V_{oo}$. Equation (57) becomes

$$\frac{d}{dt}\{cm^4(m^2 - 1)(V - X) - (12a^2/h^2)a^4k^4V\} = 0 \tag{62}$$

or, integrating, since $c = 0$ before the pressure was applied,

$$cV - cX = d(V - V_{oo}) \tag{63}$$

where

$$d = (12a^2/h^2)a^4k^4/m^4(m^2 - 1) \tag{64}$$

Averaging eqn (59) gives

$$t^* \, d/dt(m^2 V - V - cX) = (B/B_R) \int_0^L \{X \sin^2 kz + (3X^3 \sin^4 kz)/2$$

$$+ (X^5 \sin^6 kz)/8\} \, dz / \int_0^L \sin^2 kz \, dz$$

$$= (B/B_R)(X + 9X^3/8 + 5X^5/64) \text{ if } k > 0$$

$$(65)$$

where B is constant since the temperature is uniform.

In particular, the initial solution just after the pressure is applied comes by considering the limit of no creep, $B = 0$; i.e.

$$cX_0 = (m^2 - 1)(V_o - V_{oo}) \tag{66}$$

Combining with eqn (63) gives

$$V_o = \left(\frac{m^2 - 1 + d}{m^2 - 1 - c + d} \right) V_{oo} \tag{67}$$

$$X_0 = \left(\frac{m^2 - 1}{m^2 - 1 - c + d} \right) V_{oo} \tag{68}$$

Eliminating V in eqn (65) gives, putting $B = B_R$,

$$t^{***} \, dX/dt = X + 9X^3/8 + 5X^5/64 \tag{69}$$

where the new time constant is

$$t^{***} = \{(m^2 - 1)c/(c - d) - c\}t^* \tag{70}$$

with initially (at time $t = 0+$) $X = X_0$. In particular, for $d = 0$, $k = 0$, $t^{***} = t^{**}$ and the analysis reduces formally to the long-tube case. If $t^{***} < 0$, i.e. if $d > c$, the model predicts *stress relaxation*, i.e. X decreases; while if $d = c$ exactly, t^{***} is infinite, i.e. X remains equal to X_0.

Equation (70) is of the same form as eqn (29) and is solved in the same way. Consider, for example, a moderately long shell with $L = 10a$, i.e. aspect ratio (length/diameter) equal to 5. The most dangerous axial k is π/L, since this gives the smallest t^{***}. Then, with the numerical values assumed earlier, $d = 11 \cdot 03/(m^6 - m^4)$, so that $d = 0 \cdot 230$ with $m = 2$, and $d = 0 \cdot 0170$ with $m = 3$. So, since $c = 0 \cdot 2181$, the $m = 2$ mode relaxes, i.e. the most dangerous mode is $m = 3$. With $V_{oo} = 0 \cdot 6037$ (corresponding to $W_{oo} = 0 \cdot 1525$ in), $V_o = 0 \cdot 6206$ and $X_0 = 0 \cdot 6193$. This gives $J(X_0) = 0 \cdot 5347$ (from eqn 29, with $P = 9/8$, $Q = 5/64$). $t^{***} = 8 \cdot 459 \, t^* = 234 \, h$ and so the collapse time t_C

is 125 h from eqn (35). This compares with the previously calculated collapse time of 32 h for a very long shell with the same maximum out-of-roundness amplitude. So the end support results in a large increase in lifetime even for the apparently large aspect ratio of 5.

A common practical form of tube support is to have built-in ends, with $u = v = w = dw/dz = 0$. The same basic analysis applies, but is algebraically more complex. The extra end stiffness increases the collapse times. Alternatively, the temperature may be varying along the tube so that the cooler ends provide the support. This case is probably best solved by finite-difference methods.

A partial check on the validity of the approximations used is obtained by comparison with elastic linearised-buckling theory. Equation (67) predicts that elastic buckling for the mode considered occurs when $c = c_b = m^2 - 1 + d$. A comparison with the classical buckling solution [10] (again neglecting the end thrust effects) shows fairly good agreement for aspect ratios $L/2a$ greater than unity; e.g. for $a/h = 10$ eqn (64) gives $c_b = 19\cdot3,\ 8\cdot70,\ 4\cdot88$ and $3\cdot59$ as compared with Flügge's values $20\cdot6,\ 9\cdot84,\ 5\cdot19$ and $3\cdot84$ for aspect ratios of 1, 2, 3 and 4 and with $\nu = 0$. The corresponding wavenumbers are 3, 3, 2 and 2. The classical result is underestimated because of axial and shear bending energies omitted from U_{eff}.

DISCUSSION

The sandwich approximation is based on equalising the elastic bending stiffness of tube wall and sandwich model. If the creep rates dominate the elastic strain rates, it should be more accurate to equalise instead the creep rate bending stiffnesses†. For large n the situation is analogous to perfect plasticity and so h_s equals $h/2$ as compared with $h/3^{1/2}$ if the elastic stiffnesses are made equal. This implies that t^* is reduced by a quarter. More generally, t^* can be replaced by $r_c t^*$ in the analysis, where

$$r_c = 3(n/2n + 1)^{(2n/n+1)} = 0\cdot81 \text{ for } n = 5 \tag{71}$$

and lies between 1 (for $n = 1$ when the sandwich approximation is exact and the two stiffnesses coincide) and 3/4 in the large-n limit. This result comes from a straightforward calculation of the bending

†I am grateful to Mr W. R. Hodgkins for pointing this out.

stiffness in creep and assumes that the strain rate varies linearly across the tube wall (otherwise shear strain rates would arise comparable in magnitude with the in-plane direct strain rates, a contradiction). The effect of replacing t^* by $r_c t^*$ is to reduce the time taken to reach any given deformation by the factor r_c, i.e. to accelerate the collapse.

The best stiffness to use may change as the calculation proceeds and creep effects become more important. The 'best' reduction factor r to use may therefore decrease towards its limiting factor r_c as the calculation proceeds. From eqn (71) the *maximum* reduction in lifetime due to this effect should here be 19%. However, it is impossible to be completely definite because of the effect of the all-round hoop thrust which changes the plastic neutral axis and because of the thin-shell approximations used.

The long-tube problem described has been analysed independently by Harper [3], using the finite-element method. His interim calculations allowed for geometry changes as the shell deformed, but ignored the geometric stiffness terms due to the hoop thrust. These terms speed up the collapse, but the effect is relatively small for the present problem (up to 7% for the uniform temperature case). The calculations were stopped periodically so as to update the geometry and restarted. For the non-uniform temperature case ($T_{min} = 280°C$) his predicted collapse time is about half the present value. Table 4 gives his results for the creep ΔD_c along the axis of symmetry. Over 50 restarts were used in the finite-element calculations, which used a 12×2 mesh of eight-noded isoparametric elements with quadratically varying internal displacements. His initial ΔD (including elastic effects) was 0.326 in. For the uniform temperature case his calculations gave a lifetime of about 28 h (using 40 restarts) as compared with about 36 h using 23 restarts and to 31 h according to the present calculations.

Both calculations assumed plane strain (if axial creep is allowed, all collapse times are reduced by a factor $(3/4)^3 = 0.42$). The finite-difference program had been rigorously checked to verify that it did

TABLE 4
Finite-element results for long tube ($T_{max} = 520°C$, $T_{min} = 280°C$)

t (h)	0	20	40	60	80	100	200
ΔD_c (in)	0	0·015	0·033	0·055	0·080	0·107	0·172

indeed solve the governing eqn (27), by back-substitution of results into eqn (37) (with $d^2Z/d\theta^2$ discretised in the normal way). The difference between the two calculations is therefore presumably due to the different physical approximations used. The finite-element calculation can represent the actual stress variation across the tube wall more accurately than can the sandwich model. The 'normals remain normal' assumption used by the thin-shell sandwich model constrains and so stiffens the shell (see, e.g., ref. [11]). This is in addition to the extra stiffness introduced by equalising the elastic stiffnesses of tube wall and sandwich. Finally, the partial neglect of geometry changes may also contribute to the discrepancy, although there is good agreement between linerarised-buckling theory (as used here) and experiment for elastic buckling of tubes under lateral pressure [3].

It would be interesting to rework the uniform temperature and non-uniform temperature calculations for long tubes by a full finite-element analysis that includes the geometric stiffness terms, and to see whether this confirms the stress relaxation at the crown predicted by the sandwich model in the non-uniform temperature problem. For this problem the finite-element calculations apparently accumulated excessive strain at the crown because the internal stresses were not transferred across the restarts, thus underestimating lifetimes.

Some experimental confirmation of the sandwich theory is available [2], but unfortunately the initial irregularity in Bargmann's experiments was not measured accurately, so the confirmation is essentially qualitative. Finally, to put the accuracy required for engineering purposes into perspective, creep test data normally shows a very large scatter. A range of creep lifetimes of around ten in laboratory tests on nominally identical specimens is not uncommon.

CONCLUSIONS

(1) Hoff's 'sandwich approximation' method for analysing the creep buckling of a long tube under uniform all-round pressure and temperature has been generalised to cover varying temperatures and moderately long tubes.

(2) An improved form of analytical solution is given for the classical (uniform temperature) version of the problem. The improvement comes from using Fourier rather than collocation

averaging methods. An analytical solution is also given for the buckling of a closed tube under all-round pressure and having aspect ratio (length/diameter) of at least unity.

(3) End support greatly increases creep lifetimes even for apparently large aspect ratios. For the specific reheater tube problem analysed here, an aspect ratio of 5 increases the predicted lifetime fourfold.

(4) The analysis is based on a two-part variational principle which represents separately the equilibrium equations and compatibility equations. The advantage of the principle is that independent approximations to the displacements and bending moment distributions may be fed in, each involving a few unknown parameters. Equations for these unknown parameters then emerge. For a long tube, the equations derived by the variational method agree with those derived by direct methods.

(5) The analysis takes account of geometry changes and of the destabilising effect of the initial hoop thrust, but only to the same extent as classical elastic linearised-buckling theory does.

(6) Unpublished calculations by Harper [3] using the finite-element method indicate that the sandwich model systematically overestimates creep lifetimes. The error seems to be caused by a combination of the sandwich approximation, the thin-shell approximations, and the partial neglect of geometry changes. When stress relaxation occurs the finite element calculations probably underestimate lifetimes.

(7) The sandwich approximation is based on equalising the elastic bending stiffness of tube wall and sandwich. If instead the creep rate bending stiffnesses are equalised, the bending stiffness and, hence, the predicted collapse time is reduced by 25% in the perfectly plastic limit, and by 19% when the secondary creep index is 5.

ACKNOWLEDGEMENTS

This work was carried out at the Central Electricity Research Laboratories and is published by permission of the Central Electricity Generating Board. The author thanks J. V. Parker of NPC (Risley) and his own colleagues, Y. C. Wong, D. A. Morse, S. Schofield and P. G. Harper, for useful discussions during the course of this work.

REFERENCES

1. HOFF, N. J., JAHSMAN, W. F. and NACHBAR, W., A study of creep collapse of a long circular cylindrical shell under uniform external pressure, *J. Aero/Space Sci.*, **26**, 1959, 663–669.
2. BARGMANN, H., The lifetime of a long cylindrical shell under external pressure at elevated temperature, *Nucl. Eng. Design*, **22**, 1972, 51–62.
3. HARPER, P. G., Private communication.
4. GERARD, G., *Introduction to Structural Stability Theory*, McGraw-Hill, 1962.
5. TIMOSHENKO, S. P. *Theory of Elastic Stability*, McGraw-Hill, 1936.
6. EWING, D. J. F., 'The reduced deflection method for calculating structural buckling loads', This volume, pp. 345–372.
7. ZIENKIEWICZ, O. C., *The Finite Element Method in Engineering Science*, McGraw-Hill, 1971.
8. ZIENKIEWICZ, O. C. and CORMEAU, I. C., Visco-plasticity, plasticity and creep in elastic solids—a unified numerical solution approach, *Int. J. Num. Meth. Eng.*, **8**, 1974, 821.
9. SANDERS, J. L., MCCOMB, H. G. and SCHLECHTE, F. R., 'A Variational Theorem for Creep with Applications to Plates and Columns', NACA Report 1342, 1958.
10. FLUGGE, W., *Stresses in Shells*, Springer–Verlag, 1973.
11. HINTON, E., OWEN, D. R. J. and SHANTARAM, D., Dynamic transient linear and nonlinear behaviour of thick and thin plates, *2nd I.M.A. Conf. on the Finite Element Method*, (MAFELAP 1975) (Ed. J. R. Whiteman), Academic Press.

9

Creep Analysis of Turbine Structures Using the Finite-Element Energy and Extended Energy Methods

C. PATTERSON

University of Sheffield

AND

D. HITCHINGS

Imperial College of Science and Technology

SUMMARY

A typical high-temperature turbine blade problem is analysed using the 'initial strain' technique for both the finite-element energy and extended energy methods. The basic analysis is performed using a steady state law but the effects of time-hardening are evaluated. A critical comparison of the results is given and the usefulness of the extended energy method for creep problems discussed.

As a validity check on the coding and on the use of the extended energy method in creep problems the stationary creep of a thick-walled tube is analysed and the results compared with the exact solution. Simple planar configurations are also examined.

INTRODUCTION

A general finite-element extended energy formulation for state law creep processes is presented, with explicit details for the plane stress problem. Coding for the energy and extended energy methods under plane stress conditions is validated using simple test configurations, principal among which is the thick uniform cylinder subjected to internal pressure under steady state creep conditions, for which an exact solution is available. As a typical design study, a turbine blade idealisation is then analysed using an incomplete mesh. It is found that, for a given mesh, the solutions computed using the energy and extended energy methods are usually in close mutual agreement; that, in any given problem, there is a maximum time-stepping increment before onset of instability; and that incompleteness in the mesh can give rise to cumulative errors.

In the energy method the best approximate solution among the test fields is that which minimises the potential energy functional [1, 2]. Similarly, in the extended energy method the approximate solution is the minimiser of an extended functional obtained by adding an appropriate term to the potential energy. Interest in the extended energy method arises because the assured quality of convergence is better than in the energy method and also because spatial derivatives of stress and strain quantities converge to their exact counterparts, while such convergence is not guaranteed in the energy method. The added term is determined to a scale factor which is arbitrary in an exact, but not in a numerical, analysis. Previous experience [3, 4] indicates that it should be chosen so that the energy and added terms, in the linear algebraic equations, have roughly the same magnitude; this is again borne out here.

EXTENDED ENERGY FOR CREEP PROCESSES

The general formal theory of the extended energy method as applied to creep processes is presented in this section. In the next, as an example of the method, the formal quantities are given explicitly for the case of plane stress.

Linear elasto-static problems require the simultaneous satisfaction of the appropriate displacement and traction boundary conditions and the three basic sets of relations [1, 2]:

(1) the strain–displacement relations

$$\boldsymbol{\epsilon} = \mathbf{H}\mathbf{u} \tag{1}$$

(2) the linear (Hookean) constitutive relations

$$\boldsymbol{\sigma} = \mathbf{D}\boldsymbol{\epsilon}_e \tag{2}$$

(3)   the equations of static equilibrium

$$\mathbf{N\sigma} = \mathbf{f} \tag{3}$$

Here $\mathbf{u}$, $\boldsymbol{\epsilon}$, $\boldsymbol{\sigma}$ and $\mathbf{f}$ are displacement, strain, stress and load vectors, respectively; $\boldsymbol{\epsilon}_e$ is the vector of elastic strains; $\mathbf{D}$ is the familiar Hookean matrix; and $\mathbf{H}$ and $\mathbf{N}$ are appropriate matrix differential operators of the first order. Where initial strains $\boldsymbol{\epsilon}_t$ and $\boldsymbol{\epsilon}_c$, arising from thermal expansion and creep, respectively, are present, the total strain assumes the form

$$\boldsymbol{\epsilon} = \boldsymbol{\epsilon}_e + \boldsymbol{\epsilon}_t + \boldsymbol{\epsilon}_c \tag{4}$$

Equations (1) and (4) may be combined to give the field equations

$$\mathbf{ND}(\mathbf{Hu} - \boldsymbol{\epsilon}_t - \boldsymbol{\epsilon}_c) = \mathbf{f} \tag{5}$$

In a finite-element analysis a finite-element family of fields is introduced as a set of trial functions for the displacements

$$\mathbf{u} = \boldsymbol{\Phi\delta} \tag{6}$$

in which $\boldsymbol{\delta}$ is a vector of adjustable parameters, usually displacements at nodes, and $\boldsymbol{\Phi}$ describes the assumed unit functions. When the energy method is applied, the potential energy (i.e. the strain energy minus twice the external work done against the applied loads in establishing a trial displacement distribution) is defined over the set of trial fields. That displacement field which minimises the potential energy is the approximate solution to the problem. Where initial strains are present, as in eqn (4), the potential energy becomes

$$F(\mathbf{u}) = \frac{1}{2} \int \boldsymbol{\epsilon}_e \boldsymbol{\sigma} \, \mathrm{d}V - \int \mathbf{u} \cdot \mathbf{f} \, \mathrm{d}\bar{V} \tag{7}$$

and the stationary condition is

$$\mathbf{K\delta} = \boldsymbol{\phi}_o + \boldsymbol{\phi}_t + \boldsymbol{\phi}_c \tag{8}$$

where

$$\mathbf{K} = \int \boldsymbol{\Phi}^\mathrm{T} \mathbf{H}^\mathrm{T} \mathbf{DH}\boldsymbol{\Phi} \, \mathrm{d}V$$

$$\boldsymbol{\phi}_o = \int \boldsymbol{\Phi}^\mathrm{T} \mathbf{f} \, \mathrm{d}V \tag{9}$$

$$\boldsymbol{\phi}_{t/c} = \int \boldsymbol{\Phi}^\mathrm{T} \mathbf{H}^\mathrm{T} \mathbf{D}\boldsymbol{\epsilon}_{t/c} \, \mathrm{d}V$$

$\mathbf{K}$ is the familiar stiffness matrix, $\boldsymbol{\phi}_0$ the generalised load vector associated with the mechanical loads and $\boldsymbol{\phi}_t$ and $\boldsymbol{\phi}_c$ the corresponding generalised thermal and creep loads deriving from the initial strains. Under well-defined restrictive conditions the trial field which solves eqns (8) is a valid approximate solution to the given problem which converges in the sense of energy convergence to the exact solution [1, 2]. This means that the strain energy of the error displacement field tends to zero as the mesh is indefinitely refined.

In the extended energy method [3] the same general approach is employed, with the basic difference that the potential energy is replaced by the extended quantity $F + \lambda G$, in which

$$G(\mathbf{u}) = \int (\mathbf{ND}(\mathbf{Hu} - \boldsymbol{\epsilon}_t - \boldsymbol{\epsilon}_c) - \mathbf{f})^2 \mathrm{d}V \qquad (10)$$

The stationary condition now yields the modified equations for the approximate solution:

$$(\mathbf{K} + \lambda \mathbf{K}')\boldsymbol{\delta} = \boldsymbol{\phi}_0 + \lambda \boldsymbol{\phi}_0' + \boldsymbol{\phi}_t + \lambda \boldsymbol{\phi}_0' + \boldsymbol{\phi}_c + \lambda \boldsymbol{\phi}_c' \qquad (11)$$

in which λ is an arbitrary parameter and
where

$$\mathbf{A} = \mathbf{NDH\Phi} \qquad (12)$$

$$\mathbf{K}' = \int \mathbf{A}^{\mathrm{T}} \mathbf{A} \, \mathrm{d}V$$

$$\boldsymbol{\phi}_0' = \int \mathbf{A}^{\mathrm{T}} \mathbf{f} \, \mathrm{d}V \qquad (13)$$

$$\boldsymbol{\phi}_{t/c}' = \int \mathbf{A}^{\mathrm{T}} \mathbf{ND}\boldsymbol{\epsilon}_{t/c} \, \mathrm{d}V$$

The remaining quantities are defined as in the energy method. Provided that the elements used have at least quadratic polynomial freedoms, the solution to eqn (11) is a valid approximate solution to the problem, which has improved assured convergence qualities when compared with the energy method. In the energy method the stresses and strains converge in the mean to the exact values but the spatial derivatives of those quantities need not converge at all. In contrast, the extended energy method assures stronger local convergence qualities in stresses and strains and, in addition, their spatial derivatives converge. The increased 'overhead' in the extended method is that the quantities defined in eqns (13) must be evaluated. Furthermore, spatial derivatives of the initial strains must be available in order to compute the extended load vectors.

APPLICATION TO PLANE STRESS

In the course of the investigation the coding was written for axisymmetric, plane stress and plane strain conditions. Explicit forms of the required formal quantities are given here for the case of plane stress only; the others are readily evaluated.

For a material of Young's modulus E and Poisson's ratio ν, **H**, **D** and **N** assume the forms

$$\mathbf{H} = \begin{bmatrix} \partial/\partial x & 0 \\ 0 & \partial/\partial y \\ \partial/\partial y & \partial/\partial x \end{bmatrix}$$

$$\mathbf{D} = \frac{E}{1-\nu^2} \begin{bmatrix} 1 & \nu & 0 \\ \nu & 1 & 0 \\ 0 & 0 & (1-\nu)/2 \end{bmatrix} \tag{14}$$

$$\mathbf{N} = \begin{bmatrix} \partial/\partial x & 0 & \partial/\partial y \\ 0 & \partial/\partial y & \partial/\partial x \end{bmatrix}$$

If the coefficient of thermal expansion of the material is α and the temperature elevation above that corresponding to zero thermal strain is T, then the thermal strain vector is

$$\boldsymbol{\epsilon}_t = \begin{bmatrix} \alpha T \\ \alpha T \\ 0 \end{bmatrix} \tag{15}$$

If, in addition, the temperature field is defined in terms of nodal temperatures, $\mathbf{T}_e$ say, and a finite-element specification

$$T = \boldsymbol{\Phi}\mathbf{T}_e \tag{16}$$

then

$$\mathbf{ND}\boldsymbol{\epsilon}_T = \frac{E\alpha}{1-\nu} \begin{bmatrix} \boldsymbol{\Phi}_{,x} & \mathbf{T}_e \\ \boldsymbol{\Phi}_{,y} & \mathbf{T}_e \end{bmatrix} \tag{17}$$

(Note: , subscript x signifies partial differentiation with respect to x.)

Suppose, similarly, that the creep strains are given on an appropriate set of points with nodal vectors $\bar{\epsilon}_x$, $\bar{\epsilon}_y$ and $\bar{\gamma}_{xy}$. Then a finite-element representation is

$$\epsilon_{cx} = \mathbf{\Phi}\bar{\epsilon}_x; \quad \epsilon_{cy} = \mathbf{\Phi}\bar{\epsilon}_y; \quad \gamma_{xy} = \mathbf{\Phi}\bar{\gamma}_{xy} \tag{18}$$

and

$$\mathbf{ND}\epsilon_c = \frac{E}{1-\nu^2}\begin{bmatrix} \Phi_{,x}(\bar{\epsilon}_x + \nu\bar{\epsilon}_y) + (1-2\nu)/2\,\Phi_{,y}\bar{\gamma}_{xy} \\ \Phi_{,y}(\nu\bar{\epsilon}_x + \bar{\epsilon}_y) + (1-2\nu)/2\,\Phi_{,x}\bar{\gamma}_{xy} \end{bmatrix} \tag{19}$$

SOLUTION PROCEDURE

A 16-node isoparametric curved quadrilateral element, as depicted in Fig. 1, is used in which the physical x, y space is mapped into the mathematical ξ, η space so that the nodes lie at the intersection of the lines $\xi, \eta = (-1, -1/3, 1/3, 1)$. This is achieved using the 16 mapping functions constructed as follows:

(1) Define the four functions, $a_i(\xi)$, which vanish in turn on each

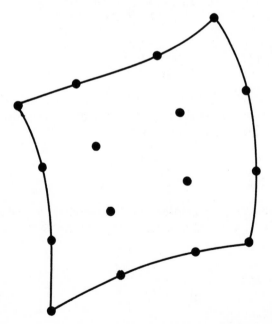

FIG. 1.   16-node element.

of the nodal lines $\xi = (-1, -1/3, 1/3, 1)$:

$$a_1 = (1 + \xi)$$
$$a_2 = (1 + 3\xi)$$
$$a_3 = (1 - 3\xi)$$
$$a_4 = (1 - \xi)$$
(20)

(2) Construct four functions, $\alpha_i(\xi)$, by combining any three of the given functions so that the resultant takes the unit value on the line for which the fourth function vanishes and is zero on the remaining lines. The required combinations are

$$\alpha_1 = -1/16 \; a_2 \; a_3 \; a_4$$
$$\alpha_2 = 9/16 \; a_3 \; a_4 \; a_1$$
$$\alpha_3 = 9/16 \; a_4 \; a_1 \; a_2$$
$$\alpha_4 = -1/16 \; a_1 \; a_2 \; a_3$$
(21)

(3) Similarly, define four functions $\beta_i(\eta)$.
(4) The 16 functions

$$\Phi_k(\xi, \eta) = \alpha_i(\xi)\beta_j(\eta) \quad (i, j = 1 \ldots 4; k = 1 \ldots 16) \quad (22)$$

obtained by taking all products of α_i with β_j in turn, take unit value at the kth node and zero value at all other nodes.

In consequence,

$$x = \Phi(\xi, \eta)\mathbf{x}_e; \quad y = \Phi(\xi, \eta)\mathbf{y}_e \quad (23)$$

in which $\mathbf{x}_e$ and $\mathbf{y}_e$ are the column vectors of the nodal coordinates of the element, define a unique mapping from the mathematical to the physical space. The inverse mapping is also unique (but not usually readily obtainable), provided that the physical element is not unreasonably distorted.

Adopting the usual approach [5], displacement fields u, v are defined in terms of their nodal values

$$u = \Phi(\xi, \eta)\mathbf{u}_e; \qquad v = \Phi(\xi, \eta)\mathbf{v}_e \quad (24)$$

Changing variables of integration from x, y to ξ, η permits the straightforward numerical evaluation of all the necessary integrals in the simple rectangular mathematical domain of an element. Six-point gaussian quadrature is employed throughout. The initial elastic problem is set up in the usual manner and solved using Cholesky factorisation, thereby providing the inverse stiffness matrix for use in the iteration procedure and the starting stress field. Throughout the initial integrations the kernel **A** (see eqn 12) is sent to file to be used later in determining the creep loads, $\boldsymbol{\phi}_c'$.

The non-linear creep problem of the extended energy method is solved using a conventional iterative procedure [6] as follows:

(1) Select a time interval Δt.
(2) Assume constant average stress $\boldsymbol{\sigma}$ over the interval and evaluate the creep strains at each gauss point.
(3) Using the kernel data on file and the creep strains, evaluate the creep loads $\boldsymbol{\phi}_c + \lambda \boldsymbol{\phi}_c'$.
(4) Obtain the corresponding displacement increments $\Delta\boldsymbol{\delta}$ using the expression

$$\Delta\boldsymbol{\delta} = (\mathbf{K} + \lambda\mathbf{K}')^{-1}(\boldsymbol{\phi}_c + \lambda\boldsymbol{\phi}_c') \tag{25}$$

and thence calculate the stress increments $\Delta\boldsymbol{\sigma}$.
(5) Take $\boldsymbol{\sigma} + \Delta\boldsymbol{\sigma}/2$ as a refined estimate of the average stress field throughout the time interval and iterate from step (2) above until the stress increment $\Delta\boldsymbol{\sigma}$ is substantially constant. Two cycles are usually adequate.
(6) The stress and displacement increments for the given time interval are $\Delta\boldsymbol{\sigma}$ and $\Delta\boldsymbol{\delta}$. Add these to the starting values for the interval to get those of the next interval and repeat from step (1) above as desired.

In the energy method the same formal procedure is adopted with λ set to zero. While the strain increments appropriate to any state creep law are easily evaluated [6], only steady state creep and time-hardening creep were used in the investigation. Furthermore, because under both these laws the creep strain rates vary rapidly as functions of stress, and therefore of position, and since each creep strain increment is a function of all 32 element degrees of freedom, the creep strain rates are evaluated at each of the 6×6 gauss points used in the integrations. In addition, the extended energy method requires the first spatial derivatives of the creep strain increments. These can be evaluated directly, but awkwardly, from the assumed creep law. Alternatively, they can be obtained approximately by taking the gauss point values as the nodal values of a subsidiary 6×6 node element and thence computing the derivatives. The latter approach is adopted because of its convenience. Furthermore, since the number of gauss points slightly exceeds the total number of degrees of freedom in the main element, there should be no substantial loss of accuracy. Formally the 36-node element is constructed in the same way as the 16-node element, six primary functions vanishing on the gauss point lines being used to construct the polynomials.

PROGRAM VERIFICATION

Few exact solutions for creep problems are available; of such solutions the problem of a long uniform thick cylinder subjected to internal pressure load, p, and obeying the steady state creep law

$$\dot{\epsilon}_c^* = C(\sigma^*)^n \tag{26}$$

is convenient as a quantitative test of the validity of the coding used. The steady state stress solution is [7]

$$\frac{\sigma_\theta}{p} = \frac{(2/n - 1)(r_0/r)^{2/n} + 1}{(r_0/r_i)^{2/n} - 1}; \qquad \frac{\sigma_r}{p} = \frac{1 - (r_0/r_i)^{2/n}}{(r_0/r_i)^{2/n} - 1}$$

$$\frac{\sigma_z}{p} = \frac{(1/n - 1)(r_0/r)^{2/n} + 1}{(r_0/r_i)^{2/n} - 1}; \qquad \frac{\sigma^*}{p} = \frac{3^{1/2}}{n} \frac{(r_0/r)^{2/n}}{(r_0/r_i)^{2/n} - 1} \tag{27}$$

where σ_θ, σ_r, σ_z and σ^* are the hoop, radial, axial and von Mises stresses, respectively, and r_i and r_0 are the internal and external radii of the cylinder. The solution is obtained under the assumption that all creep displacements are radial, so that the initial axial strain and displacement, given by the elastic Lamé solution for the thick cylinder, are fixed. The axial strain, from which the axial displacement is easily determined, is

$$\frac{\epsilon_z}{p} = \frac{1}{E} \frac{1 - 2\nu}{(r_0/r_i)^2 - 1} \tag{28}$$

where E is Young's modulus and ν is Poisson's ratio.

This solution was used as the primary test of the coding and, incidentally, as a test of the validity of the results given by the extended method in a creep context. Although the coding is written for plane stress/strain as well as axisymmetric stress, this test proves the main part of the coding, including the iteration procedure, leaving only minor options untried. These were tested using simple rectangle problems.

In addition to creep under the steady state conditions of eqn (26), the problems were examined for the time-hardening creep law

$$\epsilon_c^* = At^n(e^{\mu\sigma^*} - 1) \tag{29}$$

The configurations analysed are shown in Fig. 2. In each case a single element was used with Young's modulus 30×10^6 lbf in^{-2} and Poisson's ratio 0·3. A pressure load of 2500 lbf in^{-2} was taken for the thick cylinder and 1500 lbf in^{-2} for the rectangle problems.

Not surprisingly, a single element is not complete for the thick-cylinder problem, with the result that the radial stresses did not precisely correspond with the applied pressures even for the elastic solution. However, as the discrepancy was less than 1% of the applied pressure, it was quite acceptable from an elastic viewpoint. Furthermore, the extended energy solution was in very close agreement with the energy solution provided that the trace of $\lambda \mathbf{K}'$ was no longer than that of $\mathbf{K}$.

The effects of varying λ and Δt were investigated for the steady state creep process. As in the elastic solutions, the energy and extended energy results were always in good mutual agreement (a maximum discrepancy of 1%). However, as time elapsed, the surface radial stresses degraded slowly, giving a typical 3% error on reaching steady state. Increasing Δt had no perceptible effect for small values but eventually variations of up to 3% were observed. For such values of Δt the iteration procedure was only quasi-stable, since it was observed in a number of runs that the solution developed satisfactorily well into the steady state creep phase and then suddenly became unstable, giving very large stresses. On increasing Δt further,

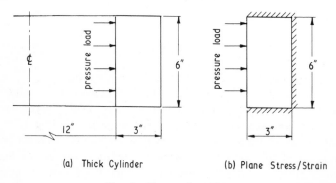

(a) Thick Cylinder   (b) Plane Stress/Strain

FIG. 2. Test configurations.

the solution process became unstable even before the steady state condition was reached. Quasi-stability was observed over a range of nominal time increments of more than one order of magnitude. With the exception of the boundary stresses, the computed steady state solution at the onset of steady state agreed with the analytical solution (eqn 27) to within 1%. The radial displacement δ, the von Mises stress σ^*, and the hoop stress σ_θ, at the loaded face for a typical run, with creep parameters $C = 1 \times 10^{-15}$, $n = 3$, are shown in Fig. 3(a).

Under the time-hardening creep law the qualitative features of the creep solutions were as expected. The initial changes in stress were similar to those in the steady state case but the time-hardening effect quickly attenuated the creep displacement rate without the onset of a steady state condition. Typical results are shown in Fig. 3(b) for time-hardening parameters $A = 1 \times 10^{-18}$, $\mu = 2 \cdot 9 \times 10^{-3}$ and $n = 0 \cdot 4$.

Since the plane stress/strain problems analysed have spatially constant stresses, the single-element model is complete. This, undoubtedly, accounts for the observation that the boundary conditions were exactly satisfied in the initial elastic solution and at all later times. Furthermore, the energy and extended energy solutions were always practically identical. In the plane stress problem (Fig. 4) under steady state creep the observed steady state condition was such that the transverse stress, σ_y, was exactly one-half the applied pressure. Under time-hardening creep the transverse stress moved towards the steady state value but the motion was inhibited by the time-hardening

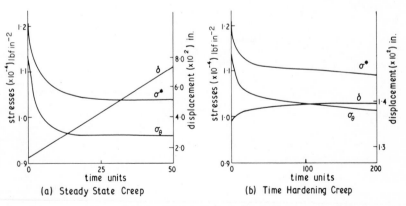

FIG. 3. Thick-cylinder behaviour.

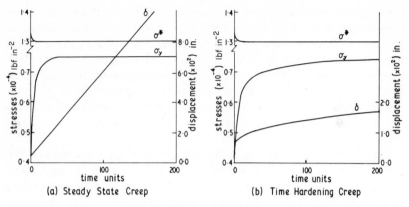

FIG. 4.   Plane stress behaviour.

behaviour. Typical results, obtained using creep parameters as for the thick cylinder, are given in Fig. 4. The plane strain response (Fig. 5) under steady state creep was also as expected, in that the results were consistent with an asymptotic approach to uniform hydrostatic pressure. Furthermore, the transverse and out-of-plane stresses (σ_y and σ_z) were equal both for the initial elastic solution and thereafter. This result follows by symmetry, since the problem examined exhibits similar plane strain in both the x, y and the x, z planes. Again, the time-hardening creep behaviour was similar to the steady state behaviour with suppressed stress changes due to hardening. Typical

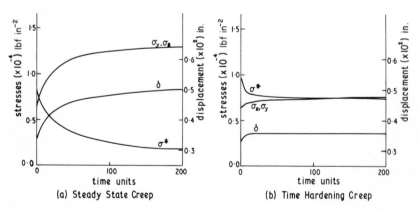

FIG. 5.   Plane strain behaviour.

results are shown in Fig. 5 for the same creep parameters as before except that the time-hardening parameter A was 1×10^{-14}.

These results validate the coding and manifest a close agreement between the energy and extended energy solutions. Important additional features are that incompleteness in the mesh, implying some error in every linear solution, can lead to cumulative errors, and that there is an upper limit of permitted time interval Δt beyond which the solution process is initially quasi-stable, then eventually unstable. Nevertheless, apart from manifest breakdown of the solution process, all variations in the generated solution were within acceptable design limits.

TURBINE BLADE PROBLEM

As a reasonable test of performance for design use, a turbine blade idealisation was examined; this is shown in Fig. 6 (the axis of rotation is in the plane of the figure). Plane stress conditions were assumed throughout and the blade and annular disc thicknesses were 0·2 and 0·4 in., respectively. Material properties were: Young's modulus, 30×10^6 lbf in^{-2}; density, 0·28 lbf in^{-3}; Poisson's ratio, 0·3. Centrifugal loading was considered with an angular velocity of 8000 rev/min and thermal effects were neglected. Consistent with design practice, a coarse mesh was used in which two elements of equal size described the blade and two elements the annular sector. Clearly, this mesh was incomplete but it was expected that it might yield adequate design data, particularly for the blade tip creep displacements. In order to improve completeness, the elements describing the blade were made unequal, in the ratio 5:1, giving a 'refined' mesh which was used in some steady state creep runs.

As before, it was found that the energy and extended energy methods gave results in close agreement. The blade root peak stress for the two meshes and tip displacements obtained for steady state creep with parameters $C = 1 \times 10^{-15}$ and $n = 3·0$ are shown in Fig. 7. As expected, the refined mesh is more capable of picking out the stress concentration but the tip displacements are relatively insensitive to this refinement. The elastic peak stresses obtained were 0·49 and $0·68 \times 10^6$ lbf in^{-2}, respectively, with corresponding steady state stresses 0·37 and $0·43 \times 10^6$ lbf in^{-2}. Blade tip displacements agreed to within 2% for both meshes and methods. The blade stress

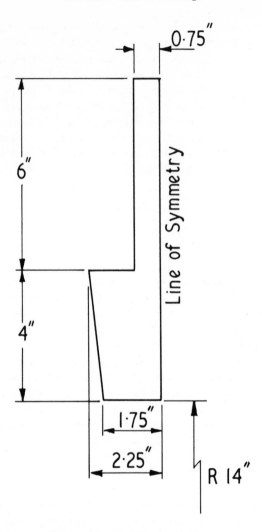

FIG. 6.   Turbine blade idealisation.

fields for both meshes were in close agreement for about five-sixths of blade length, the disparity occurring only at the localised stress concentration at the root of the blade. A slight tendency to a reduction in the rate of change of the stress gradient through an element was exhibited by the extended energy solution, which reflected the more stringent completeness requirement, but, provided that the

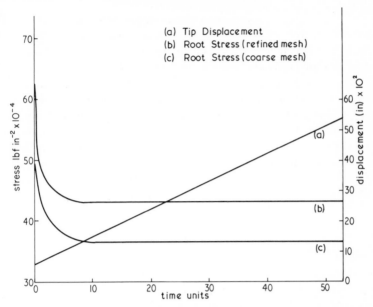

FIG. 7. Steady state creep of blade.

factor λ was taken so that the trace of the extended stiffness $\lambda \mathbf{K}'$ did not exceed that of the conventional stiffness $\mathbf{K}$, this effect was not marked. Similarly, the tip displacement decreased slightly with increasing λ but by no more than 2%. A more significant factor was that, on attaining the steady state condition, the peak stress showed a slight, but perceptible, downward trend. This undoubtedly is due to the incompleteness of mesh giving rise to an accumulating error as in the thick cylinder examination.

The effect of varying Δt was as expected from the initial tests. In the initial phase leading to steady state, it was necessary to take sufficiently small values to allow the developing field to be followed. Thereafter for suitably small values varying Δt had no significant effect; eventually the tip displacement and peak stress showed some sensitivity, but no more than 5% change, with a marked tendency for the solution procedure to become suddenly unstable well after the steady state had been established. For further increase of Δt the solution procedure was rapidly unstable even before steady state was established.

A number of runs under time-hardening creep were carried out.

Again close agreement between the methods was observed and the qualitative features of the solutions obtained were as expected.

These results demonstrate that under the adverse conditions of a typical design study the energy and extended energy methods give solutions which agree closely, not only in respect of an elastic analysis, but also under a non-linear creep regime. Such variations as do occur between the solution methods are within design tolerances. More significant are the features common to both methods: that incompleteness of mesh implies an error at each step of the iteration procedure which tends to be cumulative, and that even when steady state has been established, the time increments must be kept suitably small to avoid an instability in the iterative solution procedure.

CONCLUSIONS

An extended energy approach to the analysis of creep problems has been presented. In test problems this method gives results in close agreement with those of the energy method, even under the adverse conditions of a typical design study. In the case of the uniform thick cylinder under steady state creep, good agreement with an exact solution was observed. Significant features common to both methods are that an incomplete mesh implies slight but cumulative variations in the 'steady' stresses and that in order to maintain numerical stability the time-stepping parameter must be held sufficiently small even after the establishment of steady state conditions.

REFERENCES

1. OLIVEIRA, E. R. de A., Theoretical foundations of the finite element method, *Int. J. Solids Struct.*, **4**, 1968, 929–951.
2. PATTERSON, C., Sufficient conditions for convergence in the finite element method for any solution of finite energy, *The Mathematics of Finite Elements and Applications* (Ed. J. R. Whiteman), Academic Press, 1973, pp. 213–224.
3. PATTERSON, C., A class of functionals giving improved convergence with finite elements, *Variational Methods in Engineering* (Ed. C. A. Brebbia and H. Tottenham), University of Southampton Press, 1973, pp. 3/75–3/85.
4. PATTERSON, C., Stress intensity factors and the extended energy method, *Fracture Mechanics in Engineering Practice* (Ed. P. Stanley), Applied Science Publishers, 1977, pp. 57–68.

5. ZIENKIEWICZ, O. C., *The Finite Element Method in Engineering Science*, McGraw-Hill, 1971.
6. PATTERSON, C., 'Finite element analysis for plasticity and creep: an introduction', this Volume, pp. 125–136.
7. FINNIE, I. and HELLER, W. R., *Creep in Engineering Materials*, McGraw-Hill, 1959.

10

Biaxial Plastic Flow and Creep in Anisotropic Aluminium and Steel

D. W. A. REES AND S. B. MATHUR

Kingston Polytechnic

SUMMARY

The measured directions of a plastic strain increment vector and a creep strain rate vector for aluminium and steel in both the annealed and prestrained conditions have been compared with predictions derived from various yield functions.

For combined tension–torsion radial loading, annealed materials displayed collinearity in the direction of plastic flow and creep vectors. This indicated uniform hardening. The uniform hardening laws were not in agreement with the law derived from either the von Mises or the Tresca yield function. However, conforming predictions were obtained from the yield functions of Hill and Bailey.

The strain paths for the prestrained materials under similar test conditions displayed a rotation in the plastic flow vectors. This indicated anisotropic hardening. Tests on aluminium cylinders also showed that anisotropic hardening continued into ensuing creep with a rotation in the initial strain rate vectors. With increasing creep time strain rate vectors ultimately rotated into uniform hardening.

Plastic strain increment ratios derived from the generalised aniso-tropic yield functions of Edelman and Drucker or Yoshimura showed agreement with both uniform and anisotropic hardening. Observations on creep hardening were described by employing either one of these yield functions in a flow rule for creep.

NOTATION

A, B, C, A', B', C'	derived constants in anisotropic theory
A_0	anisotropy parameter

a, b	stress index integers
c_{ijkl}, c'_{ijkl}	tensor constants in anisotropic theory
D, K, P, Q, R	stress coefficient constants
E, G	elastic constants
F, G, H, L, M, N, S, T	anisotropy constants
f	yield function
i, j, k, l, p, q, r, s	tensor subscripts $(= 1, 2, 3)$
J'_2	second invariant of deviatoric stress (σ'_{ij})
k	yield stress in shear
$L_{p,q,r,s}$	distortion parameter
l	cylinder gauge length
m, m'	Bauschinger parameters
r, θ, z	polar coordinates
r_m	cylinder mean radius
t	time
α	rotation of axes of anisotropy
$\beta_1, \beta_2, \beta_3$	anisotropy functions
$d\epsilon^p_{ij}$	plastic strain increment tensor
$d\epsilon^p_{zz}, d\gamma^p_{\theta z}$	plastic axial and shear strain increments
$\dot{\epsilon}_{ij}$	creep strain rate tensor $(= d\epsilon^p_{ij}/dt)$
$\dot{\epsilon}_{zz}, \dot{\gamma}_{\theta z}$	axial and shear creep strain rates
$\epsilon^p_{ij}, \epsilon^p_{kl}$	plastic prestrain tensors
ϵ^p_0, γ^p_0	axial and shear prestrains
$\bar{\epsilon}^p_M$	von Mises equivalent plastic strain
$\delta_{ij}, \delta_{kl}, \delta_{ik}, \delta_{jl}$	Kronecker deltas
$d\lambda$	scalar multiplier
$\sigma_{ij}, \sigma_{kl}, \sigma_{pq}, \sigma_{rs}$	stress tensors
$\sigma'_{ij}, \sigma'_{kl}$	deviatoric yield stress tensors
$\sigma_{zz}, \tau_{\theta z}$	axial and shear stresses
$\bar{\sigma}_M$	von Mises equivalent stress
ϕ	angle of twist

Definition of Stress and Strain

Nominal stresses and engineering strains have been used throughout in the analysis of results. The plastic strain components were calculated from

$$\epsilon^p_{zz} = \frac{dl}{l} - \frac{\sigma_{zz}}{E}$$

$$\gamma^p_{\theta z} = \frac{r_m \phi}{l} - \frac{\tau_{\theta z}}{G} \tag{1}$$

The creep strain components were calculated from

$$\epsilon_{zz} = \frac{dl}{l}$$

$$\gamma_{\theta z} = \frac{r_m \phi}{l}$$

(2)

where dl and ϕ in eqns (1) and (2) refer to the measured displacements due to initial loading and subsequent creep, respectively.

INTRODUCTION

Isotropic theory of plasticity and creep relates to polycrystalline metallic materials where imposed strains occur over many randomly orientated crystal grains. The assumed isotropy is a serious limitation in the theory when the flow behaviour in a material is known to be a function of previous strain history. Uniaxial deformation, for example, would change the tensile and compressive strengths in all other directions, and a Bauschinger effect would also be evident when the direction of deformation is reversed. Many of the load-bearing components used in engineering practice would be anisotropic by virtue of the method of their manufacture. To account for such behaviour it would be necessary to employ a suitable anisotropic theory. A number of theories for anisotropic plasticity and creep deformation have therefore been advanced.

Yield functions have been proposed [1–4] for anisotropic–plastic deformation in metals. Some experimental work [5–9] has confirmed these yield functions by a curve-fitting technique on initial and subsequent yield loci for particular stress conditions and plastic strain history. Experimental work on the verification of the stress–plastic strain relations associated with each yield function is sparse. Indeed studies of this kind [8, 9] have been carried out only for Hill's yield function [1]. In creep deformation the isotropic Marin–Soderberg theory [10] has often failed to predict experimental creep rates produced under biaxial stress conditions. This has led to the development of empirical stress–strain rate relationships for aniso-tropic creep behaviour [11–13]. These relationships, however, are not fully descriptive of either the cause or the effect of anisotropy in creep.

In the present work an investigation of anisotropic stress–strain behaviour has been made for materials of known strain history. The experimental technique was to apply chosen plastic prestrains to thin-walled cylinders of commercially pure annealed aluminium and subsequently test in combined tension–torsion at room temperature. The loading path was radial, i.e. the ratio of shear to tensile stress was maintained constant. Anisotropy so produced has been examined in a plastic strain increment vector for loading and in a strain rate vector for ensuing creep. The direction of each vector then provided the basis for comparison with anisotropic theory.

THEORETICAL BACKGROUND

A theoretical strain ratio for plasticity and creep is generally derived from a particular yield function and a corresponding flow rule. In what follows yield functions (f) are first expressed in terms of the applied combined tension–torsion stress system (σ_{zz}, $\tau_{\theta z}$) and are then used in the flow rule to derive a strain increment ratio for plasticity and a strain rate ratio for creep.

Plastic Flow Theories

Plastic strain increment components are defined from the flow rule,

$$d\epsilon_{ij}^p = \frac{\partial f}{\partial \sigma_{ij}} d\lambda \tag{3}$$

where f is a plastic potential [1]. For example, a plastic strain increment ratio for combined tension–torsion, based upon the von Mises yield function, is derived in the following manner.

In general,

$$f = J_2' = \frac{1}{2} \sigma_{ij}' \sigma_{ij}' \tag{4}$$

which, for the applied stress system, becomes

$$f = \frac{1}{3} \sigma_{zz}^2 + \frac{1}{2} (\tau_{\theta z}^2 + \tau_{z\theta}^2) \tag{5}$$

From eqns (3) and (5) the plastic strain increments are

$$d\epsilon_{zz}^{P} = \frac{\partial f}{\partial \sigma_{zz}} d\lambda = \frac{2}{3}\sigma_{zz} d\lambda$$

$$d\epsilon_{\theta z}^{P} = \frac{\partial f}{\partial \tau_{\theta z}} d\lambda = \tau_{\theta z} d\lambda \qquad (6)$$

$$d\epsilon_{z\theta}^{P} = \frac{\partial f}{\partial \tau_{z\theta}} d\lambda = \tau_{z\theta} d\lambda$$

Noting that $d\gamma_{\theta z}^{P} = d\epsilon_{\theta z}^{P} + d\epsilon_{z\theta}^{P}$ and that $\tau_{\theta z} = \tau_{z\theta}$, it follows from eqn (6) that the plastic strain increment ratio is

$$\frac{d\gamma_{\theta z}^{P}}{d\epsilon_{zz}^{P}} = 3\frac{\tau_{\theta z}}{\sigma_{zz}} \qquad (7)$$

It can be shown from eqn (3) that the ratio associated with the Tresca yield function (i.e. $f = k^2 = \frac{1}{4}\sigma_{zz}^2 + \frac{1}{2}(\tau_{\theta z}^2 + \tau_{z\theta}^2)$) is

$$\frac{d\gamma_{\theta z}^{P}}{d\epsilon_{zz}^{P}} = 4\frac{\tau_{\theta z}}{\sigma_{zz}} \qquad (8)$$

The derivation of the plastic strain increment ratios associated with each of the following yield functions can be found in Ref. [14].
Hill [1]
For anisotropic materials with principal axes 1, 2, 3 the yield function given by Hill [1] has the quadratic form

$$6f = F(\sigma_{11} - \sigma_{22})^2 + G(\sigma_{22} - \sigma_{33})^2 + H(\sigma_{11} - \sigma_{33})^2 + 6(L\sigma_{12}^2 + M\sigma_{23}^2 + N\sigma_{13}^2)$$

where σ_{11}, σ_{22}, etc., are the resolved components of an applied stress system. For combined tension–torsion, where $\sigma_{11} = \sigma_{zz}$ and $\sigma_{12} = \tau_{\theta z}$ as shown in Fig. 1, then

$$\frac{d\gamma_{\theta z}^{P}}{d\epsilon_{zz}^{P}} = \left[\frac{6L}{F + H}\right]\frac{\tau_{\theta z}}{\sigma_{zz}} \qquad (9)$$

which reduces to eqn (7) for $F = H = L = 1$.

Hu [15] extended Hill's theory to include rotation of the principal axes of anisotropy caused by deformation. For a rotation, α, produced by combined tension–torsion deformation (see Fig. 1) Hu derived the ratio

$$\frac{d\gamma_{\theta z}^{P}}{d\epsilon_{zz}^{P}} = \frac{\beta_2 + \beta_3(\tau_{\theta z}/\sigma_{zz})}{\beta_1 + \beta_2(\tau_{\theta z}/\sigma_{zz})} \qquad (10)$$

where,

$$\beta_1 = (F + H)\cos^4\alpha + (F + G)\sin^4\alpha - 2F\sin^2\alpha\cos^2\alpha + \frac{3L}{2}\sin^2 2\alpha$$

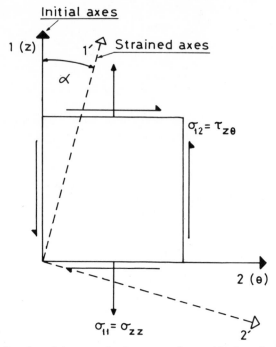

FIG. 1. Rotation of the axes of anisotropy under combined tension–torsion.

$$\beta_2 = (F + H) \cos^2 \alpha \sin 2\alpha - (2F + G) \sin^2 \alpha \sin 2\alpha + F \cos^2 \alpha \sin 2\alpha$$
$$\quad - 3L \cos 2\alpha \sin 2\alpha$$
$$\beta_3 = (4F + H + G) \sin^2 2\alpha + 6L \cos^2 2\alpha$$

For small rotations $\beta_1 \approx F + H$, $\beta_2 \approx 0$ and $\beta_3 \approx 6L$. Then eqn (9) approximates to eqn (10).

Edelman and Drucker [2]

For anisotropy produced by a prior plastic strain Edelman and Drucker [2] formulated a generalised function which described yielding in terms of an interaction between the prior strains and yield stress components in the form

$$f = \frac{1}{2} C_{ijkl} (\sigma'_{ij} - m\epsilon^p_{ij})(\sigma'_{kl} - m\epsilon^p_{kl}) \qquad (11)$$

For a tensile or compressive prestrain, ϵ^p_0, and subsequent combined tension–torsion, the associated plastic strain increment ratio is

$$\frac{d\gamma_{\theta z}^{p}}{d\epsilon_{zz}^{p}} = \frac{A\left(\sigma_{zz} - \dfrac{3m\epsilon_0^{p}}{2}\right) + B\tau_{\theta z}}{C\left(\sigma_{zz} - \dfrac{3m\epsilon_0^{p}}{2}\right) + A\tau_{\theta z}} \tag{12}$$

For a torsional prestrain, γ_0^{p}, and subsequent combined tension–torsion, the associated plastic strain increment is

$$\frac{d\gamma_{\theta z}^{p}}{d\epsilon_{zz}^{p}} = \frac{A\sigma_{zz} + B\left(\tau_{\theta z} - \dfrac{m\gamma_0^{p}}{2}\right)}{C\sigma_{zz} + A\left(\tau_{\theta z} - \dfrac{m\gamma_0^{p}}{2}\right)} \tag{13}$$

Yoshimura [3]

A further generalised yield function for describing anisotropic yielding in prior-strained material was formulated by Yoshimura [3]. This described yielding in terms of the prior strains producing a translation of the yield locus in the form

$$f = \frac{1}{2}C'_{ijkl}\sigma'_{ij}\sigma'_{kl} - m'\epsilon_{ij}^{p}\sigma'_{ij} \tag{14}$$

The associated plastic strain increment ratio for a tensile or compressive prestrain is given by

$$\frac{d\gamma_{\theta z}^{p}}{d\epsilon_{zz}^{p}} = \frac{A'\sigma_{zz} + B'\tau_{\theta z}}{C'\sigma_{zz} + A'\tau_{\theta z} - 9m'\epsilon_0^{p}} \tag{15}$$

and for a torsional prestrain by

$$\frac{d\gamma_{\theta z}^{p}}{d\epsilon_{zz}^{p}} = \frac{A'\sigma_{zz} + B'\tau_{\theta z} - 9m'\gamma_0^{p}}{C'\sigma_{zz} + A'\tau_{\theta z}} \tag{16}$$

Williams and Svensson [4]

The yield function most recently proposed [4] is an anisotropic version of the von Mises J_2' function. It expresses aggregate yield stresses and aggregate prior strains in terms of anisotropy, Bauschinger and distortion parameters in the form

$$f = (I_{ijkl} + A_0\epsilon_{ij}^{p}\epsilon_{kl}^{p})(\sigma_{ij} + \alpha_{ij})(\sigma_{kl} + \alpha_{kl}) \tag{17}$$

In eqn (17)

$$I_{ijkl} = \delta_{ik}\delta_{jl} - \frac{1}{3}\delta_{ij}\delta_{kl}$$

where δ_{ik}, δ_{jl}, δ_{ij} and δ_{kl} are Kronecker deltas, and

$$\alpha_{ij} = \epsilon_{ij}^p(L_{pqrs}\sigma_{pq}\sigma_{rs} - m)$$

For a tensile prestrain and subsequent combined tension–torsion loading the authors quote [5] an associated plastic strain increment ratio of

$$\frac{d\gamma_{\theta z}^p}{d\epsilon_{zz}^p} = \left[6\tau_{\theta z} + 6\left\{1 + \frac{3}{2}A_0(\epsilon_0^p)^2\right\}\left\{\sigma_{zz} + \frac{3}{2}\epsilon_0^p(L_{3333}\sigma_{zz}^2 + 2L_{2323}\tau_{\theta z}^2 - m)\right\}\right.$$

$$\left.\times(2\epsilon_0^p L_{2323}\tau_{\theta z})\right]\Big/\left[2\left\{1 + \frac{3}{2}A_0(\epsilon_0^p)^2\right\}\left\{\sigma_{zz} + \frac{3}{2}\epsilon_0^p(L_{3333}\sigma_{zz}^2\right.\right.$$

$$\left.\left.+ 2L_{2323}\tau_{\theta z}^2 - m)\right\}(1 + 3\epsilon_0^p L_{3333}\sigma_{zz})\right] \qquad (18)$$

and for a torsional prestrain they quote [6]

$$\frac{d\gamma_{\theta z}^p}{d\epsilon_{zz}^p} = \left[3\left\{1 + \frac{A_0(\gamma_0^p)^2}{2}\right\}\left\{\tau_{\theta z} + \frac{\gamma_0^p}{2}(L_{3333}\sigma_{zz}^2 + 2L_{2323}\tau_{\theta z}^2 - m)\right\}\right.$$

$$\left.\times(2 + 4\gamma_0^p L_{2323}\tau_{\theta z})\right]\Big/\left[2\sigma_{zz} + 6\left\{1 + \frac{A_0(\gamma_0^p)^2}{2}\right\}\left\{\tau_{\theta z} + \frac{\gamma_0^p}{2}(L_{3333}\sigma_{zz}^2\right.\right.$$

$$\left.\left.+ 2L_{2323}\tau_{\theta z}^2 - m)\right\}(2 + 4\gamma_0^p L_{2323}\tau_{\theta z})\right] \qquad (19)$$

From the condition of constancy of volume during plastic deformation it was shown that L_{3333} in eqn (18) and L_{2323} in eqn (19) were both zero.

All the foregoing theoretical plastic strain increment ratios are compared with the results of the present tests on aluminium and the published results [16, 17] for steel.

Creep Theories

Creep strain rate components are obtained from the time derivative form of eqn (3), i.e.

$$\dot{\epsilon}_{ij} = \frac{d\epsilon_{ij}^p}{dt} = \frac{\partial f}{\partial \sigma_{ij}}\frac{d\lambda}{dt} \qquad (20)$$

If the potential functions f in eqns (3) and (20) are identical, then it follows that the associated plastic strain increment ratio derived from eqn (3) would be identical with the associated strain rate ratio derived from eqn (20). For example, in the Marin–Soderberg isotropic creep theory $f = J_2'$ in eqn (20). The derived strain rate ratio in combined tension–torsion would then be identical with the right-hand side of eqn (7), i.e.

$$\frac{\dot{\gamma}_{\theta z}}{\dot{\epsilon}_{zz}} = 3 \frac{\tau_{\theta z}}{\sigma_{zz}} \tag{21}$$

Strain rates associated with the Tresca yield function have been used by Wahl [13] to describe creep in rotating discs. In combined tension–torsion it follows from eqn (8) that the associated strain rate ratio is

$$\frac{\dot{\gamma}_{\theta z}}{\dot{\epsilon}_{zz}} = 4 \frac{\tau_{\theta z}}{\sigma_{zz}} \tag{22}$$

The form of empirical equation used by Johnson [12] to describe creep anisotropy in 0·17% carbon steel at 450°C would derive from eqn (20) with f equated to Hill's yield function. For combined tension–torsion it then follows from eqn (9) that the strain rate ratio associated with this yield function is

$$\frac{\dot{\gamma}_{\theta z}}{\dot{\epsilon}_{zz}} = \left[\frac{6L}{F+H} \right] \frac{\tau_{\theta z}}{\sigma_{zz}} \tag{23}$$

An isotropic creep theory was proposed and used by Bailey [11] as an alternative to the Marin–Soderberg theory. The corresponding function f in eqn (20) was later defined by Davies [18]. For combined tension–torsion this has the form

$$f = \frac{1}{3} \sigma_{zz}^{a-2b+1} + \frac{1}{2} (\tau_{\theta z}^{a-2b+1} + \tau_{z\theta}^{a-2b+1}) \tag{24}$$

where $a - 2b$ is an odd integer. The associated strain rate ratio from eqn (20) is

$$\frac{\dot{\gamma}_{\theta z}}{\dot{\epsilon}_{zz}} = 3 \left[\frac{\tau_{\theta z}}{\sigma_{zz}} \right]^{a-2b} \tag{25}$$

which reduces to eqn (21) for $a - 2b = 1$.

It is interesting to note that Bailey's anisotropic version of his theory would lead to a strain rate ratio of the form

$$\frac{\dot{\gamma}_{\theta z}}{\dot{\epsilon}_{zz}} = \left[\frac{6L}{F+H} \right] \left[\frac{\tau_{\theta z}}{\sigma_{zz}} \right]^{a-2b} \tag{26}$$

which reduces to eqn (23) for $a - 2b = 1$.

The creep strain rates associated with the anisotropic yield functions of Edelman and Drucker, Yoshimura, and Williams and

Svensson have not appeared in the literature. In the present work the strain rate ratios associated with these and all the foregoing yield functions are compared with the measured ratio during creep in annealed and prestrained aluminium.

MATERIALS, INSTRUMENTATION AND PROCEDURE

Materials and Testpieces

All testpieces were machined from a 45 mm diameter 'Hiduminium 1A' extruded bar 9 m long. This is the commercial name for 99·8% pure aluminium containing the following percentage impurities by weight: Cu, 0·005; Fe, 0·04; Hg, 0·005; Mn, 0·01; Ni, 0·015; Si, 0·04; Ti, 0·015; Zn 0·02.

Blank tubes and miniature tensile testpieces were machined from the bar, annealed at 450°C for $2\frac{1}{2}$ h and furnace cooled. Subsequent testing of the miniature testpieces showed near-axial and transverse tensile isotropy. The tubes were light machined into the finished testpieces of Fig. 2 by a sequence of operations designed to ensure concentricity of gauge diameters, square registers and threaded ends. The gauge length profile was reproduced from a template used in conjunction with a lathe copy attachment and a single-point tool. This ensured repeatability in gauge length dimensions. Finally, each finished testpiece was given a light stress relief anneal (150°C for $1\frac{1}{2}$h).

Air plug gauge and micrometer measurements of each finished testpiece showed a variation in bore diameter from 25·392 to 25·415 mm and a variation in wall thickness from 1·496 to 1·506 mm.

Combined Tension–Torsion Machine

The testing machine was a modified version of a combined tension–torsion machine originally designed by Mathur and Alexander [19]. The modification was necessary in order to reduce the tensile load capacity from 110 kN to 40 kN and the torque capacity from 500 Nm to 150 Nm for the present test material. New load cells were manufactured which could measure these reduced capacities to within 1% accuracy.

In order that combined tension–torsion could be applied in any desired ratio, a light aluminium alloy beam was added to the test machine. The beam was supported horizontally between the original tensile and torsional loading points and carried a movable weight

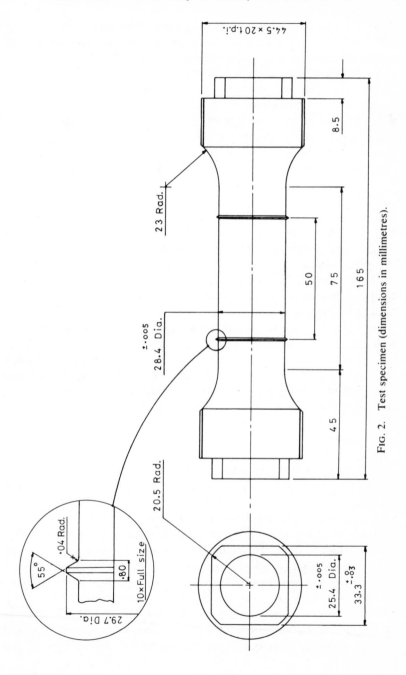

FIG. 2. Test specimen (dimensions in millimetres).

hanger on its span. Then, by applying load increments at any chosen point on the span, the ratio between the tensile and shear stress was maintained constant (i.e. radial loading in stress space).

The modified machine was calibrated by dead weight loading at various points on the span of the beam. A stress ratio $\tau_{\theta z}/\sigma_{zz}$ was calculated from the outputs of strain gauges affixed to the load cells. For each load point good agreement was found between this stress ratio and the corresponding stress ratio calculated from beam equilibrium and the known lever magnification ratios.

Full experimental details are given in ref. [14].

Extensometer

Experimental strain ratios in plasticity and creep were calculated from the continuous measurements of extension and angle of twist in a deforming testpiece. These measurements were made from two ± 2.5 mm displacement transducers mounted in an extensometer which clamped to the gauge length ridges shown in Fig. 2. Each clamping ring had spring-loaded inserts which compensated for changes in gauge diameter.

The primary and secondary windings of each transducer were incorporated into separate a.c. bridges which produced amplitude-modulated displacement–time output signals. Following amplification the original signals were restored by demodulation and filtration. They were then recorded simultaneously on a multichannel potentiometric chart recorder.

Test Procedure

Radial loading on annealed testpieces was performed for the stress ratios 0·26, 0·56, 0·80 and 1·82. For each ratio, load increments were applied at 15 min intervals to allow for creep between successive increments.

Loading was completed for a von Mises equivalent stress (i.e. $\bar{\sigma}_M = (\sigma_{zz}^2 + 3\tau_{\theta z}^2)^{1/2}$) of 37·5 N mm^{-2} and thereafter the testpiece was left to creep for 100 h. Throughout plastic and creep deformation the extension and angle of twist were continuously recorded.

Radial loading on prestrained testpieces was performed for a stress ratio of 0·95 using the same procedure but with 300–400 h of creep time after stress levels of 16 N mm^{-2} in tension and 15·2 N mm^{-2} in torsion had been achieved. This test condition was also repeated using an annealed testpiece. Prior tension, compression, forward and reversed torsion were each chosen to produce a von Mises equivalent plastic prestrain (i.e. $\bar{\epsilon}_M^p = \{(\epsilon_0^p)^2 + \frac{1}{3}(\gamma_0^p)^2\}^{1/2}$) of 2% in initially annealed testpieces. Tensile and compressive prestrains (i.e. $\epsilon_0^p = \pm 2\%$) were achieved using a strain rate of $1 \cdot 67 \times 10^{-4} \, \text{s}^{-1}$ in a 250 kN Instron machine. Torsional prestrains (i.e. $\gamma_0^p = \pm 3 \cdot 46\%$) were achieved using a strain rate of $2 \cdot 75 \times 10^{-4} \, \text{s}^{-1}$ in a 15 000 kg cm Avery machine. Adaptors and alignment plugs were used in the jaws of each machine to ensure axiality of loading.

RESULTS AND DISCUSSION

Plastic Flow and Creep in Annealed Materials

The shear strain–axial strain paths for the four stress ratios on annealed aluminium are shown in Fig. 3. Each strain path consists of the instantaneous strains and the 15 min creep strains produced by loading together with the 100 h creep strains. All strains are superimposed in the order in which they occurred. The loading strains for each stress increment are shown between successive symbols. Completion of loading is designated by 0 h creep time and thereafter the ensuing creep strains are shown for the hourly times indicated. With the exception of the strain path $\tau_{\theta z}/\sigma_{zz} = 1 \cdot 82$, where torsional buckling occurred after 25 h of creep, a common feature in Fig. 3 is the linearity of each strain path. This indicated that the annealed material hardened uniformly throughout the deformation. The gradients of each strain path have been used to obtain the following uniform hardening law in annealed aluminium:

$$\frac{d\gamma_{\theta z}^p}{d\epsilon_{zz}^p} = \frac{\dot{\gamma}_{\theta z}}{\dot{\epsilon}_{zz}} = 2 \cdot 6 \frac{\tau_{\theta z}}{\sigma_{zz}} \qquad (27)$$

The plastic strains were calculated from eqn (1) with elastic constants $E = 68 \cdot 3 \, \text{kN mm}^{-2}$ and $G = 24 \cdot 2 \, \text{kN mm}^{-2}$ obtained from the stress–strain loading plots.

Uniform hardening in annealed En24 and En25 has been observed by Rogan and Shelton [16] for similar test conditions. The law for En24 was

$$\frac{d\gamma_{\theta z}^p}{d\epsilon_{zz}^p} = \frac{\dot{\gamma}_{\theta z}}{\dot{\epsilon}_{zz}} = 3 \cdot 5 \frac{\tau_{\theta z}}{\sigma_{zz}} \qquad (28)$$

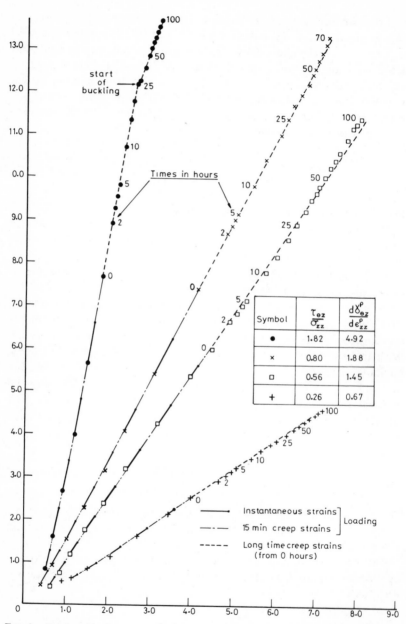

FIG. 3.   Strain paths for annealed aluminium; ordinate: shear strain, $\gamma_{\theta z}$ (%), abscissa:
axial strain, ϵ_{zz} (%).

and the law for En25 was

$$\frac{d\gamma_{\theta z}^{p}}{d\epsilon_{zz}^{p}} = \frac{\dot{\gamma}_{\theta z}}{\dot{\epsilon}_{zz}} = 3\cdot 8 \frac{\tau_{\theta z}}{\sigma_{zz}} \qquad (29)$$

Predictions of Uniform Hardening in Plasticity and Creep

It can be seen that uniform hardening laws based upon the yield functions of von Mises (i.e. eqns 7 and 21) and Tresca (i.e. eqns 8 and 22) only approximate to the experimental laws in eqns (27)–(29). This observation is consistent with the tests of Taylor and Quinney [20], who first showed discrepancies between the experimental yield loci for annealed metals and the von Mises and Tresca loci. A truly isotropic von Mises material is thus very difficult to achieve in practice. A satisfactory theoretical verification of each experimental law in eqns (27)–(29) can be obtained from the law derived from Hill's anisotropic yield function (i.e. eqns 9 and 23). In annealed aluminium, for example, F, H and L are then constants for uniform hardening in plasticity and creep such that $6L/(F + H) = 2\cdot 6$. The constants could be solved from a knowledge of two further experimental ratios, i.e. $d\epsilon_{\theta\theta}^{p}/d\epsilon_{zz}^{p}$ and $d\epsilon_{rr}^{p}/d\epsilon_{zz}^{p}$ (in plasticity), for which the theoretical solutions are $-F/(F + H)$ and $-H/(F + H)$, respectively [1].

Theories which include Hill's function as a special case will therefore describe uniform hardening in the annealed material, e.g. Bailey's eqn (26) for $a - 2b = 1$ and the yield functions of Edelman and Drucker and Yoshimura, which both reduce to a generalised form of Hill's quadratic function (i.e. $f = \frac{1}{2}C_{ojkl}\sigma_{ij}'\sigma_{kl}'$) when $\epsilon_{ij}^{p} = \epsilon_{jk}^{p} = 0$ identifies an annealed material. Hu's theory will only describe uniform hardening for $\alpha = 0$ in eqn (10) over the range of strain in Fig. 3; the theory could also describe a situation where larger strains destroy uniform hardening behaviour in annealed material. Axes of anisotropy, which would rotate with progressive deformation, would then need to be identified within the material.

Those theories which are based on the von Mises J_2' yield function cannot describe the uniform hardening law for each annealed material.

For example, Bailey's eqn (25) for $a - 2b = 1$ and the Williams and Svensson's eqns (18) and (19) for $\epsilon_0^{p} = \gamma_0^{p} = 0$ reduce to the uniform hardening law for a von Mises material only (eqn 7).

Plastic Flow and Creep in Prestrained Material

The shear strain–axial strain paths for $\tau_{\theta z}/\sigma_{zz} = 0.95$ on prestrained aluminium are shown in Fig. 4. The presentation is similar to that previously described in Fig. 3, but with longer creep strains shown in this case.

For comparison the corresponding strain path for annealed aluminium, which again conformed to the uniform hardening law of eqn (27), is also shown in Fig. 4. The strain paths in prestrained aluminium show anisotropic hardening followed by uniform hardening. Anisotropic hardening in loading is shown on the enlarged strain scales in Fig. 5, from which the elastic components have been subtracted (eqn 1). The von Mises and Tresca uniform hardening behaviours from eqns (7) and (8), respectively are also shown for comparison. The rotations in the $d\gamma_{\theta z}^p/d\epsilon_{zz}^p$ vectors of instantaneous plasticity indicate that anisotropic hardening is a function of both stress level and prestrain history. A common feature for all prestrains in Fig. 5 is that with increasing stress the $d\gamma_{\theta z}^p/d\epsilon_{zz}^p$ vector rotated towards the direction of uniform hardening in the annealed material, but remained fixed in direction during the 15 min interval of creep.

In contrast, however, the $d\gamma_{\theta z}^p/d\epsilon_{zz}^p$ vector produced by the final stress increment did not always remain fixed in direction during the 300–400 h period of ensuing creep (see Fig. 4), e.g. rotations occurred during the first 10 h in material prestrained 2% in compression and 3·46% in reversed torsion.

This showed that anisotropic hardening in plasticity continued into ensuing creep. The ultimate behaviour of each prestrained material in creep is uniform hardening as indicated by the linear $\gamma_{\theta z}$ versus ϵ_{zz} portion of each path in Fig. 4. A comparison of the gradients shows that uniform hardening in prestrained material only approximates to that in the annealed material. The fact that the gradients are different indicates that prestrain has a lasting effect on creep.

Rotating $d\gamma_{\theta z}^p/d\epsilon_{zz}^p$ vectors (see Fig. 5) are a consequence of the strengthening or weakening effect a prestrain has on the subsequent path. A 2% compressive prestrain, for example, had weakened the annealed material to tension (Bauschinger effect) by increasing the $d\epsilon_{zz}^p$ components. To a lesser extent this prestrain had also weakened the material to torsion (cross-effect) by increasing the $d\gamma_{\theta z}^p$ components. Thus, the $d\gamma_{\theta z}^p/d\epsilon_{zz}^p$ ratio was always less than that in the annealed material. With increasing stress the weakening effects grew progressively more severe, resulting in the observed upward rotation

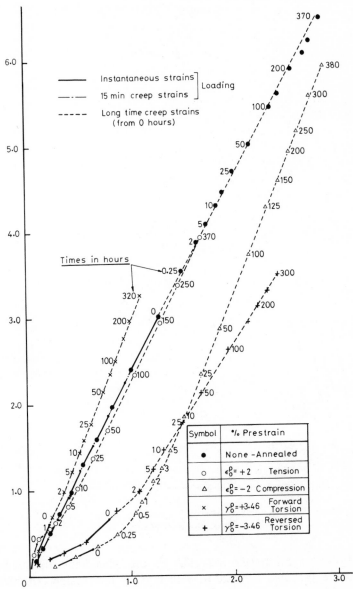

FIG. 4. Strain paths for prestrained aluminium; ordinate: shear strain, $\gamma_{\theta z}$ (%), abscissa: axial strain, ϵ_{zz} (%).

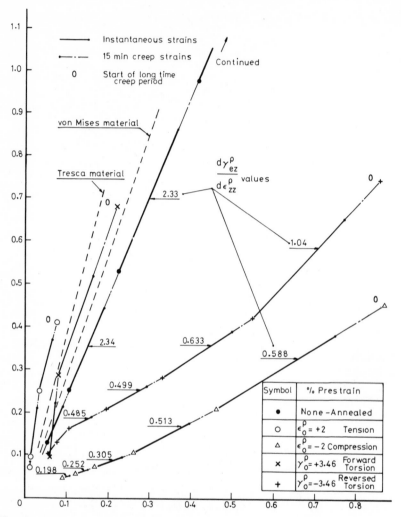

FIG. 5.   Plastic strain paths for prestrained aluminium; ordinate: plastic shear strain, $\gamma_{\theta z}$ (%), abscissa: plastic axial strain, ϵ^{p}_{zz} (%).

in $d\gamma^{p}_{\theta z}/d\epsilon^{p}_{zz}$. A 2% tensile prestrain, on the other hand, had strengthened the annealed material to both tension and torsion. The interaction and progressive deterioration between each strengthening effect with increasing stress resulted in the observed downward rotation in $d\gamma^{p}_{\theta z}/d\epsilon^{p}_{zz}$. The rotations in $d\gamma^{p}_{\theta z}/d\epsilon^{p}_{zz}$ for material prestrained 3·46% in forward and reversed torsion can be similarly explained from the strengthening or weakening effect of the prestrain in torsion and its associated cross-effect in tension. These rotations indicate that a greater cross-effect developed for torsional prestrains than for either a tensile or a compressive prestrain.

Rogan and Shelton [17] observed similar trends on rotating plastic strain increment vectors in a combined tension–torsion yield loci investigation on annealed En25 steel prestrained in tension and torsion. For example, their strain paths, reproduced in Fig. 6 for

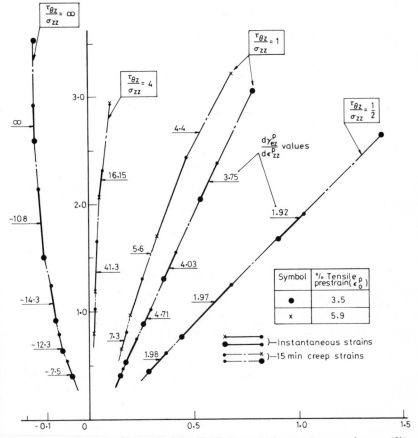

FIG. 6.   Strain paths for prestrained En25 steel; ordinate: shear strain, $\gamma_{\theta z}$ (%), abscissa: axial strain, ϵ_{zz} (%). After Rogan and Shelton [17].

stress ratios of $\frac{1}{2}$, 1, 4 and ∞, showed downward rotations in $d\gamma^p_{\theta z}/d\epsilon^p_{zz}$ for tensile prestrains of 3·5 and 5·9%.

Predictions of Anisotropic Hardening During Loading

Since anisotropic hardening was observed (see Fig. 5) to be a function of both prestrain history and subsequent stress level, a restriction is imposed on the choice of anisotropic theory which could describe it. Only the theories of Edelman and Drucker, Yoshimura, and Williams and Svensson are considered for providing a possible solution.

Edelman and Drucker [2]

Equations (12) and (13) must both reduce to the uniform hardening law of eqn (27) in annealed aluminium for $\epsilon^p_0 = \gamma^p_0 = 0$, i.e.

$$\frac{d\gamma^p_{\theta z}}{d\epsilon^p_{zz}} = \frac{A\sigma_{zz} + B\tau_{\theta z}}{C\sigma_{zz} + A\tau_{\theta z}} \tag{30}$$

Comparing eqns (27) and (30), it follows that $A = 0$ and $B/C = 2\cdot6$. Then for annealed aluminium, prestrained in tension or compression, eqn (12) becomes

$$\frac{d\gamma^p_{\theta z}}{d\epsilon^p_{zz}} = \frac{2\cdot6[\tau_{\theta z}/\sigma_{zz}]}{1 - \dfrac{3m\epsilon^p_0}{2\sigma_{zz}}} \tag{31}$$

and for annealed aluminium, prestrained in torsion, eqn (13) becomes

$$\frac{d\gamma^p_{\theta z}}{d\epsilon^p_{zz}} = 2\cdot6\left[\frac{\tau_{\theta z}}{\sigma_{zz}} - \frac{m\gamma^p_0}{2\sigma_{zz}}\right] \tag{32}$$

It will be noted that $d\gamma^p_{\theta z}/d\epsilon^p_{zz}$ in eqns (31) and (32) is a function of prestrain and stress level.

Yoshimura [3]

Comparing eqns (15) and (16) for $\epsilon^p_0 = \gamma^p_0 = 0$ with eqn (27), it again follows that $A' = 0$, and $B'/C' = 2\cdot6$. Equation (15) then becomes

$$\frac{d\gamma^p_{\theta z}}{d\epsilon^p_{zz}} = \frac{2\cdot6[\tau_{\theta z}/\sigma_{zz}]}{1 - \dfrac{9m'\epsilon^p_0}{C'\sigma_{zz}}} \tag{33}$$

and eqn (16) becomes

$$\frac{d\gamma^p_{\theta z}}{d\epsilon^p_{zz}} = 2\cdot6\frac{\tau_{\theta z}}{\sigma_{zz}} - \frac{9m'\gamma^p_0}{C'\sigma_{zz}} \tag{34}$$

It can be seen that for identical prestrains in each theory the respective equations have the same form. A comparison between the theoretical predictions of eqns (31) and (33) and the experimental $d\gamma_{\theta z}^{p}/d\epsilon_{zz}^{p}$ values from Fig. 5 for a 2% compressive prestrain is made in Table 1 for the stress levels shown. The m and m'/C' terms in eqns (31) and (33), respectively, have been calculated as positive constants in deformation by fitting each equation to the first experimental $d\gamma_{\theta z}^{p}/d\epsilon_{zz}^{p}$ value. Both equations then become

$$\frac{d\gamma_{\theta z}^{p}}{d\epsilon_{zz}^{p}} = \frac{2\cdot262}{1 + \dfrac{126\cdot3}{\sigma_{zz}}} \tag{35}$$

where the exact values of $\tau_{\theta z}/\sigma_{zz}$ and ϵ_{0}^{p} (with its sign) have been taken into account.

Table 1 shows that eqn (35) does predict the experimental trend of an upward rotation in $d\gamma_{\theta z}^{p}/d\epsilon_{zz}^{p}$ with increasing stress, but underestimates the measured values. Equations (31) and (33) can be made to predict the exact rotation for this prestrain when the Bauschinger parameters m and m' vary with deformation. Then, for the stress levels and experimental $d\gamma_{\theta z}^{p}/d\epsilon_{zz}^{p}$ values of Table 1, m and m'/C' in eqns (31) and (33), respectively, can be calculated to fit the rotation. In following this procedure the upper curve in Fig. 7 then shows the dependence of each parameter upon stress. Since eqn (33) is similar in form to eqn (31), (they are identical for $C' = 6$), the stress dependence of m and m'/C' is plotted as one graph but with suitably adjusted ordinate scales. The decrease in each parameter with increasing stress is consistent with the observed progressive deterioration in the

TABLE 1
*Comparison of experimental and theoretical $d\gamma_{\theta z}^{p}/d\epsilon_{zz}^{p}$
for a 2% compressive prestrain*

Axial stress σ_{zz} (N mm^{-2})	Plastic strain increment ratio, $d\gamma_{\theta z}^{p}/d\epsilon_{zz}^{p}$	
	From Fig. 5	From eqn (35)
12·10	0·198	0·198
13·10	0·252	0·213
14·10	0·305	0·227
15·10	0·513	0·242
16·10	0·588	0·256

D. W. A. Rees and S. B. Mathur

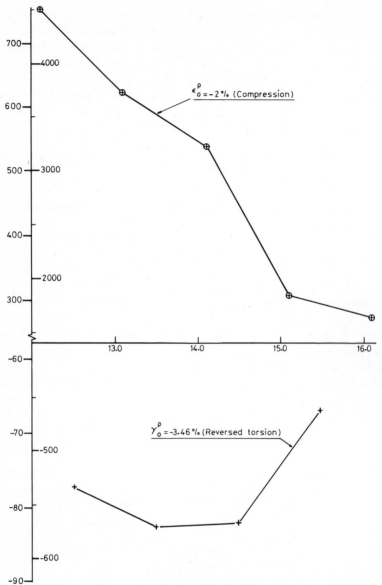

Fig. 7. Variation of Bauschinger parameter with stress and prestrain; inner ordinate scale: m, outer ordinate scale: m'/C', abscissa: axial stress, σ_{zz} (N mm^{-2}).

softening effects for this prestrain. Both the Bauschinger effect in tension and the cross-effect in torsion are described by the parameter m (or m') when each theory is applied in this way. Furthermore, the anisotropic hardening behaviour in Fig. 5 for a 3·46% reversed shear prestrain is exactly described by eqns (32) and (34) with parameters m and m' which vary with stress as shown in the lower curve in Fig. 7. Here, however, the stress dependence is less clearly defined than it was for the compressive prestrain and it is interesting to note that the interaction between the Bauschinger effect in torsion and the cross-effect in tension is described by a negative parameter in each equation.

Equations (31)–(34) could similarly be applied to describe anisotropic hardening for tensile and positive shear prestrains in Fig. 5. Since a Bauschinger effect does not then arise, m and m' should be regarded as parameters which are descriptive only of the hardening effects of such prestrains. As an example of this the two theories are applied to a wider range of test conditions in the strain paths of Rogan and Shelton (see Fig. 6). It follows from the uniform hardening law of eqn (29) that for annealed En25 prestrained in tension eqn (31) in the Edelman and Drucker theory becomes

$$\frac{d\gamma_{\theta z}^p}{d\epsilon_{zz}^p} = \frac{3 \cdot 8[\tau_{\theta z}/\sigma_{zz}]}{1 - \dfrac{3m\epsilon_0^p}{2\sigma_{zz}}} \tag{36}$$

and eqn (33) in the Yoshimura theory becomes

$$\frac{d\gamma_{\theta z}^p}{d\epsilon_{zz}^p} = \frac{3 \cdot 8[\tau_{\theta z}/\sigma_{zz}]}{1 - 9m'\epsilon_0^p/C'\sigma_{zz}} \tag{37}$$

With m and m' as positive decreasing parameters with increasing σ_{zz}, both eqn (36) and eqn (37) are consistent with Fig. 6 for the non-infinite stress ratios and two prestrains shown. The strain path for $\tau_{\theta z}/\sigma_{zz} = \infty$ showed that compressive plastic strains were initially produced from torsional loading on material prestrained in tension. A theoretical solution can be obtained by rearranging eqns (36) and (37) and putting $\sigma_{zz} = 0$. Equation (36) then becomes

$$\frac{d\gamma_{\theta z}^p}{d\epsilon_{zz}^p} = \frac{-3 \cdot 8\tau_{\theta z}}{\dfrac{3m\epsilon_0^p}{2}} \tag{38}$$

and eqn (37) then becomes

$$\frac{d\gamma^p_{\theta z}}{d\epsilon^p_{zz}} = \frac{-3\cdot 8\tau_{\theta z}}{\dfrac{9m'\epsilon^p_0}{C'}} \tag{39}$$

With m and m' as positive parameters which successively diminish to zero with increasing shear stress ($\tau_{\theta z}$), both eqn (38) and eqn (39) would predict the $d\gamma^p_{\theta z}/d\epsilon^p_{zz}$ values shown for $\tau_{\theta z}/\sigma_{zz} = \infty$ in Fig. 6, i.e. from small negative through large negative to infinity.

In replacing $d\gamma^p_{\theta z}/d\epsilon^p_{zz}$ by $\dot{\gamma}_{\theta z}/\dot{\epsilon}_{zz}$ in eqns (31)–(39), both the Edelman and Drucker and Yoshimura theories would correctly describe a plastic strain increment vector which remained fixed in direction for the 15 min creep interval allowed between load increments. For example, eqn (31) would be written

$$\frac{\dot{\gamma}_{\theta z}}{\dot{\epsilon}_{zz}} = \frac{2\cdot 6\tau_{\theta z}/\sigma_{zz}}{1 - \dfrac{3m\epsilon^p_0}{2\sigma_{zz}}} \tag{40}$$

which would describe the creep during loading in annealed aluminium prestrained in either tension or compression (see Fig. 5). It should be noted that both m and σ_{zz} in eqn (40) are constants for the creep interval.

Williams and Svensson [4]

It has been shown that this theory cannot account for the uniform hardening laws of eqns (27)–(29). An attempt to employ the theory to provide a solution to anisotropic hardening in Figs 5 and 6 would therefore be invalid because the theory applies only to a prestrained von Mises material.

Williams and Svensson successfully applied their theory to describe the observed effects of both tensile and torsional prestrain on an initial von Mises locus for aluminium. In this respect it appears that this recent theory is a refinement on the previous two theories in that it allows the Bauschinger effect to be separated from the cross-effect.

Prediction of Anisotropic and Uniform Hardening in Ensuing Creep

The Edelman and Drucker and Yoshimura yield functions can also be employed to describe the observed 'long time' creep behaviour of prestrained material in Fig. 4. For example, the theoretical creep rate ratio for annealed aluminium prestrained 2% in compression is given by eqn (40) from the Edelman and Drucker yield function. Initial

anisotropic hardening (the rotation) and subsequent uniform hardening in creep must now be described by m alone. The rotation, which
is a continuation of the diminishing Bauschinger and cross-effects
from loading, would then appear in eqn (40) with m a positive
parameter which decreases with increasing creep time. Subsequent
prevailing uniform hardening in creep would appear with m a
constant in this equation. The creep behaviour for all other prestrains
in Fig. 4 could be predicted by a similar application of either yield
function.

The yield functions of Hill and Bailey would be less satisfactorily
employed to describe anisotropic hardening in view of the absence of
terms associated with Bauschinger and allied effects in a prestrained
material. The application of either yield function would be greatly
simplified in describing the region of uniform hardening in Fig. 4.
Then a, b, F, H and L in the associated strain rate ratio eqns (23) and
(26) would be constants for a particular prestrain.

CONCLUSIONS

(1) Combined tension–torsion radial loading paths on annealed
     aluminium produced uniform hardening throughout plasticity
     and throughout ensuing creep. Published work also showed
     uniform hardening behaviour during the plastic deformation of
     annealed En24 and En25 steels under similar test conditions. It
     has been shown that the uniform hardening law in each
     annealed material was satisfactorily described with a law
     derived from the anisotropic yield functions of Hill and Bailey
     and from the generalised yield functions of either Edelman
     and Drucker or Yoshimura. This confirmed the validity of each
     yield function as a potential function in a flow rule for the
     plasticity and creep of annealed material.

     The uniform hardening law derived from the yield functions
     of von Mises and Tresca could only approximate to the law in
     each annealed material.

(2) A combined tension–torsion radial loading path on prestrained
     aluminium produced anisotropic hardening during plastic
     deformation.

     Published work also showed anisotropic hardening during
     radial loading of prestrained En25 steel. Anisotropic hardening

was observed to be a function of stress level and prestrain history and was attributed to the effect each prestrain had on the subsequent stress path. For each prestrain type the hardening behaviour was such that with increasing stress the plastic strain increment vector rotated towards the direction of the uniform hardening vector for the annealed material. The present work on prestrained aluminium showed that the rotation continued into the initial ensuing creep period before the material settled into a stable uniform hardening creep behaviour.

Of the yield functions considered, those proposed by Edelman and Drucker and Yoshimura were found to be the most consistent with overall hardening behaviour in prestrained material. In describing anisotropic hardening during plasticity the Bauschinger parameter of each function varied with stress during loading and remained constant for the time interval allowed between loads. In describing the observed creep behaviour the Bauschinger parameter of each function varied with time during anisotropic hardening and finally remained constant in uniform hardening. Both yield functions were then valid as potential functions in a flow rule for the plasticity and creep of prestrained material.

REFERENCES

1. HILL, R., *The Mathematical Theory of Plasticity*, Oxford University Press, 1950.
2. EDELMAN, F. and DRUCKER, D. C., Some extensions to elementary plasticity theory, *J. Franklin Inst.*, **251**, 1951, 581–605.
3. YOSHIMURA, Y., 'Hypothetical Theory of Anisotropy and the Bauschinger Effect Due to a Plastic Strain History,' Aero. Res. Inst. University of Tokyo, Report No. 349, 1959.
4. WILLIAMS, J. F. and SVENSSON, N. L., A rationally based yield criterion for work hardening materials, *Meccanica*, **6**, 1971, 104–110.
5. WILLIAMS, J. F. and SVENSSON, N. L., Effect of tensile prestrain on the yield locus of 1100-F aluminium, *J. Strain Anal.*, **5**, 1970, 128–139.
6. WILLIAMS, J. F. and SVENSSON, N. L., Effect of torsional prestrain on the yield locus of 1100-F aluminium, *J. Strain Anal.*, **6**, 1971, 263–272.
7. SHAHABI, S. N. and SHELTON, A., The anisotropic yield, flow and creep behaviour of prestrained En24 steel, *J. Mech. Eng. Sci.*, **17**, 1975, 93–104.
8. MEHAN, R. L., Effect of combined stress on yield and fracture behaviour of Zircaloy-2, *Trans. ASME, J. Basic Eng.*, 1961, 499–512.

9. JOHNSON, K. R. and SIDEBOTTOM, O. M., Strain history effect of isotropic and anisotropic plastic behaviour, *Expl. Mech.*, **12**, 1972, 264–271.

10. SODERBERG, C. R., Plasticity and creep in machine design, *Trans ASME*, **58**, 1936, 733–743.

11. BAILEY, R. W., The utilization of creep test data in engineering design, *Proc. Inst. Mech. Eng.*, **131**, 1935, 131–349.

12. JOHNSON, A. E., *Complex Stress Creep, Relaxation and Fracture of Metallic Alloys*, HMSO, Edinburgh, 1962.

13. WAHL, A. M., Analysis of creep in rotating discs based on the Tresca criterion and associated flow rule, *J. Appl. Mech.*, **23**, 1956, 231–238.

14. REES, D. W. A., 'Biaxial Creep and Plastic Flow in Anisotropic Aluminium,' Ph.D. Thesis, CNAA, Kingston Polytechnic, 1976.

15. HU, L. W., Studies on plastic flow of anisotropic metals, *J. Appl. Mech.*, **23**, 1956, 444–450.

16. ROGAN, J. and SHELTON, A., Yield and subsequent flow behaviour of some annealed steels under combined stress, *J. Strain Anal.*, **4**, 1969, 127–137.

17. ROGAN, J. and SHELTON, A., Effect of pre stress on the yield and flow of En25 steel, *J. Strain Anal.*, **4**, 1969, 138–161.

18. DAVIES, E. A., The Bailey flow rule and its associated yield surface, *J. Appl. Mech.*, **28E**, 1961, 310.

19. MATHUR, S. B. and ALEXANDER, J. M., 'A complex creep testing machine,' Proc. Jt. Conf. on Test Machines and Material Testing. I. Mech. E. and Soc. Env. Eng., 1965, pp. 275–288.

20. TAYLOR, G. I. and QUINNEY, H., The plastic distortion of metals, *Phil. Trans. Roy. Soc.*, **230(A)**, 1931, 323–362.

11

Redistributed Stresses as a Basis for Design for Structures Which Creep

J. DANKS
Lanchester Polytechnic

AND

A. B. LOMAX
Derby College of Art and Technology

SUMMARY

Estimation of the creep deformation of structures which operate under creep conditions can be made to a reasonable degree of accuracy by use of the redistributed stresses alone. The accuracy, when compared with intrinsic material scatter and uncertainty of the operating conditions, is quite sufficient. This is illustrated by examination of three structures, viz. two parallel bars, a beam under pure bending and a beam under pure bending with an axial load. Results from the approximate method have been compared with test results for the parallel bar structure and 'exact' theoretical solutions on all three structures, using the Graham and Walles creep equation and the time-hardening hypothesis.

NOTATION

A	area
b	breadth of beam
d	depth of beam
e	strain
e_c	creep strain
e_N	strain at neutral axis of beam
e_{NC}	creep strain at neutral axis of beam
e_o	strain at outside edge of beam
e_{oc}	creep strain at outside edge of beam

e_{oe}	elastic strain at outside edge of beam
e_s	standard strain in the Graham and Walles equation
$\dot{e}_{c_{min}}$	minimum creep strain rate
k	curvature
K	constant in the Norton law
L	length
M	bending moment
n	index of stress in the Norton law
P	load
t	time
T	temperature
$t_{1/3}, t_1, t_3$	constants in the Graham and Walles equation
y	distance from neutral axis
y_s	position of skeletal point
σ	stress
σ_a	axial stress
σ_b	bending stress
σ_N	stress at the neutral axis of beam after redistribution
σ_o	stress at outside of beam after redistribution

Subscripts

a, b	bars in parallel bar structure

INTRODUCTION

Designing components in which creep is the most probable cause of deformation is a difficult problem. Very often the exact operating conditions are unknown and little may be known about the material. In this situation a simple direct design method may well be adequate. The appropriateness of various types of hypothesis (e.g. 'equation of state', strain history, superposition, reference stress) to describe creep behaviour in situations where the stresses vary has been the subject of recent investigations by the authors [1–3] and several others [4–8]. One generally agreed conclusion is that, owing to the effects of intrinsic material scatter, it is not necessary to resort to hypotheses more complex than those based on the mechanical equation of state.

The authors consider, therefore, that the correct design philosophy for structures which creep is to develop simple approximate prediction methods, in which the designer can utilise the basic creep data available. It is also important that some comparison be available between the likely scatter in the material creep behaviour and the difference between the 'exact' and simple approximate method.

GENERAL PHILOSOPHY

All structures have an initial stress distribution which is in equilibrium with the applied loading and which is obtained using the conventional methods of elasticity and plasticity. In a structure for which creep is possible, the stresses begin to change and redistribute themselves as soon as the load is imposed. The redistribution is caused by the interaction of the time-dependent strains and the elastic–plastic time-independent strains, in conformity with some physically reasonable strain constraint. For example, in a beam where plane sections are assumed to remain plane, creep at the initial stress distribution would cause warping of the section, and redistribution of the stress is necessary to prevent this. Some time after loading, a state is reached when redistribution ceases and the stresses remain steady; under such conditions the rate at which the strain is changing all through the structure is such as to maintain the strain constraint without changing the stresses. The resulting stresses are considered to be the fully redistributed ones.

To determine the steady state stresses for a structure under creep three conditions are required:

(1) an equilibrium condition,
(2) a compatibility condition of strain rates and
(3) a material relationship between strain rate and stress; that used in this work is the Norton law [15],

$$\dot{e}_{c_{min}} = K\sigma^n \qquad (1)$$

The approximate method suggested is based on the redistributed stress values and assumes that these stresses have been acting the whole time. For structural components (or regions) in which the stress is known to increase, this method overestimates the strains and thus errs on the safe side.

The basis of the approximate method for design purposes is as follows:

(1) Determine the mean redistributed stress values based on compatibility of strain rates within the structure.

(2) Assuming that these stresses have been acting throughout, determine the deformation which is relevant to that structure. An adjustment can then be made for the elastic strain increment.

(3) Superimpose the appropriate scatter bands where this information is known.

This method implies that the redistributed stress values can be determined sufficiently accurately from steady state creep rates alone, but that the determination of the resultant creep behaviour must include the primary effects. The elastic strain increment mentioned in (2) above may be ignored if the creep strain is very much larger than the change in elastic strain. Care must be exercised, however, if structural deformation is to be calculated in a region where the stress decreases. In this case the elastic change, which will be a decrement, should be added to the estimated creep strain; otherwise the resulting strain will be underestimated.

This approach is illustrated by reference to three structures:

(1) A parallel bar model (Fig. 1a). (The compatibility condition is that the two bars must have the same extension; they can, however, have different areas, lengths and temperatures.)

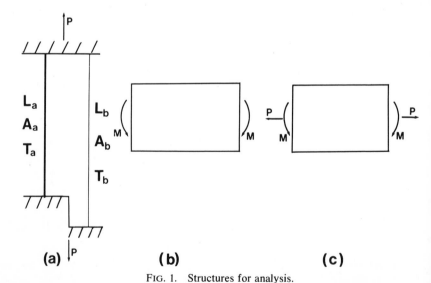

FIG. 1.   Structures for analysis.

(2)   A beam under pure bending (Fig. 1b).

(3)   A beam under pure bending with an axial tension (Fig. 1c).

'Exact' analyses are available for all three structures. These have been carried out using the time-hardening hypothesis. (Previous work [2, 3] has shown that for a structure containing regions of both increasing and decreasing stress the time-hardening and strain-hardening hypotheses give virtually identical results. The former is therefore preferred because of the resulting computational advantages.)

Tests have been conducted on a variety of parallel bar models using the rig described in a subsequent section.

MATERIALS

The two materials used in the work reported in this paper were 11% chromium ferritic steel (C, 0·16; Mn, 0·7; Si, 0·3; Cr, 11·6; Mo, 0·6; Nb, 0·27; V, 0·3; Fe balance) and Nimonic 90 (C, 0·09; Si, 0·45; Fe, 0·86; Mn, 0·05; Cr, 19·3; Ti, 2·51; Al, 1·33; Co, 16·8; Ni balance). The steel had been heat treated as follows: 0·5 h at 1150°C, air cool, 3 h at 640°C, air cool; the heat treatment for the Nimonic 90 was 8 h at 1080°C, air cool, 16 h at 700°C, air cool.

An extensive range of creep tests had been performed on these materials [9, 10], and the results from these had been cast into a form suitable for use in the creep function of the Graham and Walles equation [11]:

$$e_c = e_s \left[\left(\frac{t}{t_{1/3}} \right)^{1/3} + \left(\frac{t}{t_1} \right) + \left(\frac{t}{t_3} \right)^3 \right] \tag{2}$$

This equation was used in the present work, not because it is claimed to be generally better than other functions, but because with the data available it represented the complete creep behaviour of the two materials reasonably accurately. Also, because of the number of tests performed to obtain the constants in the function, an accurate estimate of the scatter bands for the material creep behaviour could be obtained. This scatter is predominantly intrinsic material scatter since, in the testing and analysis, other forms of scatter were reduced as much as possible. The constants ($t_{1/3}$, t_1 and t_2) in the equation are held in the form of 'cross-plots'. A typical cross-plot for t_1 is shown in

Fig. 2, in which the mean line through the points is marked M; the lines marked U and L encompass all the results which are within ±2 standard deviations of the mean. By use of the constants for a given stress the set of creep curves shown in Fig. 3 is produced.

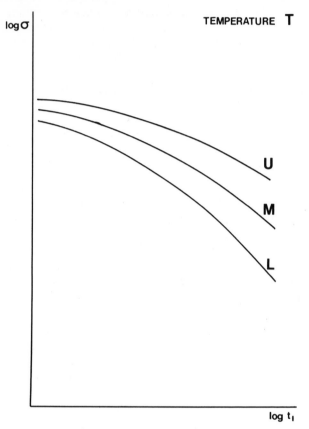

FIG. 2.   Cross-plots for use with Graham and Walles creep equation.

PARALLEL BAR STRUCTURE

Test Rig

The test rig for the creep testing of the parallel bar structure consists of two Denison T45 creep machines [12] (Fig. 4). These are so arranged that the two poise weight shafts can be linked together as

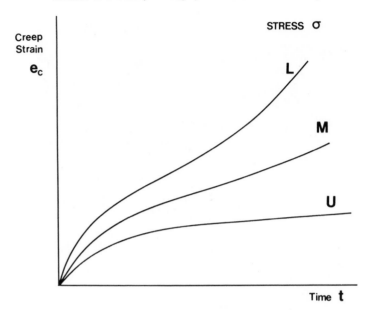

FIG. 3.   Constant stress curves showing scatter bands.

shown in Fig. 5. With a specimen in each machine the total load on the parallel bar structure can be fixed. The distribution of load between the two specimens is controlled by comparing their extensions. As creep takes place, the control circuit produces a signal which drives the common poise weight shaft, through a servomotor and gearbox, causing an increase in the load on the specimen with the smaller extension and a decrease in the load on the other. Thus, by interaction of the elastic–plastic and creep strains the extensions of the two bars are kept the same. The temperatures used were 600°C, 650°C and 700°C. The specimen extensions were measured with displacement transducers using telescopic extensometer legs attached to ridges on the gauge length of the specimens. The loads on the specimens were recorded on a chart recorder, from a potentiometric device.

Results
   A typical set of 'exact' results (i.e. results derived from the Graham and Walles equation with the time-hardening hypothesis) for the parallel bar structure is shown diagrammatically in Fig. 6. The limits

FIG. 4.   Test rig.

of expected extensions are given by the LL and UU combinations of the component bar creep curves. The limits of stress behaviour are given by the UL and LU curves [2].

The approximate method suggested depends upon the ability to predict accurately the value of the redistributed stresses. A comparison is shown in Table 1 of the predicted values of the redistributed stresses with experimental results for a wide range of area ratios, but for the same length ratio (1:1) and same temperature (600°C). It is interesting to note their insensitivity to area ratio; this was seen also in a later series of tests on parallel bars of differing lengths and areas in which the redistributed stresses appeared to depend only upon the temperatures.

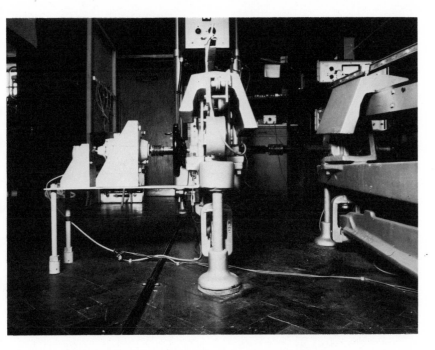

FIG. 5.   Test rig detail showing common shaft linking the two creep machines.

For the parallel bar structure the compatibility condition is given by

$$e_a L_a = e_b L_b \tag{3}$$

Taking the time derivative of eqn (3) and introducing eqn (1) gives the following as the steady state stress in bar 'a':

$$\sigma_a = \left[\frac{L_b}{L_a}\right]^{1/n}\left[\frac{P - \sigma_a A_a}{A_b}\right] \tag{4}$$

The steady state stresses for the parallel bar model are readily obtained from eqn (4). Assuming that these have been acting the whole time and applying the elastic correction, an approximate solution for the creep of this structure, as shown in Fig. 7, is obtained. Also shown in Fig. 7 are the likely scatter bands as predicted from the basic creep data, and the experimental results. It can be seen that there is no point in resorting to more complex methods, since the results from the approximate method are reasonably accurate when

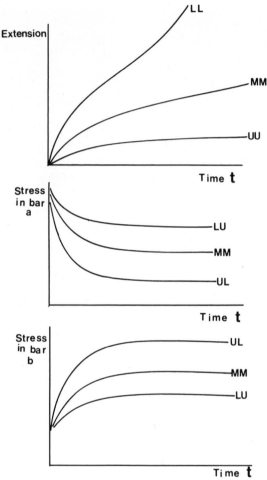

FIG. 6.   Typical extension and stress versus time graphs for the parallel bar structure
incorporating scatter.

compared firstly with the width of the scatter bands and secondly
with the exact conditions experienced by a component, as represented in a carefully controlled test. The elastic strain increment in this
case was so small as to make only an insignificant change in what was
already a satisfactory approximate solution. It was found that the
method always predicted the behaviour of the parallel bar structure
with reasonable accuracy.

TABLE 1
Redistributed stresses from a range of parallel bar structures

Area ratio	Mean stress	Predicted σ_a	Exptl. σ_a	Predicted σ_b	Exptl. σ_b
1/4	216·5	515	519	140·5	139·0
1/4	216·5	515	544	140·5	132·8
1/3	233·5	515	519	140·5	143·5
1/3	233·5	515	549	140·5	125·0
1/2	261·5	509	513	137·4	137·4
2/3	284·5	505	505	135·9	132·8
1	318·5	503	519	134·2	112·8
3/2	351·0	498	495	129·8	122·0
2	374·0	496	504	129·8	105·0
3	402·0	494	502	128·0	98·8
1/4	216·5	515	525	140·5	140·5
2/3	284·5	505	521	135·9	125·0
1	318·5	503	499	134·2	117·2
1	318·5	503	519	134·2	114·2
3	402·0	494	503	128·0	91·0

Note: Stress units are MN m^{-2}.

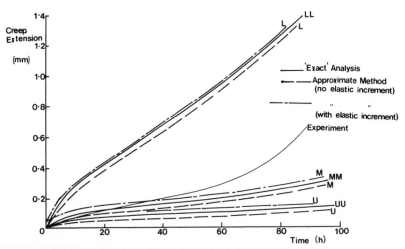

FIG. 7. Extension versus time graph for a parallel bar structure showing 'exact' and approximate analyses, and experimental result.

BEAM UNDER PURE BENDING

Beam sections loaded in pure bending are frequently referred to in the literature. The 'exact' analyses performed were for beams made from the 11% chromium ferritic steel. The basic compatibility condition has been checked practically [13] and found to be reasonable; the condition is

$$e = \frac{2y}{d} (e_{oc} + e_{oe}) \tag{5}$$

for a beam of rectangular section, depth d and width b. Applying Norton's law to the time derivative of eqn (5) and using the condition of equilibrium

$$M = \int_{-d/2}^{d/2} \sigma b y \, dy \tag{6}$$

the steady state stress at any position through the beam section is obtained in the form

$$\sigma = \left[\frac{2y}{d}\right]^{1/n} \sigma_b \left[\frac{2n + 1}{3n}\right] \tag{7}$$

For Norton index values greater than 6, the steady state stress distribution derived from eqn (7) is almost constant. The outer surface stress after redistribution is shown in Fig. 8 as a function of n. An estimation of the beam curvature may be obtained by assuming that the stress at position y had been acting the whole time. The curvature is then given by

$$k = \frac{e_c}{y} \tag{8}$$

where e_c is the creep strain resulting from the steady state stress at y; this can be calculated from eqn (1).

This analysis will also generate the 'skeletal point', where the steady state stress is equal to the initial elastic stress; its position (y_s) is given by

$$y_s = \left[\frac{2n + 1}{3n}\right]^{n/n-1} \frac{d}{2} \tag{9}$$

Figure 9 shows the steady state stress distribution obtained from eqn (7) compared with an 'exact' analysis. The stresses are seriously

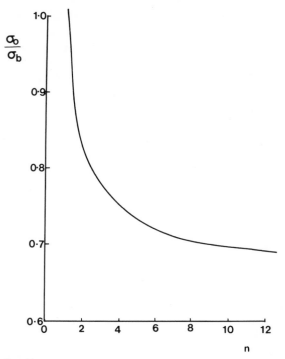

FIG. 8. Relationship between σ_o/σ_b and stress index for the beam under pure bending.

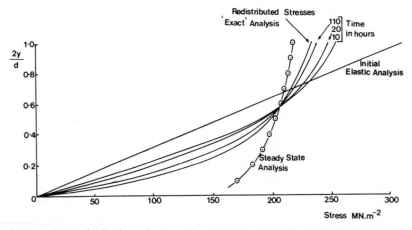

FIG. 9. Stress distribution using 'exact' and steady state analyses for the beam under pure bending.

overestimated in the lower-stressed regions, which is not necessarily a cause for concern. The curvature obtained from the approximate method (eqn 8) is shown in Fig. 10. The stress at the skeletal point has been used to calculate the curvature and there is therefore no need to apply the elastic strain increment. This would have been necessary had the stress at another point been chosen.

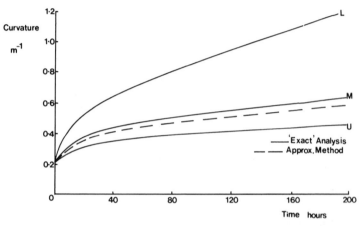

FIG. 10.   Curvature versus time graph showing 'exact' analysis incorporating scatter, and the approximate analysis, for the beam under pure bending.

The stress distribution in Fig. 9 shows a point on the 'exact' analysis where the stresses have remained approximately constant; this point does not lie on the line of the initial elastic analysis. It was observed from other analyses that the distance of this constant stress point from the elastic stress line varied with the size of the bending moment. The material considered (11% chromium ferritic steel) does not have a linear minimum creep rate–stress relationship, i.e. the creep index does not remain constant. The variation of the constant stress point from the elastic line is interpreted as an effect of this variation in creep index, the variation being greater the greater the bending moment or range of stress in the beam. The relevance of the variation in creep index and its dependence upon stress has already been noted by Anderson, Gardner and Hodgkins [14].

BEAM UNDER PURE BENDING WITH AN AXIAL TENSION

In this case a further equilibrium condition is necessary in addition to that of eqn (6), viz.

$$P = \int_{-d/2}^{d/2} \sigma b \, \mathrm{d}y \tag{10}$$

As there are two independent loads, the stress and strain distributions must be related to two known stresses or strains. Therefore, the compatibility condition is written in terms of the strain at the outer edge of the beam and the strain at the neutral axis (see Fig. 11):

$$e = e_{\mathrm{N}} + \frac{2y}{d}(e_0 - e_{\mathrm{N}}) \tag{11}$$

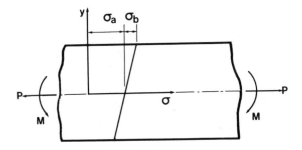

Initial Elastic Conditions

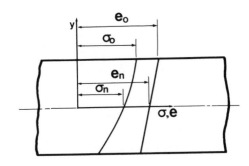

Steady State Conditions

FIG. 11. Initial and steady state stresses and strains in the beam with pure bending and an axial tension.

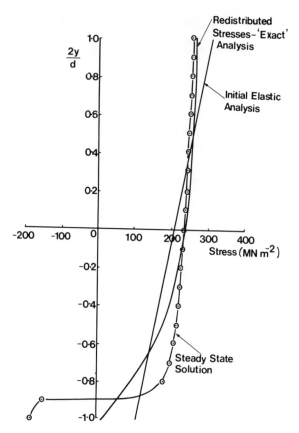

FIG. 12.   Stress distribution using the 'exact' and steady state analyses for the beam with
pure bending and an axial tension.

Substituting eqn (1) into the time derivative of eqn (11), the following function for the stress distribution is obtained:

$$\sigma = \left[\sigma_N^n + \frac{2y}{d} (\sigma_o^n - \sigma_N^n) \right]^{1/n} \tag{12}$$

By substitution of eqn (12) into eqns (10) and (6) the following two
further equations are produced:

$$\sigma_o = \frac{2P(n+1)(1-\hat{R})}{b\,dn} \left[\frac{1}{1-(2\hat{R}-1)^{(n+1)/n}} \right] \tag{13}$$

and

$$1 = \frac{3\bar{R}}{(1-\hat{R})}\left[\frac{1}{1-(2\hat{R}-1)^{(n+1)/n}}\right]$$
$$\times\left[\frac{n+1}{2n+1} - \hat{R} - \left[\frac{n+1}{2n+1}\right](2\hat{R}-1)^{(2n+1)/n} + \hat{R}(2\hat{R}-1)^{(n+1)/n}\right] \tag{14}$$

where $\hat{R} = (\sigma_N/\sigma_o)^n$ and $\bar{R} = \sigma_a/\sigma_b = Pd/6M$.

Values of σ_N and σ_o and, hence, the steady state stress distribution can be obtained from these equations. The stress distribution obtained for a beam with an $\bar{R}$ value of 2, where no stress exceeds the yield stress, is shown in Fig. 12. The stress distribution is well described by the steady state analysis over the more highly stressed regions, but very poorly over the region of low stress. If the assumption is made that this stress distribution has been acting all the time, then by using the following equation an estimate of the curvature can be made:

$$k = \frac{2}{d}(e_{oc} - e_{NC}) \tag{15}$$

where e_{oc} and e_{NC} are the creep strains due to σ_o and σ_N.

Curvature values calculated from eqn (15) are shown in Fig. 13; clearly the results are not very satisfactory. As mentioned earlier, it is preferable to deal with a region where the stress increases rather than one in which the stress decreases. It is clear in this case that the region of decreasing stress is much better defined and gives a more suitable basis for curvature estimation. In these circumstances it is not unlikely that the curvature will be underestimated (curve A, Fig. 13). In this case, although the elastic increment is negative, it should be added to the estimated curvature. When this is done, curve B is obtained, which is a more reasonable estimate of the beam curvature.

CONCLUSIONS

The simple technique of determining the stress distribution after redistribution has taken place, using a simple Norton law, and then using this distribution to determine the deformation is seen to be adequate, when compared with the possible scatter in material creep properties and the uncertainty of the conditions which the structure experiences.

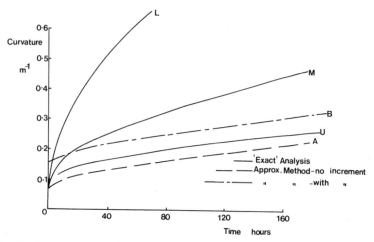

FIG. 13.   Curvature versus time graph showing the 'exact' analysis incorporating scatter and the approximate analysis, for the beam with pure bending and an axial tension.

An elastic strain increment should always be added to the estimated creep strain even though in practice this increment is negative. It is preferable to use as a basis for creep strain estimation a region of increasing stress.

REFERENCES

1. LOMAX, A. B., The influence of scatter on some simple variable stress creep predictions, *J. Strain Anal.*, **8**, 1973, 10–18.
2. LOMAX, A. B., Creep characteristics of an experimental model composite, *J. Strain Anal.*, **9**, 1974, 238–246.
3. DANKS, J., 'Realistic Creep Predictions for Structures', M.Phil. Thesis, CNAA, 1977.
4. ELLISON, E. G. and WEBSTER, G. A., Creep deformation of a composite beam subjected to combined axial and bending loads, *J. Mech. Eng. Sci.*, **9**, 1967, 98–106.
5. BARNES, J. R., CLARKE, J. M. and CLIFTON, T. E., An experiment in stress redistribution caused by creep, *J. Strain Anal.*, **2**, 1967, 280–289.
6. BULLARD, J. B. and CLIFTON, T. E., Further experiments in stress redistribution caused by creep, *J. Strain Anal.*, **5**, 1970, 277–283.
7. MARRIOTT, D. L., Approximate estimation of strain hardening creep deformation, *J. Strain Anal.*, **4**, 1968, 297–303.

8. MARRIOTT, D. L. and PENNY, R. K., Strain accumulation and rupture during creep under variable uniaxial tensile loading, *J. Strain Anal.*, **8**, 1973, 151–159.
9. WALLES, K. F. A., 'Analysis of creep data from a collaborative programme for an 11% chromium steel', N.G.T.E. Note No. NT.759, 1969.
10. WALLES, K. F. A., 'Random and systematic factors in the scatter of creep data', ARC Current Paper 935, April 1966.
11. GRAHAM, A. and WALLES, K. F. A., 'Regularities in creep and hot fatigue data', Parts I and II ARC Current Papers CP 379 and CP 380, 1958.
12. LOMAX, A. B., 'A model system to simulate creep of composites', *Metals Mater.*, **1**, 1967, 116–120.
13. MacCULLOUGH, G. H., An experimental and analytical investigation of creep in bending, *Trans. A.S.M.E.*, **55**, 1933, 55–60.
14. ANDERSON, R. G., GARDNER, L. R. T. and HODGKINS, W. R., Deformation of uniformly loaded beams obeying complex creep laws, *J. Mech. Eng. Sci.*, **5**, 1963, 238–244.
15. NORTON, F. H., *The Creep of Steel at High Temperatures*, McGraw-Hill, 1929.

12

Uniaxial and Biaxial Properties of a Material for Modelling Creep

H. Fessler and T. H. Hyde
University of Nottingham

SUMMARY

A lead–antimony–arsenic alloy in chill-cast form has been calibrated by constant load, room temperature tests in uniaxial tension, combined internal pressure and torsion applied to thin cylinders, and internal pressure applied to hemispherical-ended pressure vessels. The results are presented as constitutive equations and compared with prototype material data.

A uniaxial tensile rig for flat strip specimens and the combined pressure–torsion arrangement are described.

INTRODUCTION

Creep is important in engineering structures which have to operate in a temperature range around $0·5T_m$, where T_m is the absolute melting temperature of the prototype material. Prototype operating temperatures may be well above ambient temperature. At high temperatures, precision strain measuring and recording for tests of these materials over long periods of time is difficult and expensive. Because prototype materials creep at inconveniently high temperatures, model materials which creep at room temperature are useful. For lead alloys $0·5T_m$ occurs near ambient temperature, making them attractive as model materials because resistance strain gauges can be used on models and specimens, and it is simple to provide the temperature-controlled environment for model and specimen tests.

For the prediction of prototype response the important properties are the strains, rupture times and rupture ductilities produced by different stresses. The magnitudes of the three principal stresses control the response. This paper is limited to the strains produced by constant stresses; the effect of variable stresses is being studied at present.

All the work presented in this paper was carried out at $20 \pm \frac{1}{2}°C$ but tests at other temperatures between 0°C and 80°C are in progress to allow prototype temperature variations to be modelled.

For creep most material testing is performed with uniaxial tensile stresses (i.e. ratio of principal stresses 1:0:0). As the onset of plasticity and the creep rate are usually assumed to depend on the effective stress and the deviatoric stresses at the point considered, it is important to carry out tests with other stress ratios to test these assumptions for the model material. Stress ratios from equibiaxial (1:1:0) to pure shear (1:0:−1) have been investigated here by testing hemispherical-ended pressure vessels under internal pressure and thin tubes under combined torsion and internal pressure as well as uniaxial tensile specimens.

The model material is cast in the laboratory to facilitate the making of models of complicated shape. A 1·6% antimony, 0·16% arsenic lead alloy is chill-cast to produce a fine-grained structure. Details of the pressure vessels (P) and cylinders (C) used in this investigation are given in Table 1. The casting procedures will be described in a future paper.

UNIAXIAL TESTING

The uniaxial specimens, shown in Fig. 1 as item A, were cut from the pressure vessels and cylinders. It is most important [1] that the tensile force acts along the axis of the (square cross-section) gauge length. The profile of the square ends of the specimens and errors in the position of the clamping stud holes (item B) do not affect the alignment, because the circular knife edges (item C) are aligned with the sides of the gauge length in a special symmetrical jig before the clamping studs (item D) are tightened. Circular knife edges form the contacts with the flat faces of the loading links (item E). The width of the loading links is equal to that of the clamping plates; this, together with the use of setting-up shims between the loading links and the clamping plates, ensures that the flat faces of the loading links are perpendicular to the axis of the gauge length. Shims are used to set up

TABLE 1
Initial strains (microstrain)

Stress ratio	Specimen	Composition % Sb	% As	σ^*_{VM} (N mm^{-2})	Outside Total ϵ_1	ϵ_2	Outside Elastic ϵ_1	ϵ_2	Outside Plastic ϵ_1	ϵ_2	Inside Total ϵ_1	ϵ_2	Inside Elastic ϵ_1	ϵ_2	Inside Plastic ϵ_1	ϵ_2
1:1:0	P14	1·67	0·16	5·17	146	115	125	125	21	−10	122	125	125	125	−3	0
	P10	1·54	0·15	6·89	475	458	167	167	308	291	560	591	167	167	393	424
	P16	1·49	0·15	6·89	491	408	167	167	324	241	262	262	167	167	95	95
	P12	1·53	0·14	9·64	1 150	1 211	234	234	916	977	1 229	1 298	234	234	995	1 064
1:0·47:0	C9/3	1·34	0·14	6·89	311	−46	271	10	40	−56						
	P14	1·67	0·16	8·96	1 818	127	352	13	1 466	114	1 900	−128	352	13	1 548	−141
	P10	1·54	0·15	11·94	8 630	260	407	15	8 223	245	11 154	−790	407	15	10 747	−805
	P16	1·49	0·15	11·94	5 921	310	407	15	5 514	295	6 816	−209	407	15	6 409	−224
	C18/1	1·31	0·14	13·77	2 004	−68	542	20	1 462	−48						
	C19/5	1·45	0·15	13·77	3 707	152	542	20	3 165	132						
	P12	1·53	0·14	16·71	2·06%	—	657	24	1·99%	—	—	—	657	24	—	—
1:0·25:0	C20/2	1·48	0·15	13·77	1 360	531	588	−125	772	−406						
1:0:0	C20/1	1·48	0·15	6·89	450	−309	298	−131	152	−178						
	C19/3	1·45	0·15	13·77	1 453	−1 090	595	−262	858	−828						
1:0:0	P7	1·56	0·15	Various												
1:0:0	P9	1·66	0·15	Various												
1:0:−0·5	C19/4	1·45	0·15	13·77	1 528	−1 262	275	−212	1 253	−1 053						
1:0:−1	C9/1	1·34	0·14	6·89	333	−337	247	−247	86	−90						
	C18/5	1·31	0·14	10·33	647	−642	371	−371	276	−271						
	C19/2	1·45	0·15	13·77	6 583	−7 743	494	−494	6 089	−7 349						

σ^*_{VM} is the von Mises effective stress.

N.B. Inside strains were not measured on the C specimens.

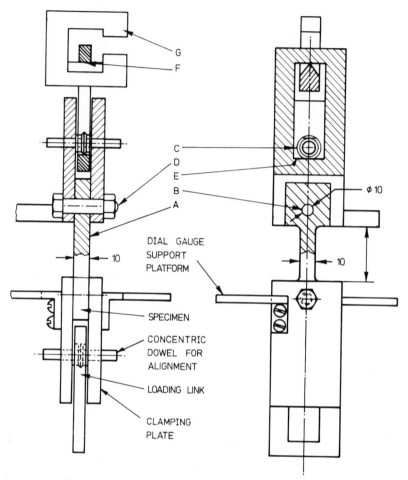

F$_{IG}$. 1. Uniaxial specimen and 'jaws' (dimensions in mm).

the knife edges (item F) of the open links at the outer ends of the loading links. The open links (item G) connect the assembly to the loading frame as shown in Fig. 2.

With the sides of the load hangers parallel to the sides of the clamping plates and the open links placed in the centre of the load hangers, the system should pull axially. The loading is applied to the uniaxial specimen through an 8:1 lever arrangement. The three knife edges on the lever arm are in line and the lever arm is stiff enough to ensure that the three knife edges stay in line when the load is applied;

FIG. 2.   Uniaxial tensile loading frame.

this ensures that the lever ratio remains constant while the lever rotates. With this loading arrangement, which incorporates six knife edges, tests carried out on an aluminium uniaxial specimen showed that the load is applied axially to within 0·03 mm.

Strains up to 2% were measured with pairs of electrical resistance strain gauges attached to opposite faces of the gauge length. It has been shown [2] that the reinforcing effect of gauges on lead alloy is negligible. Large strains were measured with dial gauges, shown in Fig. 2, arranged symmetrically to avoid bending.

Most of the creep strains were less than 1%, so that the results from these constant force creep frames may be considered as constant stress data.

UNIAXIAL TENSILE STRAINS

The total strain, ϵ_t, is a function of stress (σ), time (t) and temperature (T). For the isothermal tests dealt with here it is convenient to omit the temperature functions. The instantaneous strain consists of an elastic part (suffix e) and a plastic part (suffix pl), leading to a total strain, ϵ_t, given by

$$\epsilon_t = (\epsilon_e + \epsilon_{pl}) + \epsilon_c \tag{1}$$

where suffix c refers to creep.

The best fit to the first 200 h lead alloy data from 53 tests was obtained with

$$\epsilon_t = \frac{\sigma}{E} + \left[\frac{\sigma}{B}\right]^q + A \sinh \left[\frac{\sigma}{H}\right] t^{m(\sigma)} \tag{2}$$

The Young's modulus, E, was obtained from unloading, typically after 200 h of creep. The constants B and q were obtained from the instantaneous plastic strains of all specimens of a particular casting presented in Fig. 3(a). The time exponent, m, varied with stress. The empirical function

$$m = a + b\sigma + c\sigma^2 \tag{3}$$

has no physical justification but, together with the other coefficients, fits the data well, as shown in Fig. 4. For the range of stress considered m varied between 0·26 and 0·50. The values of the constants in eqns (2) and (3) are shown in Table 2 for lead alloys with three slightly different amounts of antimony.

If the creep strain is divided into primary (suffix pr) and secondary (suffix s) parts, it has been shown [3–6] that a relationship of the form

$$\epsilon_c = \epsilon_{pr}[1 - \exp(-\alpha t)] + \dot{\epsilon}_s t \tag{4}$$

gives a reasonably accurate fit to creep data for some materials, where the factor α is generally stress-dependent. For the lead alloy tested at 20°C α does not depend on the applied stress over the range $0·18 < \sigma/\sigma_u < 0·85$, where σ_u is the ultimate tensile strength. There is evidence [3–5] to suggest that the primary and secondary components of creep can be described by simple power relations, so that

$$\epsilon_c = \left[\frac{\sigma}{C}\right]^n [1 - \exp(-\alpha t)] + \left[\frac{\sigma}{D}\right]^r t \tag{5}$$

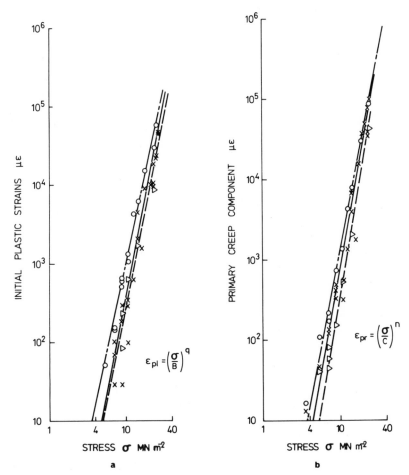

FIG. 3.   a: Instantaneous uniaxial plastic strains. b: Primary uniaxial creep strains.
————×, P7; —·—○, P9; ----▷, C19 and C20.

Equation (5) fits the lead alloy data well after more than 10 h of creep; this may be seen in Fig. 3(b) for C and n, in Fig. 4 for the exponential time function and in Fig. 5 for D and r. Values for these constants are also shown in Table 2.

Introducing further terms, such as an anelastic creep term, improves the fit to primary data but, in the absence of detailed knowledge of anelastic strains, a more complex primary creep function cannot be justified.

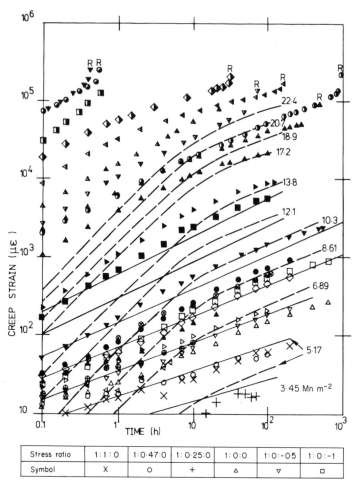

FIG. 4. Uniaxial creep strains from casting P7. ———, Prediction based on $\epsilon_c = A \sinh (\sigma/H)t^{m(\sigma)}$; — — —, prediction based on $\epsilon_c = \epsilon_{pr}[1 - \exp(-\alpha t)] + \dot{\epsilon}_s t$. R indicates that the specimen ruptured.

Stress ratio	1:1:0	1:0·47:0	1:0·25:0	1:0:0	1:0:-0·5	1:0:-1
Symbol	X	O	+	△	▽	□

BIAXIAL TESTING

Test specimens must contain a significant region of uniform stress to avoid plastic and creep redistribution of stresses and they must not fail at the grips. They should also allow a wide range of stress ratios to be used and should be simple to make. Cruciform plates [7] and

TABLE 2
Material behaviour parameters

Casting Ref.	Antimony content %	Instantaneous			Creep									
		Elastic	Plastic		Empirical					Primary			Secondary	
		E (GN $\times$ m^{-2})	B (MN $\times$ m^{-2})	q —	A —	H (MN $\times$ m^{-2})	a —	b (m$^2 \times$ MN^{-1})	c (m$^2 \times$ MN^{-1})^2	C (MN $\times$ m^{-2})	n —	p (h^{-1})	D (MN $\times$ m^{-2})	r —
P7	1·56	22·8	50·7	5·00	3·90	2·44	0·172	0·032	$-6·89 \times 10^{-4}$	38·4	5·28	0·060	138	4·92
P9	1·66	23·4	45·5	4·54	5·56	2·27	0·070	0·069	$-31·19 \times 10^{-4}$	37·4	4·98	0·058	433	3·44
C19 C20	1·45 1·48	23·4	82·9	4·95	3·75	2·73	0·175	0·027	$-6·51 \times 10^{-4}$	40·1	5·65	0·054	567	3·49

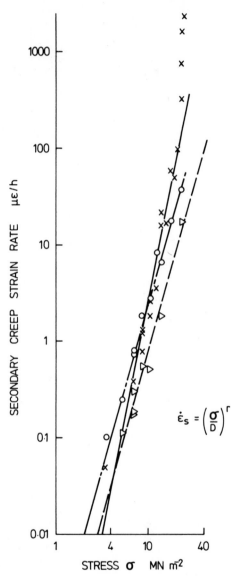

FIG. 5.   Uniaxial secondary creep rates; symbols as defined in Fig. 3(a).

thin-walled tubes [8] have been used; the former were rejected as too expensive and too liable to buckle in compression. Thin tubes are simple to make and can be subjected to tension, pressure or torsion. Combinations of internal pressure and torsion permit stress ratios from $1:0\cdot5:0$ to $1:0:-1$. This range, together with $1:1:0$ from thin hemispheres, was considered adequate.

The wall thickness of the tubular specimens must be large enough for the grain size of the material to be unimportant and to allow accurate machining; it should be small, relative to diameter and cylinder length, to minimise the difference between inner and outer surface stresses. These considerations led to tubes of 3 mm wall thickness, 50 mm inside diameter. The length of the tubes was determined by finite-element calculations using PAFEC 70+ [9] to give, under internal pressure, a uniformly stressed length equal to one diameter when both ends were rigidly clamped. (The finite-element method was used with the original intention of extending these calculations.) These elastic stresses are shown in Fig. 6. It was assumed that in torsion the end effects would be more localised.

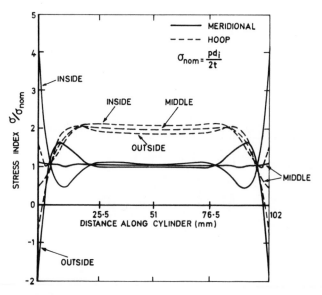

FIG. 6.   Elastic stress distribution in thin cylinder under internal pressure.

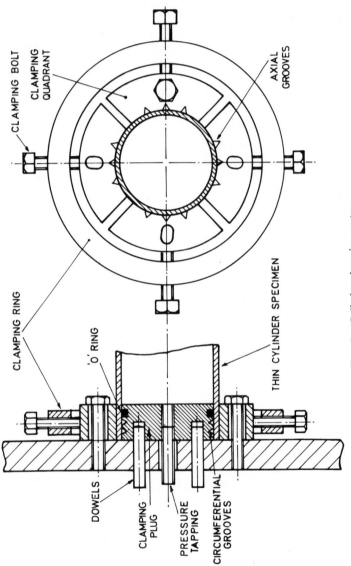

Fig. 7. Cylinder clamping system.

The hemispheres were end-closures of pressure vessels of constant wall thickness, the cylindrical portion of which had the same thickness/diameter ratio as the tubes and was one diameter long.

The tubes were machined from solid bars (called C), whereas the pressure vessels (called P) were cast to finished size.

Biaxial Loading Equipment

The clamping system is shown in Fig. 7. Grip of the thin cylinder is achieved by tightening the four clamping bolts, causing the lead to be extruded into the circumferential grooves in the sealing and clamping plugs and the axial grooves in the clamping quadrants. 'O'-rings are used to prevent oil leakage when the cylinder is pressurised. With this clamping system a pressure of 4 N mm^{-2} and a torque of 250 Nm have been applied without any leakage. A solid cylinder is placed inside the specimen to reduce the compressibility of the system.

The torsion system is shown in Fig. 8. The load application straps and the upper pulley support straps are made of thin steel strips (0·3 mm thick) which offer very little resistance to bending. As the upper pulley support straps are long compared with the change of length of the thin cylinders, the upper pulley is free to move in the axial direction without applying any axial restraint to the cylinder. The upper pulley is mounted on a roller bearing and it is balanced so that resistance to rotation is negligible. Bending caused by the weight of the lower pulley is eliminated during operation by a balance weight attached to the load-applying bar. This introduces a small additional torque on the cylinder. A knife edge fixed in the centre of the load-applying bar equalised the tension in the straps. Before applying the torque, the total weight (i.e. weights, load hanger, load-applying bar and balance weight) is supported by two levers mounted on a platform (shown in Fig. 9). During this time the lower pulley is balanced by additional balance weights. The application of the torque requires the simultaneous lowering of the load-applying bar and raising of the additional set of balance weights.

Pressure is supplied by a continuously running pump to a hydraulic accumulator which absorbs pressure surges; it is controlled by a relief valve and the seal-less piston shown at the right of Fig. 9, above the valve which is opened to load the specimen.

Small strains were measured with strain gauges mounted on the outside in the middle of the specimens. A symmetrical, multi-half-bridge system, continuously energised by a 4 V stabilised supply,

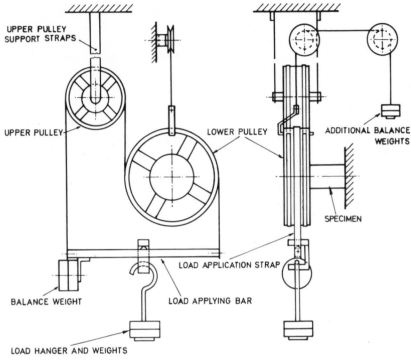

FIG. 8.   Cylinder torsion system.

maintained at $20 \pm \tfrac{1}{2}$°C, showed no more than $\pm 10\,\mu\epsilon$ drift in 300 h; the latter was measured separately and deducted from all readings. Large strains were obtained from displacement measurements. Twist was determined by a dial gauge from the vertical movement of the load-applying bar with a sensitivity of $6\,\mu\epsilon$ per division; the arrangement is shown in Fig. 9. Another dial gauge (also shown in Fig. 9) measured the axial movement of the lower pulley and, hence, the axial strain with a sensitivity of $25\,\mu\epsilon$ per division. Hoop strain was obtained with a sensitivity of $17\,\mu\epsilon$ per division from the diametral expansion measured by two pairs of opposing linear capacitance transducers which replaced the other dial gauges shown in Fig. 9. To eliminate the effect of out-of-roundness of the specimen, the transducers must be in contact with the same point on the specimen surface throughout a test, i.e. they must rotate during torsion. They are mounted in a concentric ring which surrounds the specimen and is

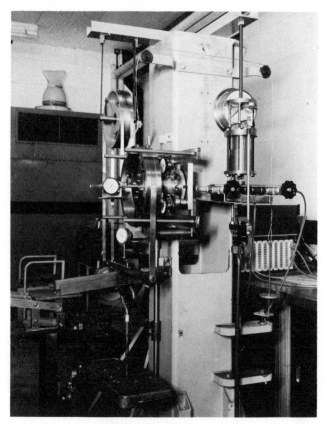

FIG. 9.   Cylinder test rig.

supported on four roller bearings, two of which are also visible in Fig. 9. A small projection (cemented with a plastic adhesive to the specimen surface) touches one of the transducers and provides the minute torque required to rotate the assembly.

Calibration with a strain-gauged aluminium specimen confirmed the finite-element results and showed that no bending moments were applied to the specimen.

BIAXIAL STRAINS

Multiaxial stress systems can be defined in terms of the principal stresses ($\sigma_1, \sigma_2, \sigma_3$) or a hydrostatic stress ($\sigma_{kk}/3 = (\sigma_1 + \sigma_2 + \sigma_3)/3$) and the deviatoric components $(S_{ij} = \sigma_{ij} - \delta_{ij}\sigma_{kk}/3$, where δ_{ij} is the Kronecker delta).

As in the uniaxial case, the components of total strain can be divided into instantaneous and creep components, i.e.

$$(\epsilon_{ij})_t = (\epsilon_{ij})_e + (\epsilon_{ij})_{pl} + (\epsilon_{ij})_c \tag{6}$$

The elastic strain components are given by

$$(\epsilon_{ij})_e = E_{ijkl}\sigma_{kl} \tag{7}$$

where E_{ijkl} are the usual functions of Young's modulus (E) and Poisson's ratio (ν). From unloading tests performed on two thin-cylinder specimens elastic values of Poisson's ratio of 0·439 and 0·435 were determined. For subsequent calculations a Poisson's ratio of 0·44 is assumed and the Young's modulus used is the mean of the values given in Table 2, which were obtained from the uniaxial tests. Table 1 gives the total initial strains obtained from the biaxial tests together with the calculated elastic and plastic strains. The third component of plastic strain is obtained by assuming that plastic deformation is a constant volume process.

For a monotonically loaded, statically determinate specimen the onset of yielding and the plastic strains depend on an effective stress, σ^*; the usual criteria are due to Tresca and von Mises (suffix VM). Using the von Mises criterion, Fig. 10 shows the effective initial plastic strains plotted against the effective stresses for various stress ratios. Also shown in Fig. 10 are the straight lines fitted to the uniaxial data in Fig. 3(a). Although the scatter of the biaxial data is greater than that for the uniaxial data and the effective strains in the tension-tension quadrant are generally greater than those in the tension-compression quadrant, the agreement is reasonably good. The flow rule associated with the von Mises effective stress criterion is the Prandtl–Reuss flow rule. Figure 11 shows the directions of the initial plastic strain vectors plotted on the π plane. For the Prandtl–Reuss flow rule to be applicable the strain vectors should be normal to the von Mises effective stress contours. For the tests which require only a single load application (i.e. internal pressure or torque) this was nearly the case. For the other tests (stress ratios $1:0:0$, $1:0·25:0$ and $1:0:-0·5$) the simultaneous application of internal pressure and torque to the thin cylinders was difficult and the strain vectors for these data were not as close to the normals. However, since most of the plastic strain vectors are reasonably close to the normals, the flow

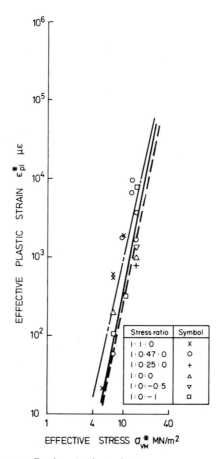

FIG. 10. Instantaneous effective plastic strains:

$$\epsilon_{pl}^* = \frac{2^{1/2}}{3}[(\epsilon_1 - \epsilon_2)^2 + (\epsilon_2 - \epsilon_3)^2 + (\epsilon_3 - \epsilon_1)^2]^{1/2}$$

$$\sigma_{VM}^* = \frac{1}{\sqrt{2}}[(\sigma_1 - \sigma_2)^2 + (\sigma_2 - \sigma_3)^2 + (\sigma_3 - \sigma_1)^2]^{1/2},$$

Full line, P7; chain dotted line, P9; dotted line, C19 and C20 combined.

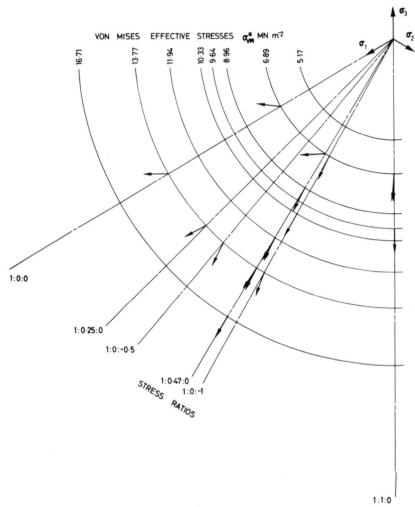

FIG. 11. Directions of initial plastic strains, presented on a π plane.

rule associated with the von Mises effective stress (i.e. the Prandtl–
Reuss flow rule) is applicable. Since Figs 10 and 11 show that the von
Mises effective stress and the Prandtl–Reuss flow rule are reasonably
accurate, the plastic term in eqn (2) can be generalised to

$$(\epsilon_{ij})_{pl} = \frac{3S_{ij}}{2\sigma_{VM}^*} \left(\frac{\sigma_{VM}^*}{B} \right)^q \tag{8}$$

Similarly, the creep strain vectors at 1, 10 and 100 h and the creep strain rate vector at 100 h have been shown to obey the Prandtl–Reuss flow rule.

Having established that the Prandtl–Reuss flow rule is applicable, noting that the only assumption made is that creep is a constant volume process, this has been used in conjunction with the von Mises effective stress criterion to predict biaxial creep strains for all the tests in Table 1. All these data are available [10]; typical examples are shown in Figs 12 and 13. When it is considered that creep usually has a great deal of associated scatter, Figs 12 and 13 and the other results in Ref. [10] show that the von Mises effective stress criterion gives reasonable predictions. Using it in conjunction with the Prandtl–Reuss flow rule, the predicted creep strains in 100 h have been compared with the measured creep strains in 100 h in Fig. 14. Also shown in Fig. 14 are the extreme values obtained from the uniaxial tensile tests of specimens taken from the P7, P9, C19 and C20 castings. Figure 14 shows that the creep strains in the tension–tension quadrant are generally greater than those in the tension–compression quadrant. However, the results indicate that a multiaxial creep law for

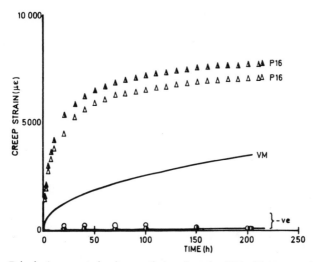

FIG. 12.   Principal creep strains from cylinder of casting P16 with stress ratio $1:0.47:0$ and $\sigma^*_{VM} = 11.94 \, \text{MN} \times \text{m}^{-2}$. The full lines are the predictions of principal creep strains based on the von Mises effective stress criterion and the Prandtl–Reuss flow rule. $\blacktriangle$, $\mathbb{O}$: Creep strains measured on inside surface. $\triangle$, $\bigcirc$: Creep strains measured on outside surface.

the creep term in eqns (2) can be written in the form

$$\epsilon_c = \frac{3AS_{ij}}{2\sigma^*_{VM}} \sinh\left[\frac{\sigma^*_{VM}}{H}\right] t^{m(\sigma^*_{VM})} \qquad (9)$$

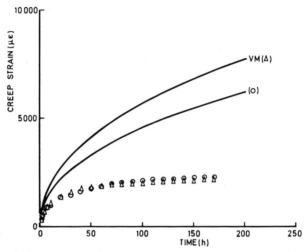

Fig. 13. Principal creep strains from casting C19/4 with stress ratio $1:0:-\frac{1}{2}$ and $\sigma^*_{VM} = 13.77 \, MN \times m^{-2}$. The full lines are the predictions of principal creep strains based on the von Mises effective stress criterion and the Prandtl–Reuss flow rule. $\triangle$, $\bigcirc$: Creep strains measured on outside surface.

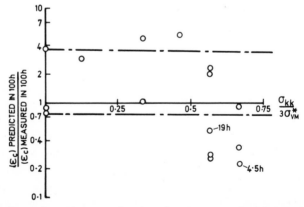

Fig. 14. Comparison of measured and predicted creep strains after 100 h. The chain dotted lines indicate the scatter obtained from uniaxial tests of specimens from the P7, P9, C19 and C20 castings.

RUPTURE

The breaking strength and ductility are material properties which are related to the times required for rupture to occur due to different, constant stresses and to the associated strains. Most of the data (for model and prototype materials) is obtained from uniaxial tests. At 20°C the lead alloys have an ultimate tensile strength of about 28 MN m^{-2}. Their elongation at fracture is between 15 and 30%, independent of rupture time up to 1000 h. The rupture strength falls from the ultimate strength to about 21 MN m^{-2} at 1000 h.

It is unlikely that the model material can be used to predict prototype rupture directly, but an experimental creep rupture reference stress method which gives safe predictions for most prototype materials has been developed by the authors [11].

COMPARISON WITH PROTOTYPE MATERIALS

Although recent model techniques [11–13] based on the reference stress approach show that close correlations between model and prototype material properties are not required, it is interesting to note some similarities between the lead alloy model material and prototype material properties. Values normally used [14] for primary creep time exponents, m, lie between 0·33 and 0·5, but for most prototype materials, experimentally determined primary creep time exponents are not usually available. However, for a CMn steel, m is stress-dependent and lies between 0·2 and 0·7 [15]. For the lead alloy m was found to vary between 0·26 and 0·5. A more complex primary time function such as that in eqn (5) is sometimes used. Values of α have been quoted for various materials [3–6], the order of magnitude varying from 10^{-3} to $10\,h^{-1}$. For the lead alloy α was found to be about $6 \times 10^{-2}\,h^{-1}$. From these comparisons it can be seen that the time function for the lead alloy is similar in form to that of many other materials.

The stress function used in eqn (2) (i.e. sinh (σ/H)) is not usually used for engineering calculations, the simpler power stress function (i.e. σ^n) being preferred.

Evans and Wilshire [3] found that for nickel, zinc and iron the stress index for the initial strain, q, was 2 and the B values were 600, 450 and 450 MN m^{-2}, respectively. For the lead alloy $q = 5$ and

$B = 60 \, \text{MN m}^{-2}$. For all of these materials the B values are of the same order of magnitude as the tensile strengths, but the q value for the lead alloy is significantly greater than the q values for the other materials.

Creep stress indices are quoted in the literature for many materials [3–5, 16–20]. For most practical materials these lie between 2·5 and 7. For the lead alloy the creep stress indices are in the range 3·5 to 5·7, making the material suitable as a creep model material.

The rupture ductility of practical materials as well as the rupture times are important. Rupture times and ductilities for prototype materials are quoted in ref. [21]. The rupture ductility of the lead alloy is between 15% and 30% for rupture times between 0·5 and 1000 h. This ductility is less than or equal to that for most practical materials. Also, the normalised creep curves for the lead alloy (i.e. ϵ_c/ϵ_R versus t/t_R) are above those for prototype materials. It has been shown [11] that these properties, used in conjunction with the rupture reference stress method, render the material suitable as a rupture model material.

DISCUSSION

The properties presented here are for one temperature only (20°C). Work is in progress to obtain uniaxial data at several other temperatures to extend the use of the material to models subjected to steady state temperature gradients and mechanical loading.

The initial plastic strains are similar in magnitude to the primary creep strains. The stress indices for initial plastic and primary creep response, q and n, respectively, are also approximately equal; they are greater than the secondary creep stress index, r.

Material properties are usually defined as the response to constant loads. The effect of varying mechanical loads has been studied [22] and it was found that the 'strain-hardening' hypothesis gives good predictions, but a 'total plastic work' hypothesis, which appears promising, has been proposed by the authors.

The model material obeys the von Mises effective stress criterion approximately for both plastic and creep strains. The Prandtl–Reuss flow rule is also applicable. Although there are few data for prototype materials, these assumptions are usually made.

Chill-casting the material into metal moulds produced small-grained castings which were stable, i.e. the creep properties did not vary with time between casting and testing. There was slight dendritic grain growth from the surfaces but the surface grain pattern appeared uniform. As the material was not 'worked' (rolled, pressed, etc.) before testing and the grain structure was found to be homogeneous, it was assumed to be isotropic. The casting procedures were relatively simple, but it was not possible to control the antimony content sufficiently accurately to reduce cast-to-cast variations of creep properties to the scatter within one casting. Although the overall scatter was less than found in much good-quality prototype material creep-testing, the systematic variation due to antimony content was allowed for in some other work [13]; creep strains at a particular time were plotted against antimony content and this curve was used for interpolation for other compositions.

The model material behaviour is similar to that of typical prototype materials; time indices, m, and stress indices, n, vary over similar ranges and creep rupture strains are also similar.

Experiments with model materials are simpler than with prototype materials because the loads are smaller and temperatures are lower. The lower model test temperatures permit the use of electrical resistance strain gauges and model tests can therefore be much more informative than prototype tests. By using experimental reference stress methods [11, 12] the almost unattainable similarity conditions [23, 24] for valid prototype response predictions are no longer necessary. These experimental reference stress methods also make it unnecessary to formulate accurate constitutive equations for model and prototype materials (a formidable task for variable load and variable temperature response when significant plastic strains occur).

CONCLUSION

Lead–antimony–arsenic alloys at room temperature have creep properties similar to those of prototype materials in the creep range. Because they exhibit the same features, results from model tests can be used to validate the complicated, non-linear programs necessary to predict component response directly. Model tests can also be used to determine experimental reference stresses for creep deformation and creep rupture.

ACKNOWLEDGEMENTS

The work was supported by the Department of Mechanical Engineering, University of Nottingham, while Dr Hyde held an SRC studentship. The authors gladly acknowledge the help of the technicians of their Department and of the Department of Geology who analysed the composition of many castings.

REFERENCES

1. PENNY, R. K. and LECKIE, F. A., The mechanics of tensile testing, *Int. J. Mech. Sci.*, **10**, 1968, 265–273.
2. BELLAMY, R. A. 'The Development of Model Techniques for Prediction of Creep Strains Applied to Steam Turbine Casings', Ph.D. Thesis, University of Nottingham, 1973.
3. EVANS, W. J. and WILSHIRE, B., Transient and steady state creep behaviour of nickel, zinc and iron, *Trans. Met. Soc. A.I.M.E.*, **242**, 1968, 1303–1307.
4. EVANS, W. J. and WILSHIRE, B., Work-hardening and recovery during transient and steady-state creep, *Trans. Met. Soc. A.I.M.E.*, **242**, 1968, 2514–2515.
5. EVANS, W. J. and WILSHIRE, B., Transient and steady-state creep behaviour of a Copper-15 at.-% aluminium alloy, *Metal Sci. J.*, **4**, 1970, 89–94.
6. BLACKBURN, L. D., Isochronous stress–strain curves for austenitic stainless steels, *A.S.M.E.*, 1972, 15–48.
7. HAYHURST, D. R., Creep rupture under multi-axial states of stress, *J. Mech. Phys. Solids*, **20**, 1972, 381–390.
8. JOHNSON, A. E., HENDERSON, J. and KHAN, B., *Complex Stress Creep, Relaxation and Fracture of Metallic Alloys*, H.M.S.O., Edinburgh, 1962.
9. HENSHELL, R. D. (Ed.), *PAFEC 70+ Manual'*, Department of Mechanical Engineering, University of Nottingham, 1975.
10. HYDE, T. H., 'Experimental Reference Stress Techniques for the Prediction of Creep Deformations using Lead Alloy Models', Ph.D. Thesis, University of Nottingham, 1976.
11. FESSLER, H., HYDE, T. H. and WEBSTER, J. J., 'Prediction of creep rupture of pressure vessels', A.S.M.E. Energy Technology Conference, Houston, Texas, September 1977, Paper No. 77-PVP-54.
12. FESSLER, H., HYDE, T. H. and WEBSTER, J. J., Stationary creep prediction from model tests using reference stresses, *J. Strain Anal.*, **12**, 1977, 271–285.
13. FESSLER, H., HYDE, T. H. and WEBSTER, J. J. 'Experimentally determined reference stresses for the prediction of creep deformation and life of components using models', to be published (Conference on Recent Developments in High Temperature Design Methods, I. Mech. E., November, 1977).

14. PENNY, R. K. and MARRIOTT, D. L., *Design for Creep*, McGraw-Hill, 1971.
15. CUMMINGS, W. M., 'Numerical methods for the analysis of creep strain data', British Steelmakers Creep Committee, *Proc. Symposium, October, 1971*, 139–161.
16. CARLTON, R. G., POYNER, J. and TOWNLEY, C. H. A., 'Creep data requirements for the design of power plant', British Steelmakers Creep Committee, *Proc. Symposium, October, 1971*, 12–22.
17. GULVIN, T. F., HACON, J., HAZRA, L. K. and MARES, H. W., 'The creep properties of carbon steels to BS 1501–161 and 224 grades', British Steelmakers Creep Committee, *Proc. Symposium, October, 1971*, 37–60.
18. GLEN, J. and HAZRA, L. K., 'Some information of creep behaviour of low alloy steels', British Steelmakers Creep Committee, *Proc. Symposium, October, 1971*, 61–94.
19. GERRARD, J. C. and MERCER, C., 'Low strain creep behaviour of 12% CrMoVNb steels', British Steelmakers Creep Committee, *Proc. Symposium, October, 1971*, 95–113.
20. MURPHY, M. C., 'Rating the creep behaviour of heat-resistant steels for steam power plant', *Metals Eng. Q.*, February, 1973, 41–50.
21. *B.S.C.C. High Temperature Data*, Iron and Steel Institute, 1973.
22. FESSLER, H. and HYDE, T. H., 'Creep deformation of metals', to be published in *J. Strain Anal.* (Special Creep Issue).
23. FREDERICK, C. O., Model correlations for investigating creep deformation and stress relaxation in structures, *J. Mech. Eng. Sci.*, 7, 1965, 57–66.
24. FESSLER, H. and BELLAMY, R. A., 'Use of models for the prediction of creep behaviour of components', I. Mech. E. Conf. Publication 13, 1973, 170.1–170.10.

13

Cyclic-Strain-Induced Creep under Complex Stress

M. R. BRIGHT and S. J. HARVEY

Lanchester Polytechnic

SUMMARY

The permanent and cumulative plastic strains occurring in the cyclic-strain-induced creep process are examined experimentally and analytically for the case of tubes in cyclic plastic torsion with superimposed axial tension at room temperature.

The development of material properties for cyclic hardening (EN 32B) and softening (EN 19) materials is shown to be relatively independent of the elastic follow-up stresses.

The mode of deformation within a cycle is examined in detail and the use of anisotropic yield surfaces and flow rules are discussed.

A yield surface is determined, applicable to the cyclic condition, and a region of large curvature is shown to exist at the loading point. This yield surface could be used to explain the modes of deformation occurring within a cycle of plastic strain.

NOTATION

T	shear yield stress
X	tensile yield stress
γ_p	plastic shear strain
$d\gamma_p$	plastic shear strain increment
ϵ_p	axial plastic strain
$d\epsilon_p$	axial plastic strain increment
$d\bar{\epsilon}_p$	equivalent plastic strain increment
σ	axial stress
τ	shear stress
τ_{max}	maximum shear stress in any half-cycle

INTRODUCTION

If a component is undergoing cycles of plastic strain, due to mechanical or thermal loading, and at the same time is subjected to a sustained or 'follow-up' load, then an irrecoverable plastic strain increment will occur in the direction of this load with each cycle of plastic strain. The effect is cumulative and can often result in failure of a component by gross deformation before fatigue failure occurs. Design rules against this ratcheting mechanism must ensure that the accumulated permanent strain is below a certain limit. The mode of strain accumulation at low temperatures is similar to that obtained in high-temperature creep tests, with primary, secondary and tertiary stages.

This cyclic creep or incremental collapse process has been examined previously for many loading conditions, including cyclic torsion of tubes with a steady end load and cyclic push–pull or torsion with a mean strain. An excellent survey of the problems and methods of analysis at low and high homologous temperatures has been given by Krempl [1]. Confining the discussion to time-independent problems, the cumulative creep strains are induced by the cyclic plastic strains, and a number of methods have been used to predict the strain accumulation. Empirical creep relationships have been proposed [2] to predict the secondary stages of cyclic creep, but, in general, the methods of approach have been based on the use of classical plasticity theories using idealised material properties, and some general proposals have been made which allow the inclusion of isotropic, kinematic and anisotropic changes in the yield surface. Total deformation theories have been used [3], together with assumed idealised stress–strain material behaviour, to examine the problem of pipe bends subjected to internal pressure and cyclic bending. The incremental flow theories of plasticity have been used to examine cyclic torsion of solid cylinders and tubes, with superimposed steady tension, under isotropic and kinematic hardening conditions.

Schwiebert and Moyar [4] adopted an approach equivalent to kinematic hardening in conjunction with an assumed linear hardening stress–strain relationship, and were able to explain qualitatively the primary stages of cyclic creep and predict that the cumulative strains reached a limit after a few cycles. Although their theoretical curves were similar in shape to the initial stages of cyclic creep, no quantitative comparisons between theory and experiment were presented.

Moyar and Sinclair [5] have discussed the application of isotropic hardening behaviour for strain cycling, and it can be shown that this approach predicts a limit in the cyclic creep and leads to shakedown. It has been shown [6] that the cumulative strains obtained experimentally are much higher than those predicted by the kinematic hardening approach and that the condition of shakedown does not occur for cyclic strains of reasonable magnitude. Many design procedures for pressure vessel components still rely heavily on the use of plasticity approaches developed for non-cyclic conditions, often incorporating material data which are inappropriate and do not represent the true cyclic material properties.

It is considered that to base analytical methods on the assumptions of isotropic or kinematic hardening behaviour is unrealistic. During the cyclic strain process the material may harden or soften and cyclic properties are developed which may lead to anisotropic material behaviour. The cumulative plastic strains in the direction of the follow-up loads may also result in the development of some anisotropy. An effective evaluation of the cyclic creep strains can only be made when realistic material data are used, together with the appropriate plastic stress–strain flow relationships.

A research programme was undertaken in which several cyclic-strain-induced creep processes have been examined, including cyclic plastic torsion of tubes with steady axial load or internal pressure and cyclic push–pull of tubes with steady internal pressure. The effects and influence of the cyclic properties were assessed by using a range of materials which cyclically harden or soften.

This paper deals essentially with the results of a series of tests in cyclic plastic torsion with steady axial loads, and examines the use of plastic flow theories, the nature of the plastic creep strains within a cycle and the possible shape and development of the yield surface. It is hoped that a better understanding of the factors controlling the mode and magnitude of the plastic strains will follow and that this will result in more realistic design procedures being developed.

APPARATUS AND SPECIMENS

The cyclic torsion testing machine was one which had been used in a previous investigation [6] and consisted essentially of a driving head operated by a motor and a reacting head, which incorporated elec-

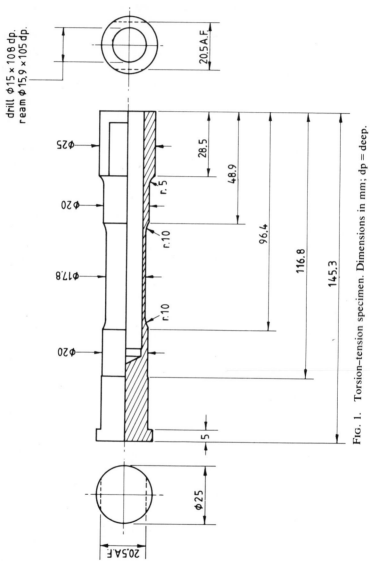

FIG. 1. Torsion–tension specimen. Dimensions in mm; dp = deep.

trical resistance strain gauge load cells to indicate the torque applied. The reacting head was fixed to a carriage, which was mounted on four linear bearings and was free to slide in the axial direction. The axial load was applied by a chain attached to the carriage which passed over a sprocket system to give a load magnification of 10 : 1. The head jaws were designed to transmit this end load to the specimen and also to impart torsion without bending. An example of the specimens used is shown in Fig. 1; the flats on the specimen were used to transmit the torque and the axial load was carried by the flange. The cyclic shear strain could be applied automatically, for fatigue investigations, or manually when the mode of deformation within a cycle was examined. The shear strain was monitored and controlled using a rotary inductive transducer over a gauge length of 0·5 in (12·7 mm). The axial extension was measured in several ways, using strain gauges, Huggenburger extensometers and displacement transducers. The electrical outputs of torque, shear strain and axial extension were monitored on a double-pen X–Y plotter.

The main test programme was carried out with a cyclic hardening carbon steel, EN32B (C1115); there was also a more limited range of tests on a cyclic softening material, EN 19 (A1S1 4140), used for high-temperature components and tested in the hardened and tempered condition. The range of axial loads used was 1070 lbf (4·76 kN), 1570 lbf (6·98 kN), 2070 lbf (9·21 kN), 2570 lbf (11·43 kN) and 3070 lbf (13·66 kN), giving initial axial stresses from 6·12 tonf in^{-2} to 17·57 tonf in^{-1} (94·6 MN m^{-2} to 271·6 MN m^{-2}). A range of cyclic shear strains up to ±6% was examined, but the results presented in this paper are for shear strains of ±3%, which represent the general cyclic creep behaviour.

EXPERIMENTAL RESULTS

The cyclic creep curves showing the cumulative plastic strains are shown in Fig. 2 for the EN32B (C1115) material. It can be seen that there is a transient region followed by an approximately constant steady state cyclic creep rate. These tests were not continued until failure and the tertiary stage was not reached. Although some axial extension does occur with zero axial load, it is negligibly small compared with that occurring when follow-up loads are present. Two typical steady state stress–strain hysteresis curves are shown in Fig.

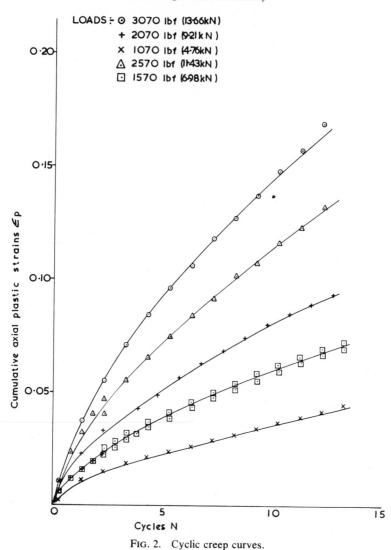

FIG. 2. Cyclic creep curves.

3. Although there are some slight differences in the shape, in general, the curves all tend towards the same peak shear stress. All the other curves lie within the two shown, and in view of the expected experimental differences, due to variations in material properties between specimens, it would be reasonable to assume that a common

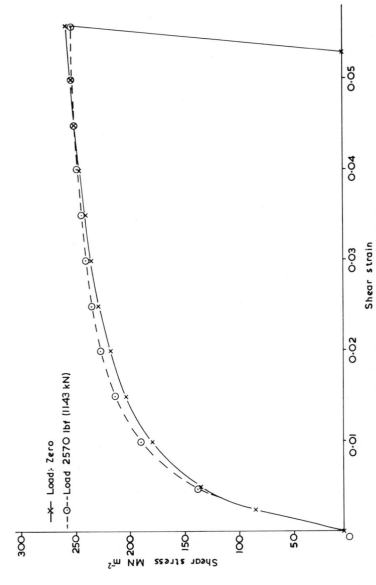

FIG. 3.   Steady state cyclic shear stress–strain hysteresis curve.

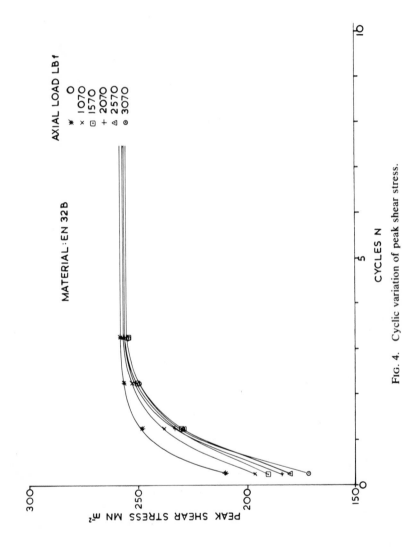

FIG. 4.   Cyclic variation of peak shear stress.

curve could be drawn which would represent the steady state cyclic behaviour, irrespective of the axial stress. The variations in peak shear stresses with cycles are shown in Fig. 4, where it can be seen that, although the expected differences exist in the initial cycles, with large axial stresses resulting in lower peak shear stresses, these differences disappear very quickly and all the specimens tend to cyclically harden towards the same steady state condition. In fact, for this material, a steady state condition was reached after about $3\frac{1}{4}$ cycles.

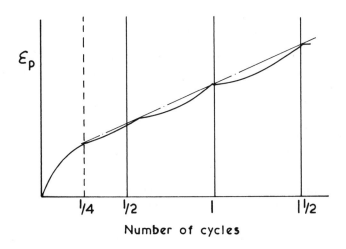

FIG. 5. Schematic development of cumulative plastic strains showing the changing pattern of strains with cycles.

The general development of the mode of deformation within a cycle is shown in Fig. 5. It is apparent that the character of the longitudinal strain changes after the first quarter-cycle. In the first quarter-cycle (i.e. the initial application of load) the axial extension is that which would be expected when isotropic hardening occurs, with the rate of extension being reduced as the shear stress increases. In subsequent cycles the opposite trend is observed and the rate of increase of extension increases towards the end of the cycle.

PLASTIC STRAIN IN FIRST QUARTER-CYCLE

The longitudinal plastic strains occurring in the first quarter-cycle were calculated using the deformation theory,

$$\epsilon_p = \frac{\sigma}{3\tau} \gamma_p \tag{1}$$

and the incremental flow theory,

$$d\epsilon_p = \frac{\sigma}{3\tau} d\gamma_p \tag{2}$$

Two general observations can be made when comparing the predictions of the two theories with the experimental results:

(1) The results show good agreement with both theories with respect to the mode of deformation and the magnitude of the plastic strain. This agreement is particularly good when the axial strains are small, i.e. when the axial stresses are small or during the initial period of deformation for the higher values of axial stress.

(2) For larger values of the axial strain the theoretical predictions begin to diverge and there is less agreement with the experimental results.

The strains which arise during the initial twist have been examined previously on numerous occasions, but perhaps not so frequently at such large values of shear strain.

In general, the predictions of the deformation theory agreed better with the experimental results. The observed trend of closer agreement at lower axial strains than at the larger axial strains is shown in Fig. 6, where the experimental and theoretical strains at the end of the first quarter-cycles are compared.

It can be easily shown that, for the case of proportional or radial loading, the incremental theory reduces to the deformation theory and the plastic strain is then a function of the current state of stress and is independent of the loading path. When the axial stresses are small, the loading path is not far from radial, so that the close agreement with both theories in this range is not surprising. Furthermore, it has been proposed by Budiansky [7] that there are ranges of loading paths other than proportional loading for which the basic postulates of plasticity theory are satisfied by deformation theories.

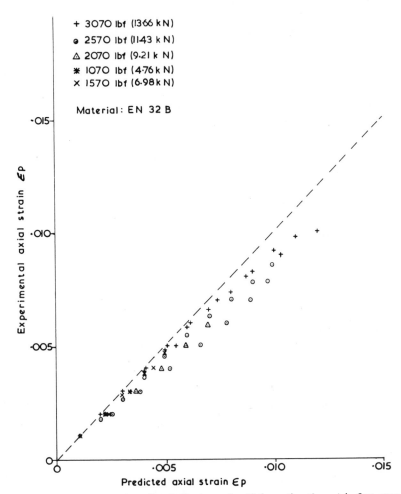

FIG. 6.   Experimental and predicted plastic strains (deformation theory) in first quarter-cycle.

The use of a deformation theory greatly simplifies the calculations of the plastic strains. However, the plastic strains cannot, in general, be independent of the loading path and there are theoretical objections to the use of the deformation theory.

With increase in axial stress the loading path becomes less radial and, consequently, the predictions given by the two theories will diverge.

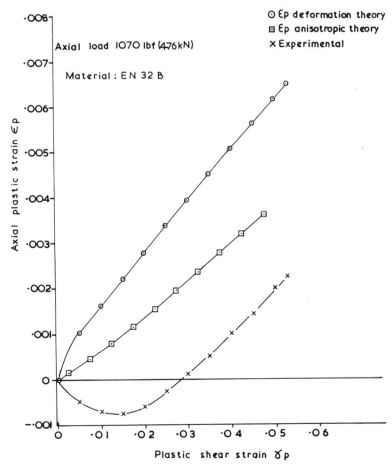

FIG. 7(a)–(e).   Experimental and predicted plastic strains during a half-cycle in the steady state cyclic conditions.

PLASTIC STRAINS IN STEADY STATE CONDITIONS

The changes in the axial strain within a half-cycle in the steady state cyclic condition are shown in Figs 7(a)–(e) for the full range of axial loads. The general deformation characteristics are apparent from these results.

At low values of the axial stress a shortening occurs on reversal of the shear stress and the tube only begins to extend in the latter part of

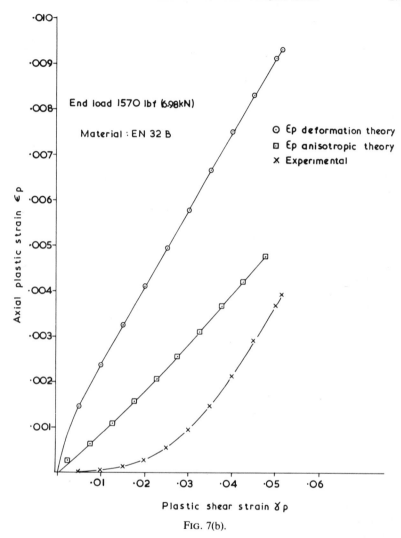

Axial plastic strain ϵ_p

Plastic shear strain γ_p

End load 1570 lbf (6.98kN)

Material : EN 32 B

⊙ ϵ_p deformation theory
□ ϵ_p anisotropic theory
× Experimental

FIG. 7(b).

this half-cycle (Fig. 7a). With increased axial stress the shortening effect disappears, but the longitudinal strain remains exceedingly low after shear stress reversal and then the rate of extension increases.

This deformation pattern cannot be explained by the deformation theory or the incremental theory, since both would predict a fall in

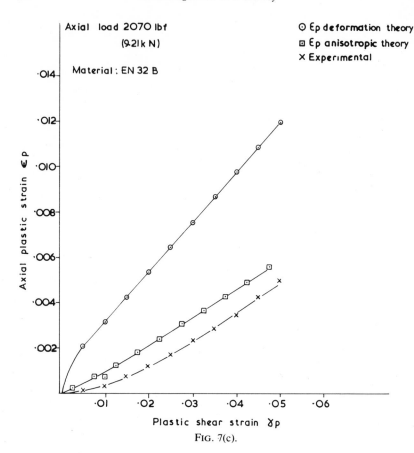

FIG. 7(c).

the rate of extension as τ increases for a constant axial stress σ, as was observed in the first quarter-cycle.

An anisotropic yield surface model has been proposed by Chandler [6] which can qualitatively predict the pattern of strains occurring within any half-cycle in this steady state region. The model takes the form of a series of expanding yield surfaces in the form of ellipses (see Fig. 8) which can be expressed in the form

$$\frac{\tau^2}{T^2} + \frac{\sigma^2}{X^2} = 1 \tag{3}$$

where T and X are the shear and tensile yield stresses, respectively.

On the basis of a series of tension tests on the cycled material it

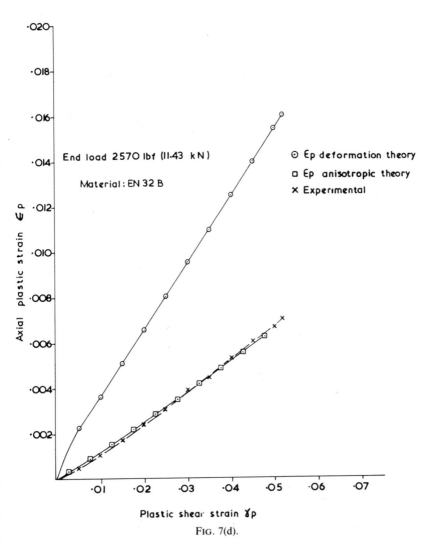

FIG. 7(d).

was concluded [6] that the tensile yield stress remained constant throughout the cycle with the value

$$X = (3^{1/2}\tau_{max} + \sigma)$$

It was also suggested that the above expression would account for some strain hardening in the axial direction due to the cumulative

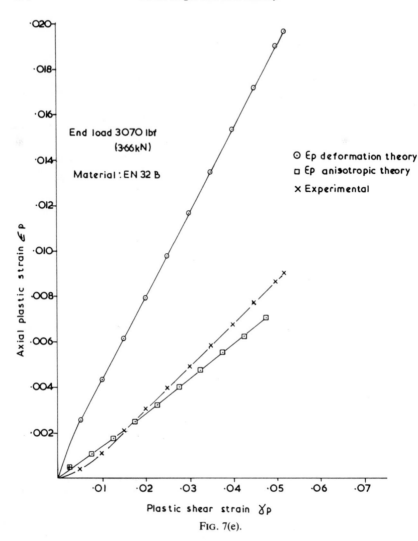

FIG. 7(e).

plastic strain. The longitudinal yield stress was defined on the basis of a 0·1% proof stress, and in view of the fact that the shape of the yield surface is extremely sensitive to the definition of yield, the range and validity of eqn (3) in describing the yield surface at the loading point is open to question and requires further investigation. It was pointed out that this anisotropic approach was not generally applicable and

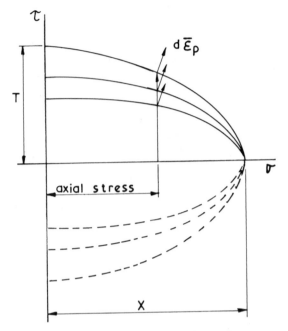

FIG. 8. Schematic anisotropic yield surface.

other forms would have to be found for other loading conditions. However, the model can predict the general strain pattern within any half-cycle.

With the strain increment vector normal to the yield surface, the axial plastic strains are given by

$$d\epsilon_p = \frac{\sigma\tau \, d\gamma_p}{X^2 - \sigma^2} \tag{4}$$

This equation can be solved either analytically or graphically when the appropriate τ–γ relationship is known. Since the steady state cyclic condition is being considered, then the steady state shear stress–shear strain curve (Fig. 3) can be considered to be a basic material property and used to evaluate eqn (4).

The strain prediction using this anisotropic theory and the deformation theory are shown in Figs 7(a)–(e) and compared with the experimental results.

The deformation theory grossly overestimates the extensions for all values of axial stress. The anisotropic approach predicts the general strain pattern, but cannot predict the shortening behaviour and certainly overestimates the plastic strains at the lower values of axial stress. However, as the value of the axial stress increases, good and sometimes excellent agreement with the anisotropic theory is obtained.

These results show that if the axial strains in each half-cycle are calculated using the deformation theory and the steady state hysteresis curve, this would lead to a considerable overestimate of the cumulative creep strains, but would constitute a simple and conservative design procedure. A general and perhaps important observation can be made regarding these results. Although large errors can occur when the anisotropic theory is used, particularly immediately after stress reversal at low axial stress, as the shear strains increase, the experimental rate of increase of the axial strain agrees remarkably well with the anisotropic prediction. We could conclude from this that eqn (4) predicts the incremental plastic strains accurately over a considerable portion of the half-cycle and eqn (3) provides a reasonable representation of the yield surface at the loading point in this range.

EFFECT OF LOAD CHANGES DURING CYCLING

The effect of changing the axial load at the end of any half-cycle in the steady state condition is shown in Figs 9 and 10. It can be seen from these results that the steady state cyclic condition, reached with a particular axial load, has a significant influence on the subsequent deformation when the loading conditions are changed. After cycling to a steady state with a load of 2570 lbf (11·43 kN), and then removing this load and recycling, the deformation pattern shown in Fig. 9 is obtained. A significant shortening of the tube occurs over several cycles before the expected zero length change conditions are reached. A similar shortening effect was measured for other axial load reductions. When the axial load is increased after cycling to a steady state, a transient period exists (Fig. 10) before a new steady state condition is established. During this transient period the rate of axial extension is higher than that occurring either before or after the axial load is increased.

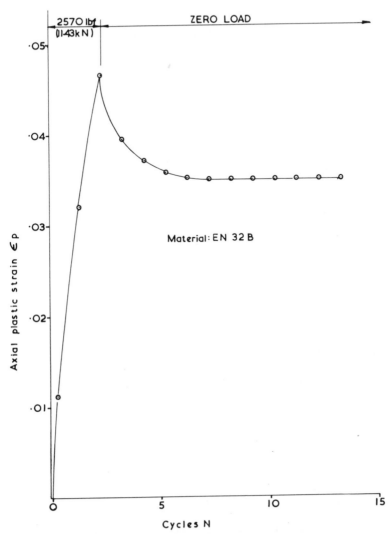

FIG. 9.   Effect of axial load reduction on subsequent plastic strains.

THE YIELD SURFACE

It was apparent from these results that a closer examination of the yield surface would be necessary if the deformation phenomena are to be satisfactorily explained. However, the accurate determination of

M. R. Bright and S. J. Harvey

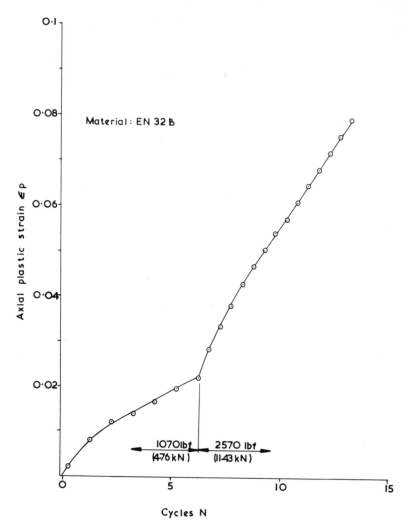

FIG. 10. Effect of axial load increase on subsequent plastic strains.

the shape of the yield surface at any stage of the cyclic process would be difficult. It is now well established that many stress 'probes' into the yield surface cannot be made without distorting and changing the shape of the surface significantly.

It was therefore decided that 'probes' into the yield surface should be restricted to a few important points from which the general shape could be inferred. For example, if the axial tension and compression stress–strain curves could be determined and the shape of the yield surface in the region of the loading point established, the general shape of the yield surface could be found. Tension and compression tests were carried out on specimens which had been cycled to a steady state with a range of axial loads. As might be expected, a Bauschinger effect was apparent, with the tensile stress–strain curve lying above the compression curve. In addition, departure from linearity starts much earlier in compression than in tension. The results show that if departure from linearity is used to define yield, then the tension and compression yield stresses are significantly smaller than $(3^{1/2}\tau_{max} + \sigma)$.

The general shape of the yield surface in the region of the loading point was determined by observing the nature of the plastic strain increments when the axial load was slightly reduced or increased during a half-cycle. The results confirmed the earlier ones (Figs 9 and 10) when the loads were changed. A small reduction in axial load resulted in a shortening of the specimen when the shear strain was reapplied. An increase in the rate of axial extension occurred when the load was slightly increased.

After an extensive range of tests it was concluded that, at the loading point, a region of large curvature existed and it was possible that a corner was present. If a yield surface is constructed on the basis of all the data obtained, then the presence of a corner is a distinct possibility. The general shape of the surface is shown in Fig. 11.

The yield surface ABC is proposed on the basis that no plastic shear strains occur during unloading to zero shear stress. If plastic shear strains do occur during this unloading period, and there are indications that they do for some conditions, then the yield surface would not enclose the origin, although it was considered that its shape would be similar to that shown in Fig. 11.

The presence of a corner or a region of large curvature at the loading point could explain the modes of deformation observed during (i) shear stress reversal and (ii) changes of axial load. The corner must develop as loading proceeds, and the axial strains on stress reversal would be small and may even be negative, resulting in the axial shortening which was observed for small axial loads. The

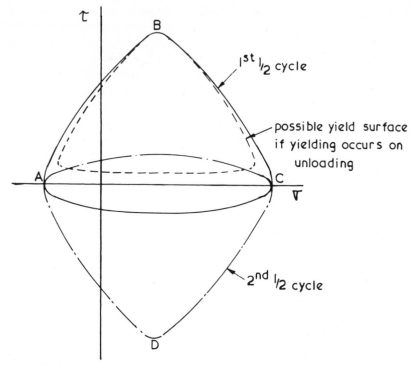

FIG. 11.   General shape of yield surface in steady state cyclic condition.

yield surface on the subsequent half-cycle would therefore develop from ABC to that shown by ACD (Fig. 11).

CYCLIC SOFTENING MATERIALS

A wide range of cyclic tests had been carried out previously [6] using EN 19 (AISI 4140) and some were repeated in this investigation. It was confirmed that the general development of the strains, both cumulative and within a cycle, followed the same pattern as observed for the cyclic hardening EN 32B (C1115). The material cyclically softened to a steady state and the no-load cyclic stress–strain curve could be used to approximate to the cyclic stress–strain curves when follow-up loads were present. The cyclic softening was relatively slow and a long period was required to reach steady state. The yield surface model proposed for the cyclic hardening material would appear to be equally applicable for this cyclic softening material.

APPLICATION TO DESIGN

The problems associated with predicting accurately the cumulative plastic strains are enormous. The problems are not so much those of computation as of having sufficient material data available at all stages of the cyclic process.

It would be necessary to know the cyclic material property for each cycle and the shape and development of the yield surface at every instant. However, it would seem that some simplifications could be made which could lead to conservative design procedures. For example, many materials will cycle to a steady state quite rapidly and it has been shown in this investigation that the steady state stress–strain hysteresis curve with follow-up loads is approximately the same as that obtained without axial loads.

It was found that the same modes of deformation occurred within a cycle for both the cyclic softening and cyclic hardening materials. The EN 19 (AISI 4140) material cyclically softened to approximately the same steady state stress–strain curve for a wide range of axial loads and cyclic strain, although it took considerably longer to reach this condition.

The steady state cyclic shear stress–shear strain curve might be considered as basic material data, since it would often be determined during the course of testing to find the low cyclic fatigue data, and could be used as a basic property for predicting the plastic strains.

If it is used in conjunction with a deformation theory, for simplicity, it would lead to an overestimate of the plastic strain in each half-cycle and therefore in the cumulative cyclic creep curve, thus providing a very conservative design procedure. If the anisotropic theory is used, then quite accurate predictions of the cumulative strains can be made, but there is a danger of underestimating these strains for large follow-up stresses.

If materials cyclically harden or soften to a steady state, then design procedures can be simplified, particularly if this steady state is reached rapidly. Many of the stainless steels used in pressure vessels do not reach a steady state condition, but continue to cyclically harden considerably, and further work needs to be done for these materials. It should be possible to use a cyclic stress–strain curve which represents average cyclic behaviour, but since this curve would give stress values larger than those obtained in the initial cycles and smaller than those relevant to the later cycles, it could lead to some errors in strain predictions at various stages of life.

CONCLUSIONS

A simple idealised stress system, amenable to plasticity analysis, has been used in this investigation of the cyclic-strain-induced creep process. It has been shown that the cyclic material property will influence and control the nature and magnitude of the cumulative plastic strains. In many practical situations such simplified stress systems will not be encountered, but the results of this investigation have given a good indication of how the yield surface develops and changes throughout a cycle of plastic strain, and this can form the basis for investigating other cyclic creep systems. For example, it has been shown that the nature and magnitude of the cumulative expansion of a tube, subjected to a steady internal pressure, is exactly the same as the strains of a tube with axial load and cyclic shear, when a common loading programme is followed [8].

Difficulties with this approach will obviously arise if there are significant and regular variations in the follow-up loads, since these will lead to continued changes in the shape of the yield surface and the cyclic material property.

The stress systems examined so far are those in which the direction and type of follow-up stress are different from those of the cyclic stress. More practical examples exist when the follow-up stress is the same as the cyclic stress: for example, a pressure vessel or tube subjected to cycles of push–pull with a sustained internal pressure. In the case of cyclic torsion–tension the axial material properties are influenced by shear on transverse sections, whereas in the push–pull internal pressure systems the material properties in the circumferential direction will be more closely related to the effects of the longitudinal cyclic stresses. Tests have been carried out for this system [8] and it can be shown that the cyclic strain pattern and the magnitude of the cumulative strains differ significantly from those in cyclic torsion–tension, and the anisotropy is developed on different planes.

Obviously there remains much more to be done, but it is considered that the approach adopted for determining the shape of the yield surface in this investigation enables a much more fundamental understanding of the cyclic creep process to be achieved. It is capable of extension to other stress systems and essentially uses realistic cyclic material data, which in many instances would be readily available.

REFERENCES

1. KREMPL, E., 'Cyclic Creep—An Interpretive Literature Survey', Welding Research Council Bulletin No. 195, 1974.
2. FELTNER, C. E. and SINCLAIR, G. M., 'Cyclic stress-induced creep of close-packed metals', *Joint Int. Conference on Creep*, I. Mech. E., 1963.
3. EDMUNDS, H. G. and BEER, F. J., Notes on incremental collapse in pressure vessels, *J. Mech. Eng. Sci.*, 3, 1961, 187–199.
4. SCHWIEBERT, R. D. and MOYAR, A. J., 'Application of Linear Hardening Plasticity Theory to Cycle and Path Dependent Strain Accumulation', Dept. of Theoretical and Applied Mech., University of Illinois, Report No. 212, 1962.
5. MOYAR, G. J. and SINCLAIR, G. M., 'Cyclic strain accumulation under complex multiaxial loading', *Joint Int. Conference on Creep*, I. Mech. E., 1963.
6. CHANDLER, H. D., 'Cyclic Strain Induced Creep and Fatigue under Biaxial Stress Conditions', Ph.D. Thesis, CNAA, Lanchester Polytechnic, 1971.
7. BUDIANSKY, B., A re-assessment of deformation theories of plasticity, *Trans. ASME, J. Appl. Mech.*, 26, 1959, 259–264.
8. BRIGHT, M. R., 'Cyclic Creep under Complex Stress', Ph.D. Thesis, CNAA, Lanchester Polytechnic, 1977.

14

Plasticity of Steels in Reversed Cycling

J. C. RADON

Imperial College of Science and Technology

AND

P. W. J. OLDROYD

The Polytechnic of Central London

SUMMARY

Earlier investigations have shown that, even in the case of metals known to have the same shape of curves for unidirectional tension and compression tests (e.g. annealed copper), there may exist no simple relationship between the phenomena of cyclic hardening (or softening) and cyclic creep. The former depends on changes of shape common to both sides of the stress–strain loop, but the latter is more sensitive to changes that may occur in the relative shapes of the two sides of the loop—particularly if these are where the curves approach the horizontal as they near the cusps.

The present investigation is concerned with the even more complex behaviour that may be expected with certain other materials (e.g. those of b.c.c. structure). Cast-iron and steel have been studied as materials displaying marked differences in tensile and compressive behaviour (at least for the first few cycles) and stainless steel as a material liable to be influenced by progressive metallurgical changes during cycling.

The latter material has also been used for an investigation into the effect of mean stress on cyclic creep in order that a comparison might be made with the results of an earlier investigation into the behaviour of annealed aluminium.

INTRODUCTION

In the study of plastic cyclic deformation between fixed limits of stress the terms 'cyclic hardening' and 'cyclic softening' are used to describe decreases and increases in strain range, respectively, and the term 'cyclic creep' to describe progressive change in strain from cycle to cycle.

Earlier investigations have shown that the relationship between these phenomena is not a simple one [1, 2]. For instance, a rapid increase in the rate of cyclic creep need not be accompanied by cyclic softening. When rapid cyclic hardening takes place during the early stage of cycling, either positive or negative creep can be observed during the same cycle—depending on whether the strain is observed after the removal of the compressive or the tensile load. This is illustrated in Fig. 1, which is an autographic record of the cyclic behaviour of austenitic cast-iron. (Tensile stresses and strains are defined as positive and compressive stresses and strains as negative.) The problem of defining strain range and creep per cycle is not a simple one and it has been fully discussed in an earlier paper [3].

Because of the large changes that are liable to occur in both strain range and creep per cycle, it is convenient to use a logarithmic scale when plotting values of these quantities. This has been done in the earlier work and the same procedure has been adopted here. In the present investigation cases occur in which the sign of the creep per cycle changes during cycling. As the graphs show only numerical values, it has been necessary to label the curves to indicate positive or negative creep. In the case of a reversal of creep there is a discontinuity in the curve.

In their earlier work the authors comment on the possible large effects that minute changes in the form of the settled cyclic stress–

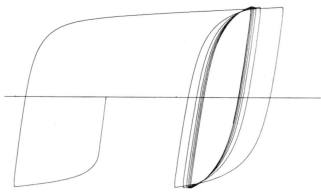

FIG. 1.   Cast-iron. Load cycling, first seven cycles. (Stress amplitude 24·25 tonf in^{-2}.)

strain loop may have on the rate of creep per cycle [3]. In studies of the fatigue-softening of low-carbon steel by the expansion of zones of plastic deformation Abel and Muir [4] have shown that the change in shape of the curves that form the settled cyclic loop is towards a steeper rise and an almost horizontal approach to the peaks of the loop. Such a tendency towards levelling off at the peak values can explain an increased rate of creep per cycle. Since an almost horizontal stress–strain curve indicates an approach to instability in tension, such changes may also account for the intermittent high values of creep per cycle observed by the present authors in tests made on cold-worked copper [3].

In addition to small changes in the form of the settled cyclic loop, some metals show large changes in the form of the curves that compose the loop in the course of the earlier cycles of deformation. The metallurgical reasons for these large changes differ from metal to metal and, in some cases, the earlier changes may introduce a considerable lack of symmetry in the tension–compression proper-ties—the effect of which may persist for many cycles—which may have a profound effect on the rate of creep per cycle during the early life of the material.

The metals chosen for the present investigation are ones in which the first or second loading is known to introduce a lack of symmetry in the first cycle. Since this lack of symmetry is eliminated by subsequent cycling, it is to be expected that it might result in gradual changes in the shape of the curves that form the loops and that the effect of such minute changes might be revealed by appreciable changes in the rate of creep per cycle.

EQUIPMENT AND SPECIMENS

Cyclic tension–compression tests were carried out between fixed limits of stress at room temperature using equipment described in Ref. [5]. The test results were recorded autographically. The stresses are nominal stresses and the strains axial strains derived from the measured changes in the waist diameter of the specimens. The form of the specimens and the end fittings have been described elsewhere [6].

The materials tested were austenitic cast-iron (2·16% C), cast mild steel (0·21% C) and austenitic stainless steel (0·06% C, 8·6% Ni and 18·2% Cr).

RESULTS

The as-recorded stress–strain loops for an austenitic cast-iron specimen tested at a nominal stress amplitude of $\pm 24 \cdot 25$ tonf in^{-2} are
shown in Fig. 1. The decrease in mean loop width shows the cyclic
hardening of the material. If the first loop is disregarded, it will be
observed that the intersection of the left-hand side of the loop with
the axis indicates positive creep and that of the right-hand side
negative creep.

Autographic records similar to that shown in Fig. 1 were used to
produce the curves in the figures that follow.

The total creep for austenitic cast-iron cycled at various stress
amplitudes between equal limits of tensile and compressive stress is
plotted against number of cycles in Fig. 2. The creep values for the
right-hand side of the loops only are shown. The apparent final
increase in the rate of (positive) creep is caused by the development of
cracks in the specimens.

Figure 3—which, like Figs 4–9, is drawn with logarithmic scales for
both axes—shows creep per cycle against number of cycles for the
specimens represented in Fig. 2; Fig. 4 shows the corresponding
curves representing axial plastic strain range against number of

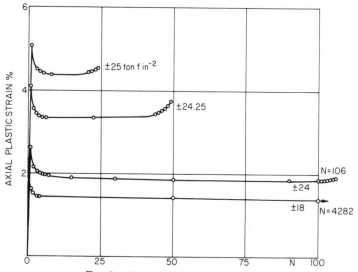

FIG. 2.   Total cyclic creep of cast-iron.

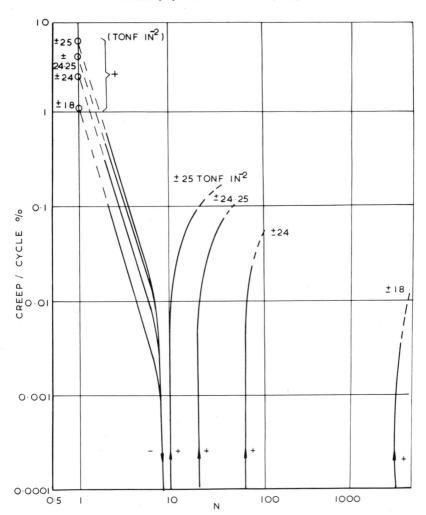

FIG. 3.   Austenitic cast-iron. Load cycling. Cyclic creep/cycle versus *N*.

cycles. Comparison of Figs 3 and 4 shows that during the main part of the life of the specimens (prior to the starting of a crack) there is no appreciable change in the hardness of the material and the creep per cycle is negligible.

Figure 5 shows the creep per cycle against number of cycles for a specimen of cast mild steel; Fig. 6 shows the corresponding curves

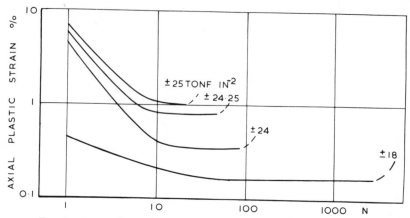

FIG. 4.   Austenitic cast-iron. Load cycling. Axial plastic strain range versus *N*.

representing axial plastic strain range. The curves representing creep recorded for both sides of the cyclic loop are shown. After the first cycle both curves show a positive rate of creep per cycle but this decreases from cycle to cycle. This decrease in the rate of cyclic

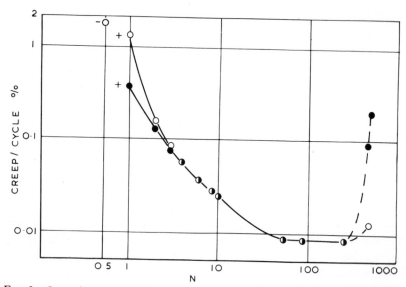

FIG. 5.   Cast mild steel. Cyclic creep/cycle. Load cycling ± 24 tonf in^{-2}. ●, Tension; ○, compression.

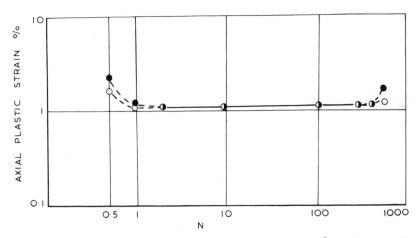

FIG. 6. Cast mild steel. Strain range/cycle. Load ± 24 tonf in⁻². ●, Tension; ○, compression.

creep is not, however, associated with an appreciable change in hardness of the material.

The behaviour of stainless steel is more complicated than that of mild steel because, after a certain amount of cyclic plastic deformation, a metallurgical change occurs which causes rapid hardening of the material. Tests made between equal strain limits show that after the initial cyclic hardening, which at small strain amplitudes is gradual, there occurs a rapid cyclic hardening of the material. The smaller the strain range used the later the onset of rapid hardening produced by changes in the lattice structure [7]. Figure 7 shows that similar behaviour occurs when the tests are made between fixed limits of cyclic stress. With a stress amplitude of ± 25 tonf in⁻² the settled value of the cyclic strain range is 0·95% and there is no change until 100 cycles, when a decrease in the strain range marks the onset of rapid hardening. This agrees well with the results obtained during cyclic tests made between fixed limits of plastic strain. With a constant strain range of 0·85% the onset of rapid hardening occurs shortly after 100 cycles [7]. It is evident that the onset of rapid cyclic hardening can be estimated from the average value of the strain range regardless of the form of the test.

In the tests that followed the limits were not equal (for either stress or strain) but, as the plastic strain range was always less than 0·5%, it was considered reasonable to suppose that rapid hardening would not

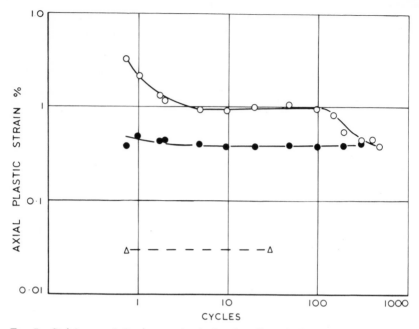

FIG. 7.　Stainless steel. Strain range/cycle. Load cycling ±25(○), ±20(●) and ±10(△) tonf in⁻².

occur before 1000 cycles. The results of the tests made on stainless steel could therefore be considered to be free from the effect of lattice structure changes induced by prolonged cycling.

Figure 8 shows the creep per cycle against number of cycles for two specimens of stainless steel, one cycled between 0 and + 25 tonf in^{-2} and the other between 0 and + 20 tonf in^{-2}. As is to be expected, apart from some random variation in the observed values, the creep per cycle for the latter specimen is always less than for the former. In both cases a rapid decrease in the creep per cycle occurred as cycling continued.

In Fig. 9 the specimen cycled between 0 and + 20 tonf in^{-2} is compared with one cycled between − 10 and + 20 tonf in^{-2}. (The initial loading was tensile.) Because of the Bauschinger effect the latter specimen shows a smaller amount of creep in the first cycle than does the former. However, the rate at which the creep per cycle decreases is less for the latter than the former and, as a result, the curves cross. This may be because the reversals of plastic deformation in the latter produce some cyclic hardening of the material.

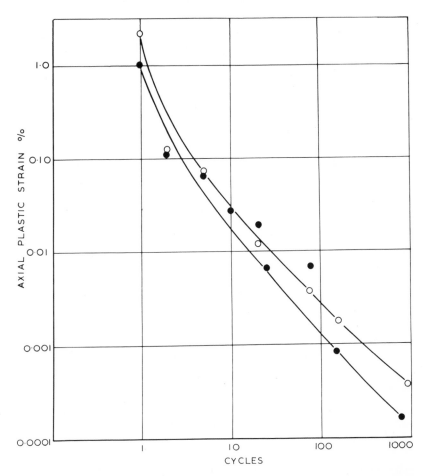

FIG. 8.   Stainless steel load cycling. Creep/cycle versus cycles. 0 to 25(○) and 0 to 20(●) tonf in^{-2}.

In an earlier paper the authors discussed the effect of a small mean stress on the rate of cyclic creep of annealed aluminium [1]. The opportunity was taken to make similar tests—though over only a limited range of mean stress—on a specimen of stainless steel. A stress range of 40 tonf in^{-2} was used and the mean stress raised in steps to 6 tonf in^{-2} in tension (no compressive values were used). As the strain range was always small, it is not thought that the results were influenced by the phenomenon of rapid cyclic hardening. Only sufficient cycles to achieve a stable loop for each mean stress value

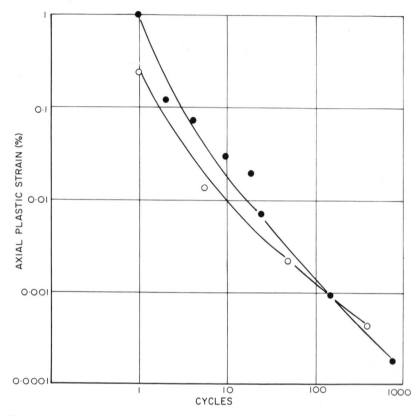

F𝚒ɢ. 9. Stainless steel load cycling. 0 to +20(●) and −10 to +20(○) tonf in⁻².
Creep/cycle versus cycles.

were applied. Figure 10 shows that, unlike aluminium, the stainless
steel gives a curve which has an increasing slope with initial increase
of mean stress.

DISCUSSION AND CONCLUSIONS

Although both cyclic hardening or softening and cyclic creep result
from changes in the shapes of the two curves that form the cyclic
loop, the phenomena are not directly related. When a mean stress is
present, cyclic creep results from the marked difference in shape of
the two curves and it follows that any changes of shape of the curves

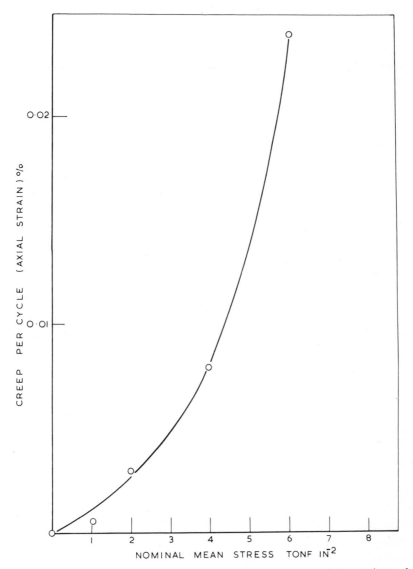

FIG. 10.   Variation of creep rate with mean nominal tensile stress: single specimen of austenitic stainless steel cycled at a stress range of 40 tonf in⁻².

that results in cyclic softening must also cause an increase in the rate of cyclic creep. If no mean stress is present, it is possible for the cyclic creep to change appreciably without there being any visible change in the hardness of the material (as represented by the loop width). Cyclic hardening or softening results from a progressive change of shape occurring in the two curves that form the loop, but cyclic creep results from the development of a difference of shape between the two curves. Under certain circumstances a small difference in shape can produce a high rate of creep per cycle. The changes in shape can be of particular importance if they occur in the part of the curve near the peaks of the loop.

Studies of the cyclic hardening of annealed metals such as copper [3], using fixed limits of strain, show that their progress towards a settled cyclic state does not depend on whether the initial loading is tensile or compressive. The progress is, however, influenced by prior cold-working of the material by initial tensile or compressive overstrain. Similar phenomena are observed when cycling between fixed limits of stress, but in this case cyclic creep can occur—even with zero mean stress—in addition to cyclic hardening and softening. The nature of this creep can be very sensitive to the sign of the initial plastic overstrain. It has been shown (unpublished work, P.W.J.O.) that annealed copper, which when cycled at zero mean stress normally has a positive rate of creep, will, if subjected to a large initial positive overstrain, have a negative rate of creep at the commencement of cycling. As cycling continues, the rate of creep decreases to zero and then the material tends to assume its normal positive rate of creep. The creep rate can be profoundly affected by initially biasing the material by giving it markedly different tension–compression properties prior to cycling. In the case of copper this bias results from the Bauschinger effect which arises from unidirectional overstrain. With the materials used in the present investigation the bias introduced during the first cycle results from a lack of symmetry in the tension–compression properties of the material.

In the case of cast-iron the carbon that fills the cavities provides help to the material during the first half-cycle in which the material has to resist compressive stresses. In the next half-cycle—during which the stresses are tensile—the cavities behave as if empty and the large stress concentrations associated with them result in the material undergoing considerable strain. It is suggested that the carbon provides less and less assistance in helping to withstand compression during subsequent cycles because unidirectional deformation has caused the walls of the cavities to lose contact with the carbon. This suggestion supports the experimental evidence that after a few cycles there is little creep and little cyclic hardening.

In the case of cast mild steel the locking of dislocations by the presence of interstitial atoms provides much of the resistance to deformation during the first loading. Provided that the strain is adequately large—as is the case in the present tests—the freeing of the dislocations will occur entirely in the first half-cycle. The Frank–Read source [8], which accounts for the generation of dislocations when the loading is in one direction, may equally well generate dislocations when the loading is in the other direction. However, the movement of dislocations cannot be supposed to be a simple to-and-fro motion along the same paths—they may, for instance, climb to new planes of movement. It is, therefore, to be expected that the nature of the pattern of dislocation movement produced in the first loading will, to some extent, influence the cyclic stress–strain behaviour during many subsequent cycles. It is suggested that this is the cause of a small gradual change in the two curves that form the loop, resulting in changes in the rate of cyclic creep long after the loop has assumed the sensibly symmetrical shape associated with the absence of cyclic hardening or softening.

It is assumed, from the evidence of cyclic tests made on stainless steel between equal limits of stress (or strain), that within the number of cycles made during the present tests there would be no appreciable cyclic hardening or softening of the material. The phenomena observed for cycling between zero and a maximum tensile stress or a small compressive and a maximum tensile stress are therefore to be associated with the small gradual changes in shape of the two curves that form the loop—as described in the case of cast mild steel. In this case the behaviour is particularly sensitive to changes occurring at the top of the tensile curve where this tends towards the horizontal. The tests made to investigate the effect of mean stress on cyclic creep in stainless steel suggest that the behaviour is more complex than is that of aluminium.

REFERENCES

1. OLDROYD, P. W. J. and RADON, J. C., Behaviour of aluminium in uniaxial stress cycling, *Proc. Symp. Mech. Behaviour of Materials, Kyoto, Japan,* 1974, pp. 251–256.

2. KLESNIL, M. and LUKAS, P., Dislocation structure associated with fracture surface of fatigued copper single crystals, *Phil. Mag.*, **17**, 1968, 1295–1298.
3. RADON, J. C., OLDROYD, P. W. J. and BURNS, D. J., Mean strain and hardness changes in uniaxial stress cycling, *Proc. Int. Conf. Mech. Behaviour of Materials, Kyoto, Japan*, 1972, pp. 285–298.
4. ABEL, A. and MUIR, H., Fatigue softening of low carbon steel, *Phil. Mag.*, **32**, 1975, 553–563.
5. OLDROYD, P. W. J., Modifications to an instron testing machine for cyclic loading in the plastic range, *Proc. I. Mech. E.*, **180** (3A), 1965/66, 229–237.
6. RADON, J. C., BURNS, D. J. and BENHAM, P. P., Push–pull low-endurance fatigue of cast irons and steels, *J. Iron St. Inst.*, **204**, 1966, 928–935.
7. OLDROYD, P. W. J., BURNS, D. J. and BENHAM, P. P., Strain hardening and softening of metals produced by cycles of plastic deformation, *Proc. I. Mech. E.*, **180** (3A),1965/66, 392–396.
8. BROOM, T. and HAM, R. K., The hardening and softening of metals by cyclic stressing, *Proc. Roy. Soc., London*, **A242**, 1957, 166–179.

15

Thermal Ratcheting of a Hollow Stepped Cylinder

V. Sagar Dwivedi†, H. Fessler, T. H. Hyde and J. J. Webster

University of Nottingham

SUMMARY

Cyclic thermal loading of components subjected to steady mechanical loads may induce plastic strain cycling and ratcheting. The life of such components may be determined from the magnitudes of the cyclic plastic strains and strain increments per cycle.

Finite-element results for the stress–strain histories of a hollow stepped cylinder and a thin cylinder (the shank portion of the stepped cylinder) subjected to steady axial loads and cyclic thermal loading resulting from axisymmetric cyclic temperature variations are given. Results for the thin cylinder are compared with results from a well-established uniaxial model for this problem. The results for the hollow stepped cylinder illustrate the effect of a stress concentration on the cyclic plastic straining and ratcheting of the component.

The work was inspired by problems which may arise in nuclear power plant but is of interest wherever high thermal loading occurs repeatedly.

NOTATION

b bore of stepped cylinder
E Young's modulus
E_p plastic modulus
s distance along the outer surface of stepped cylinder as measured from shank end

†All papers by V. Sagar Dwivedi prior to 1976 were published in the name V. Sagar.

T	temperature
T_{melt}	melting temperature
ΔT	maximum temperature difference between the surfaces of cylinder
z	distance along the axis of cylinder as measured from shank end
α	coefficient of linear expansion
ν	Poisson's ratio
ϵ_m^p	meridional component of plastic strain
ϵ_z^p	axial component of plastic strain
ϵ_θ^p	circumferential component of plastic strain
$\Delta\epsilon_z^p$	axial component of reverse plastic strain
$\Delta\epsilon_\theta^p$	circumferential component of reverse plastic strain
σ_t	thermal stress
σ_y	initial yield stress
σ_z	axial stress
σ_θ	circumferential stress
$\bar{\sigma}$	mean axial stress

INTRODUCTION

Many components in power and chemical plant are subjected to steady mechanical loads in operation and cyclic thermal loads due to plant start-ups and shut-downs. These components should be designed against failure due to excessive deformation, creep rupture and fatigue failure. Ratcheting and reverse plasticity can be eliminated for components which do not operate in the creep range, by designing them within the shakedown loads. However, such a design procedure would be unduly conservative, because many materials are capable of withstanding many plastic strain reversals, and components subjected to a few load cycles can withstand modest cyclic strain increments. The optimum design of components subjected to repeated loads in the creep regime is very complicated, because creep reduces the beneficial effects of the stresses induced by plastic deformation and also because of the complex interactive effects of creep and plasticity on material behaviour. Leckie [1] gives an excellent review of the literature and design procedures for components subjected to plastic strain cycling and ratcheting.

The authors are currently engaged on a joint experimental and theoretical investigation to validate computational methods for predicting the stress–strain histories of components subjected to steady mechanical loads and large repeated thermal shocks. Axisymmetric lead–antimony–arsenic alloy model components (Fig. 1) will be subjected to tensile axial loads, and thermal shocks will be induced by reducing and increasing the temperature of the fluid flowing along the outside surface while the temperature of fluid flowing through the bore is maintained approximately at the initial uniform temperature of 60°C. The material has a melting point of 325°C (598K) and creeps significantly at 60°C (333K) (i.e. $T/T_{melt} = 0.56$). Experimental results for the surface strain histories will be compared with finite-element solutions based on the material data and will be used to investigate the effect of, and validate various laws for, the creep and plastic behaviour of the material.

This paper reports some preliminary calculations for the cyclic elastic–plastic behaviour of the component without creep. The component is subjected to a steady axial load and is cyclically loaded by alternately quenching and reheating its outside surface. In the text odd and even 'half-cycles' refer to quenching and reheating, respectively. The shank end of the component (Fig. 1) is a hollow circular cylinder and is investigated separately. The results from axisymmetric analyses of this cylinder are compared with results from the Miller–Bree [2, 3] approximate uniaxial model for thin pressure vessels and tubes. Some results for the complete component which illustrate the effect of the stress concentration around the fillet are also given.

FINITE-ELEMENT ANALYSIS

The component was idealised with 8-noded isoparametric axisymmetric temperature and stress elements. Solutions for transient temperature and elastic stress distributions were obtained using the standard facilities of the PAFEC 70+ suite of finite-element programs [4]. Routines for elastic–plastic analysis for combined mechanical and cyclic thermal load were developed from the routines developed in the department by Dawson [5] for mechanical loading.

The elastic–plastic solutions were obtained using incremental loading and the method of successive elastic solutions. For the initial mechanical load the post-yield load was incremented in steps of approximately 0.2 of the yield load. During the thermal transients incremental solutions were obtained at time intervals corresponding

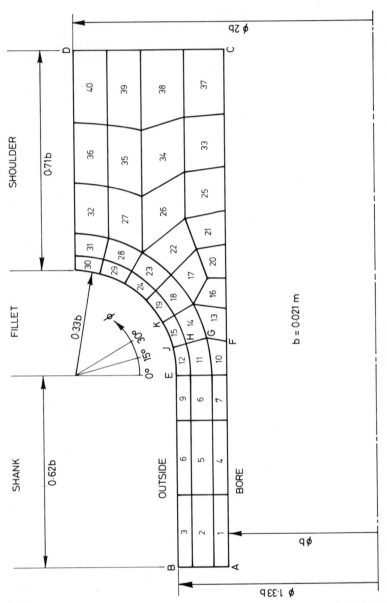

FIG. 1. Hollow stepped cylinder with finite-element mesh showing nomenclature, dimensions and important positions.

to surface thermal strain changes ($\alpha \Delta T$) of approximately one-quarter or one-half of the initial yield strain. Solutions for each load increment were obtained iteratively using methods developed by Dawson and described in Ref. [6]. Stress distributions were determined, for successive estimates of the plastic strain increments, by elastic re-solution until convergence within a specified accuracy was achieved. For the solutions reported here the criterion was that the change in successive estimates of the equivalent plastic strain increment at each gauss point should be less than 10^{-8} times the total equivalent strain at that gauss point. The determination of the plastic strain increments was based on the von Mises yield criterion and the Prandtl–Reuss flow rule. Further details of the finite-element transient temperature and elastic–plastic analyses are given in Ref. [6].

HOLLOW CIRCULAR CYLINDER

Previous Work

A hollow circular cylinder is one of the classic problems in thermal ratcheting because of its wide application and its relevance to the nuclear fuel can problem. Miller [2] considers thin-walled pressure vessels; Bree [3] considers pressurised thin tubes; and one of the present authors has investigated thick-walled circular cylinders under steady axial tension and torsion [7]. Both Miller [2] and Bree [3] formulate their problems as approximate equivalent uniaxial models. Bree [3] allows for biaxiality in determining the equivalent uniaxial thermal strain but neglects all other biaxial effects. In effect, Bree's model is a thin narrow rectangular strip, constrained to remain straight and subjected to a constant axial stress equal to the thin-tube hoop stress and an equivalent thermal strain cycle. This model is equally applicable to thin cylinders subjected to an axial load and uniform through-thickness cyclic temperature changes; the results for this model are compared with finite-element solutions for a cylinder with the same dimensions as the shank end of the component, shown in Fig. 1.

Loading

The cylinder is initially at a uniform temperature with a uniform axial tensile stress, $\bar{\sigma}$. It was found that the plastic strains in the cylinder resulting from the thermal cycle induced by changing the

temperature of the fluid flowing along the outside of the cylinder
could be modelled sufficiently accurately by incrementally imposing a
through-thickness linear temperature distribution with the same
maximum temperature difference, ΔT, through the wall of the cylinder. Most of the results for the cylinder were obtained with this
model of the thermal cycle.

Procedure

The finite-element idealisation used for the plane strain analysis of
the cylinder was one element long in the axial direction and had seven
elements of equal radial thickness. Appropriate boundary conditions
were imposed to provide plane strain conditions.

Numerical results are presented for the stress and strain histories
for various ratios of mean axial stress to initial yield stress, $\bar{\sigma}/\sigma_y$, and
for a ratio of thermal stress to initial yield stress, σ_t/σ_y, of 2·6 (note:
$\sigma_t = E\alpha\Delta T/2(1 - \nu)$). Three materials with different isotropic hardening characteristics were studied. Other relevant material properties,
assumed to be independent of temperature, are given in Table 1.

TABLE 1
Model material properties

Initial yield stress (σ_y)	7·0 MN m^{-2}
Young's modulus (E)	23·2 GN m^{-2}
Poisson's ratio (ν)	0·4
Thermal conductivity	35·1 W °C^{-1} m^{-1}
Specific heat/unit volume of material	14·3 × 10^5 kJ °C^{-1} m^{-3}
Coefficient of linear expansion (α)	0·286 × 10^{-4}°C^{-1}

Elastic–Perfectly Plastic Material

Finite-element solutions for the variation of the components of
stress and plastic strain through the thickness of the cylinder after
each of the first four half thermal cycles are shown in Fig. 2 for
$\bar{\sigma}/\sigma_y = 0·7$ and $\sigma_t/\sigma_y = 2·6$. It may be seen that there is reverse plastic
straining in the regions of both surfaces of the cylinder in successive
half-cycles, and incremental plastic straining. The cyclic behaviour of
the cylinder stabilises after a few half-cycles for the elastic–perfectly
plastic material. The stress and plastic strain distributions after successive thermal loading and unloading are similar to those shown in

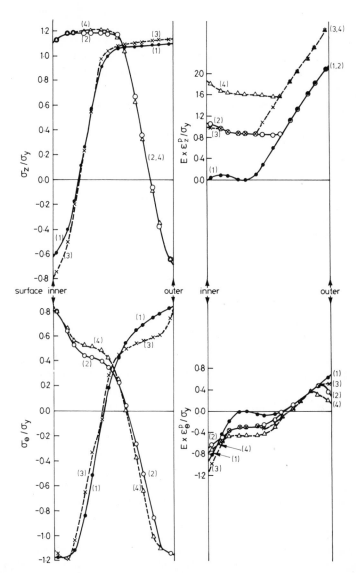

FIG. 2. Stress and strain distribution in hollow cylinder for first four half thermal cycles ($\bar{\sigma}/\sigma_y = 0.7$; $\sigma_t/\sigma_y = 2.6$): ———●———, at the end of first half thermal cycle; ———○———, at the end of second half thermal cycle; ----×----, at the end of third half thermal cycle; ----△----, at the end of fourth half thermal cycle.

Fig. 2 for the third and fourth half-cycles, respectively, except for the addition of the cyclic plastic strain increments.

Solutions were obtained for different steady axial loads and the same cyclic thermal loading ($\sigma_t/\sigma_y = 2\cdot6$). In each case thermal cycling was continued until successive cycles produced practically identical changes of strain. For the inner and outer surfaces of the cylinder, components of these stabilised reverse plastic strains are shown in Fig. 3(a) and cyclic plastic strain increments (ratchet strains) in Fig. 3(b). Values obtained from the equations derived by Bree [3] for the uniaxial model are also shown. The approximate uniaxial model accurately predicts the onset of ratcheting at a value of $\bar\sigma/\sigma_y = 0\cdot38$. However, it seriously overestimates the reverse plastic strains in the non-ratcheting region ($\bar\sigma/\sigma_y \leqslant 0\cdot38$) and the ratchet strains. It may be seen that reverse plasticity occurs at much larger values of $\bar\sigma/\sigma_y$ than predicted by the uniaxial model; however, the magnitudes of these reverse strains are generally less than those of the ratchet strains.

Linear-Hardening Material

The same yield stress as for the lead alloy was chosen with a ratio of the plastic to elastic moduli of $0\cdot15$, a reasonable value for comparison with the lead alloy (see Fig. 4).

The accumulation of the components of plastic surface strain during the first ten half-cycles of thermal load for $\bar\sigma/\sigma_y = 1\cdot4$ and $\sigma_t/\sigma_y = 2\cdot6$, where σ_y is the initial yield stress (see Fig. 4), is shown in Fig. 5. Values obtained from the equations derived by Bree [3] are also shown in Fig. 5. These results show the effect of the material hardening due to the successive cyclic plastic straining, the consequent reduction of the cyclic plastic strain increments and the monotonic convergence of the total accumulated plastic strain to a finite value. Figure 6 shows, for the same cyclic thermal load ($\sigma_t/\sigma_y = 2\cdot6$) and different steady mechanical loads ($1\cdot0 \leqslant \bar\sigma/\sigma_y \leqslant 2\cdot0$), the components of both the total accumulated plastic strain and that accumulating during the thermal cycling, i.e. after the initial application of the mechanical load. It may be seen that the accumulated plastic strain, subsequent to the initial application of the axial stress, is independent of the mean axial stress when this stress is greater than the initial yield stress. This phenomenon is predicted by the uniaxial model [3]. Results for the uniaxial model, also shown in Fig. 6, again show that this model overestimates the plastic strains.

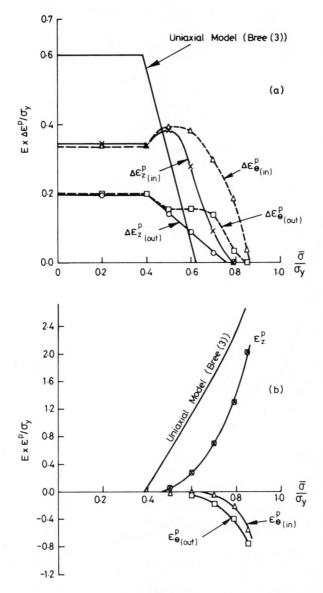

FIG. 3. (a) Cyclic reverse plastic strains at surfaces of hollow cylinder; (b) ratchet strains per cycle at surfaces of hollow cylinder. Suffixes (in) and (out) refer to inner and outer surface, respectively.

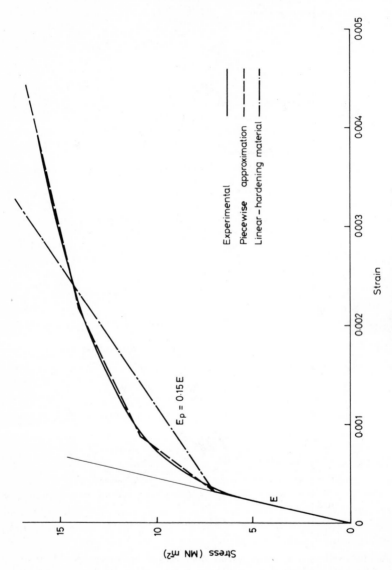

FIG. 4.   Stress–strain curve for lead alloy.

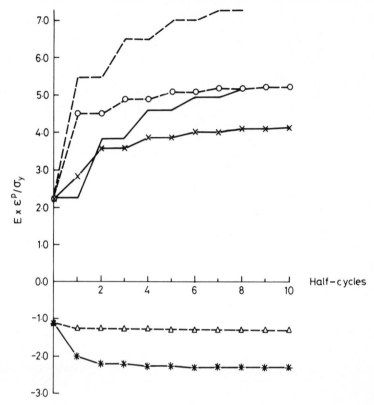

FIG. 5. Accumulation of plastic surface strains in hollow cylinder of linear-hardening material ($\bar{\sigma}/\sigma_y = 1\cdot4$; $E_p/E = 0\cdot15$): ——×——, ϵ_z^p at inner surface; ----O----, ϵ_z^p at outer surface; ——*——, ϵ_θ^p at inner surface; ----△----, ϵ_θ^p at outer surface; ————, at inner surface; --------, at outer surface; uniaxial model (Bree [3]).

Lead Alloy

The lead alloy material stress–strain curve was modelled by the piecewise linear relationship shown in Fig. 4. The behaviour of the cylinder of this material is qualitatively similar to that of the cylinder of a linear-hardening material. The variation of the accumulated plastic strain components with mean stress for $\sigma_t/\sigma_y = 2\cdot6$ is shown in Fig. 7. Comparing these results with those for the linear-hardening material (Fig. 6), it may be seen that the total accumulated plastic strains are larger in the linear-hardening cylinder at low values of $\bar{\sigma}/\sigma_y$ and the plastic strains are larger in the lead alloy cylinder at the high

310 V. Sagar Dwivedi, et al.

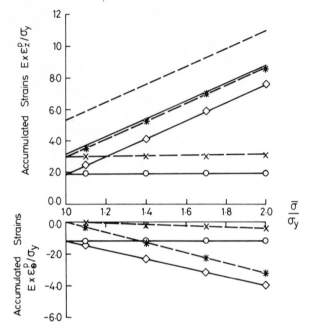

FIG. 6. Accumulated plastic surface strains in hollow cylinder of linear hardening material ($E_p/E = 0\cdot15$): ——◇——, total at inner surface; ----*----, total at outer surface; ——O——, accumulated during thermal cycling at inner surface; ---- × ----, accumulated during thermal cycling at outer surface; ————, total at inner surface; --------, total at outer surface; uniaxial model (Bree [3]).

values of $\bar{\sigma}/\sigma_y$, as would be expected from the material stress–strain curves (Fig. 4). The cross-over of the stress–strain curves for the lead alloy and linear-hardening materials (Fig. 4) occurs at a plastic strain of approximately 18×10^{-4}. This strain corresponds to the average axial component of the total accumulated strain in the linear-hardening cylinder at $\bar{\sigma}/\sigma_y \approx 1\cdot6$ and, at this value of $\bar{\sigma}/\sigma_y$, the accumulated strains in both cylinders are approximately equal (compare Figs 6 and 7). Hence, reasonably good predictions for the total accumulated plastic strains of the lead alloy cylinder would be obtained from a linear-hardening material based on the yield stress and the total accumulated plastic strain. The results for the piecewise linear-hardening material, unlike those for the linear-hardening material, show that the accumulated plastic strain, subsequent to the initial application of the axial stress, is not independent of the mean axial stress when this stress is greater than the initial yield stress.

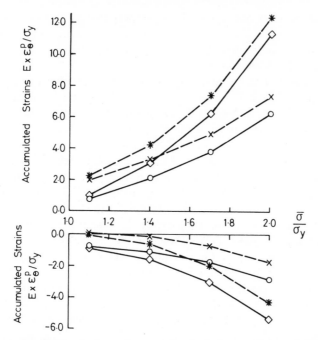

FIG. 7. Accumulated plastic surface strains in hollow cylinder of lead alloy: ——◇——, total at inner surface; ----*----, total at outer surface; ——○——, accumulated during thermal cycling at inner surface; ----×----, accumulated during thermal cycling at outer surface.

HOLLOW STEPPED CYLINDER

A hollow stepped cylinder is being used in the investigation to validate methods for predicting stress and strain histories under thermal ratcheting conditions. The portion of the component being studied in detail and the finite-element idealisation used for its analysis are shown in Fig. 1. The component is initially at a uniform temperature and subjected to an axial load equivalent to a uniform stress of $0.7\,\sigma_y$ at the shank end and $0.18\,\sigma_y$ at the shoulder end. In each thermal cycle the outside surface temperature of the component is reduced by 21 °C at a constant rate of 30 °C s^{-1}, held for a time and then increased to 21 °C at the same rate and held again. The inside surface temperature is held constant throughout and the hold times are such that the steady state conditions are effectively achieved at the end of each half-cycle. This temperature cycle is equivalent to $\sigma_t/\sigma_y = 1\cdot65$. The material is assumed to be elastic–perfectly plastic and plane strain boundary conditions are imposed at each end of the finite-element idealisation.

The extent of the plastic zone in the component $0\cdot2$ s after the start of the first thermal cycle and at the end of the first three half thermal cycles is shown in Fig. 8. Components of the plastic strains at the inside and outside surfaces of the component, after each of the first four half-cycles, are shown in Figs 9 and 10.

Analyses of the elastic stresses due to mechanical and thermal loading [8, 9] showed that the maximum stress concentration due to both mechanical and thermal loading occurred in the fillet at J, shown in Fig. 8. This feature is also exhibited in the present elastic–plastic analysis, in which the first plastic enclave appears around J (Fig. 8) during the early stages of the first cooling.

At the shank end of the component the behaviour is generally in agreement with predictions based on the uniaxial model [3] in so far as there is a modest amount of thermal strain ratcheting and no plastic strain reversal. However, the yield in the region of the inside surface in the first half-cycle would not be predicted by the uniaxial model.

The uniaxial model is not valid for the shoulder end of the component because it is based on thin-cylinder theory with a linear through-thickness temperature variation. The most severe temperature distribution at the shoulder end occurs during the transient parts of the thermal cycle, when it is very non-linear, and this produces reverse plastic straining in the region of the outside surface, as shown in Fig. 9.

In the region of the fillet there is combined reverse plastic strain cycling and ratcheting (Fig. 9). The magnitudes of the components of reverse plastic strain are similar to those occurring at the shoulder end of the component. However, the cyclic increments in the components of plastic strain are considerably greater than those occurring elsewhere in the component, showing the effect of the concentration.

CONCLUSIONS

(1)   The Miller–Bree uniaxial model provides a very good insight into the behaviour of thin cylinders subjected to steady axial

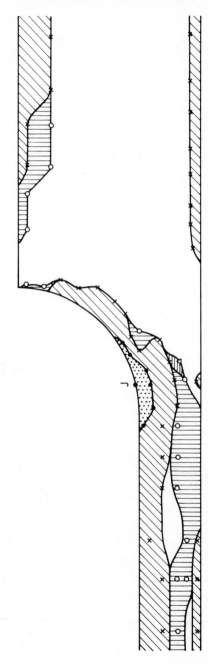

FIG. 8. Growth of plastic zones in stepped cylinder: ▓▓, after 0·2 s in first half thermal cycle; ▒▒, additional growth in first half thermal cycle; ▒▒▒, additional growth in second half thermal cycle; ▤, additional growth in third half thermal cycle; no additional growth in fourth half thermal cycle.

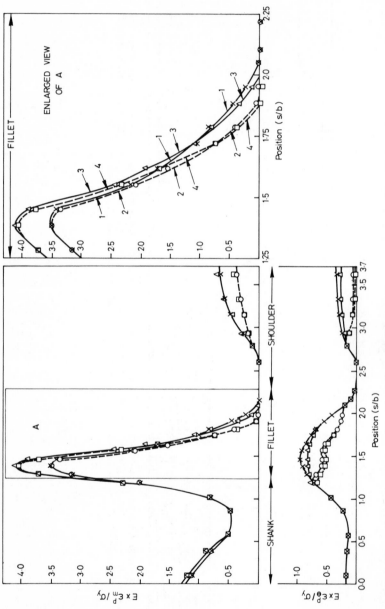

FIG. 9. Plastic strains at the outer surface of stepped cylinder: ——×——, at the end of first half thermal cycle; ----○----, at the end of second half thermal cycle; ——△——, at the end of third half thermal cycle; ----□----, at the end of fourth half thermal cycle.

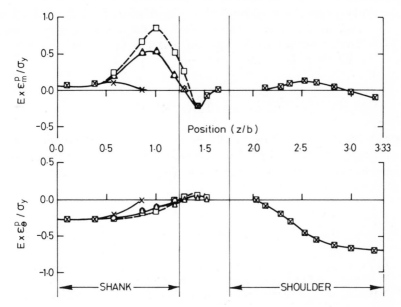

FIG. 10. Plastic strains at the inner surface of stepped cylinder: ——×——, at the end of first half thermal cycle; ----O----, at the end of second half thermal cycle; ——△——, at the end of third half thermal cycle; ----□----, at the end of fourth half thermal cycle.

load and cyclic axisymmetric thermal loads induced by through-thickness temperature variation. The model illustrates the regimes of reverse plastic straining and strain ratcheting. In general, predictions for the reverse plastic strains and ratchet strains, based on the model, are significantly larger than values obtained from a full two-dimensional analysis. However, under some loading conditions the model fails to predict reverse plastic straining.

(2) The accumulated plastic strains in a thin cylinder of a non-linear-hardening material may be predicted, with acceptable accuracy, from results for an appropriate linear-hardening material.

(3) The plastic straining in the region of the transition of a hollow stepped cylinder, subjected to a steady axial load and cyclic axisymmetric thermal loads, is significantly different from that in the two plane cylindrical portions of the component. For the case considered here the maximum reverse plastic strains in

the transition fillet were of similar magnitude to those occurring in the larger-diameter end of the component, but the ratchet strains were very much larger than those occurring anywhere else in the component.

ACKNOWLEDGEMENTS

This work is supported by an SRC grant. The authors thank the many contributors to the PAFEC finite-element scheme [4, 6] for the use of computer programs and particularly R. Dawson for the elastic–plastic mechanical loading program.

REFERENCES

1. LECKIE, F. A., 'A Review of Bounding Techniques in Shakedown and Ratchetting at Elevated Temperature', W. R. C. Bulletin 195, 1974.
2. MILLER, D. R., Thermal stress ratchet mechanism in pressure vessels, *Trans. A.S.M.E., J. Basic Eng.*, **81**, 1959, 190–194.
3. BREE, J., Elastic–plastic behaviour of thin tubes subjected to internal pressure and intermittent high-heat fluxes with application to fast-nuclear-reactor fuel elements, *J. Strain Anal.*, **2**, 1967, 226–238.
4. HENSHELL, R. D. (Ed.), *PAFEC 70+ Manual*, Department of Mechanical Engineering, University of Nottingham, 1972.
5. DAWSON, R. J., Private communication, 1975.
6. HENSHELL, R. D. (Ed.) *PAFEC 75 Theory, Results*, Department of Mechanical Engineering, University of Nottingham, 1975.
7. SAGAR, V. and PAYNE, D. J. Incremental collapse of thick-walled circular cylinders under steady axial tension and torsion loads and cyclic transient heating, *J. Mech. Phys. Solids*, **23**, 1975, 39–53.
8. DWIVEDI, V. SAGAR, FESSLER, H., HYDE, T. and WEBSTER, J. J., 1st Progress Report to S.R.C., 'Creep and Plasticity Interaction: Elastic Steady-state Stresses in Shouldered Tube', Department of Mechanical Engineering, University of Nottingham, September 1976.
9. DWIVEDI, V. SAGAR, FESSLER, H., HYDE, T., and WEBSTER, J. J., 2nd Progress Report to S.R.C., 'Elastic Transient Thermal Stresses in a Shouldered Tube', Department of Mechanical Engineering, University of Nottingham, February 1977.

16

Incremental Plastic Deformation of a Cylinder Subjected to Cyclic Thermal Loading

A. M. GOODMAN

Berkeley Nuclear Laboratories

SUMMARY

Incremental plastic deformation due to cyclic thermal loading is an increasingly important consideration in the design of advanced high-temperature plant. Present understanding has been strongly influenced by the Bree Diagram, which is now incorporated in ASME Code Case 1592. This paper examines a form of cyclic thermal loading which cannot be properly assessed for design purposes using the Bree Diagram. It is shown, however, that simple and reliable analyses are still possible using established principles of stress analysis. The approximate formulae that result are verified by comparison with detailed computation. Simple material behaviour laws for cyclic elasto-plasticity are then considered with reference to experimental data reported in the literature.

NOTATION

a cylinder radius

D flexural stiffness, $D = \dfrac{Eh^3}{12(1 - \nu^2)}$

E Young's modulus
F attenuation factor
h cylinder thickness
L length of ramp
M moment resultant
N force resultant
P axial stress
w radial displacement (positive inward)
x axial distance

α coefficient of thermal expansion

β $= \left[\dfrac{3(1 - \nu^2)}{a^2 h^2} \right]^{1/4}$

γ $= \dfrac{E}{\sigma_y a \beta}$

θ temperature
ν Poisson's ratio
σ_y yield stress (σ_{yh} hot, σ_{yc} cold)

Coordinate Subscripts
x axial
ϕ circumferential

Subscript
R   residual

Superscript
p   plastic

Dimensionless Notation

$\bar{M} = \dfrac{\gamma M}{D\beta}, \quad \bar{N} = \dfrac{N}{h\sigma_{yh}}, \quad \bar{P} = \dfrac{P}{\sigma_{yh}},$

$\bar{w} = \beta w, \quad \bar{x} = \beta x, \quad \bar{\epsilon} = \dfrac{E\epsilon}{\sigma_{yh}}, \quad \bar{\theta} = \dfrac{E\alpha \Delta\theta}{2\sigma_{yh}}$

INTRODUCTION

Previous assessments of the cyclic thermal ratcheting of a cylindrical vessel—for example, the work of Edmunds and Beer [1] and of Bree [2]—have considered the application and removal of a through thickness thermal gradient in the presence of pressure stresses. The analyses of Bree in particular are now receiving considerable attention from designers following their use in ASME Code Case 1592 [3]. If there are no mechanical loads, the cyclic thermal loading, which is symmetric, will not cause incremental plastic deformation. Assuming an elastic–ideally plastic material, Bree then showed that there are combinations of cyclic thermal and steady mechanical loads above shakedown, which do not result in ratcheting and for which the fatigue damage at the surface may be acceptable in design. However, it must be remembered that these analyses refer to a specific geometry subjected to one form of cyclic loading.

This paper examines a form of cyclic loading which cannot be assessed using the results of Bree's work. A cylindrical vessel is used to contain two fluids, which are at different temperatures and have different heat transfer coefficients with the structural steel. The surface between the two fluids then fluctuates in level, subjecting the vessel to a cyclically varying axial temperature gradient. The vessel walls also carry the self-weight of the fluids.

This problem is fundamentally different from that examined by Bree and other workers, since the cycle of thermal stresses is asymmetric. It is now possible for the thermal load to give rise to ratcheting in the absence of a mechanical load.

The objectives of the present paper are to establish the conditions under which plastic ratcheting may occur in this case, and in doing so to demonstrate both the differences between this loading and that more usually considered and the way simple concepts can still be used to assess component response. Once the designer is aware of the potential problem, there are a number of ways of 'designing out' of any difficulty. To give two examples, it may be possible to surround the vessel with graduated insulation, so that the thermal gradient is suitably less severe, or the section may be corrugated to absorb the radial expansion.

In order to ensure conservatism, it is considered that one fluid has an indefinite heat transfer coefficient while the other is a relatively good insulator. This maximises the asymmetry of the cycle of thermal loads. As the fluid with a high heat transfer coefficient advances, the walls of the vessel will change temperature rapidly, so that the axial temperature gradient becomes steeper. When this fluid retreats again, the temperature change will be slow, giving rise to a less severe axial gradient. In the limiting case considered here, the temperature history is idealised as an axial step change (or very steep ramp) travelling in one direction only, the return half-cycle being neglected. If the hotter fluid has the high heat transfer coefficient, then an advance by this fluid is described as the travel of a 'hot front' (see Fig. 1a). Traverses of both hot and cold fronts are considered.

The stress and strain histories are assessed in terms of simple formulae supplying estimates of the cyclic strain range and any incremental plastic deformation. These are supported by more detailed computations using a modified version of the CEGB program

PATAS [4, 5]. It is useful to establish first the classical thin-shell solution for a stationary axial thermal gradient.

ANALYSIS

Stationary Axial Thermal Gradient

In order to develop simple approximate formulae from closed-form analysis it is convenient to idealise the temperature profile as a step change $\Delta\theta$ in temperature. It will be assumed throughout that material properties other than the yield stress are independent of temperature. Each half of the cylinder may be analysed separately as a semi-infinite isothermal shell, the conditions at the near boundary being $M_x = 0$ and $w = 0$ (see Fig. 1). Following Timoshenko and Woinowsky-Krieger [6], the complete elastic solution for a cylinder of radius a and thickness h, subjected to a step change in temperature $\Delta\theta$ at $x = 0$ and an axial load Ph, is given by

$$w = \frac{a\alpha\Delta\theta}{a} e^{-\beta x} \cos\beta x + \frac{a\nu P}{E} - \frac{a\alpha\Delta\theta}{2} \quad (x > 0)$$

$$w = \frac{-a\alpha\Delta\theta}{2} e^{\beta x} \cos\beta x + \frac{a\nu P}{E} + \frac{a\alpha\Delta\theta}{2} \quad (x < 0) \qquad (1)$$

where

$$\beta = \left[\frac{3(1 - \nu^2)}{a^2 h^2}\right]^{1/4}$$

The hoop force resultants at the discontinuity are given by

$$(N_\phi)_{\text{peak}} = \pm E\alpha\Delta\theta\frac{h}{2} \qquad (2)$$

For analysis beyond yield the material is assumed to be elastic–ideally plastic and a dimensionless notation is introduced:

$$\bar{M} = \frac{\gamma M}{D\beta}, \quad \bar{N} = \frac{N}{h\sigma_{\text{yh}}}, \quad \bar{P} = \frac{P}{\sigma_{\text{yh}}}, \quad \bar{w} = \beta w,$$

$$\bar{x} = \beta x, \quad \bar{\epsilon} = \frac{E\epsilon}{\sigma_{\text{yh}}}, \quad \bar{\theta} = \frac{E\alpha\Delta\theta}{2\sigma_{\text{yh}}}, \quad \bar{\sigma} = \frac{\sigma}{\sigma_{\text{yh}}}$$

where

$$D = \frac{Eh^3}{12(1 - \nu^2)} \quad \text{and} \quad \gamma = \frac{E}{\sigma_{\text{yh}}a\beta} \qquad (3)$$

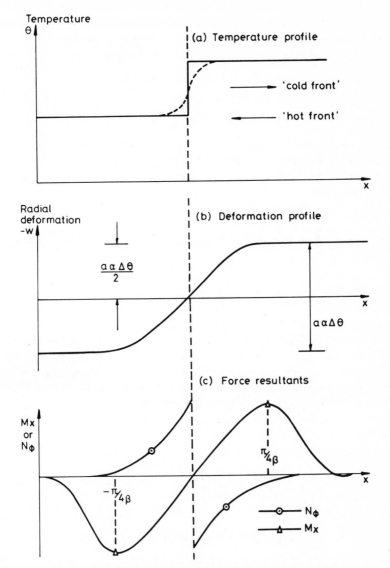

Fɪɢ. 1. Temperature, deformation, elastic bending moment and force resultant profiles.

The yield stress σ_{yh} applies at the higher temperature; the value at the lower temperature σ_{yc} may be higher.

Using this notation, the elastic equations become

$$\gamma\bar{w} = \bar{\theta}e^{-\bar{x}}\cos\bar{x} + \nu P - \bar{\theta} \quad (\bar{x} > 0)$$
$$\gamma\bar{w} = -\bar{\theta}e^{\bar{x}}\cos\bar{x} + \nu\bar{P} + \bar{\theta} \quad (\bar{x} < 0)$$

(1a)

and

$$(\bar{N}_\phi)_{\text{peak}} = (\bar{\sigma}_\phi)_{\text{peak}} = \bar{\theta}$$

(2a)

These displacement and force resultant distributions are shown schematically in Figs 1(b) and 1(c), respectively.

The ideally plastic material is assumed to follow the von Mises yield criterion,

$$\bar{\sigma}_x^2 + \bar{\sigma}_\phi^2 - \bar{\sigma}_x\bar{\sigma}_\phi = 1$$

(4)

while plastic flow is distributed according to the Levy–Mises flow rule

$$\frac{d\bar{\epsilon}_x^p}{d\bar{\epsilon}_\phi^p} = \frac{2\bar{\sigma}_x - \bar{\sigma}_\phi}{2\bar{\sigma}_\phi - \bar{\sigma}_x}$$

so that

$$\left[\frac{d\bar{\epsilon}_x^p}{d\bar{\epsilon}_\phi^p}\right]_{\bar{x}=0} = \frac{2\bar{P} - \bar{\theta}}{2\bar{\theta} - \bar{P}}$$

(5)

A post-yield computer analysis cannot follow the axial movement of a step change in temperature with reasonable efficiency. For this purpose it is better to approximate the profile as a steep linear ramp change in temperature. The analytical solution for a step thermal load $F\bar{\theta}$ will be compared with the computed solution for a ramp load $\bar{\theta}$, where F is chosen so that the elastically calculated peak hoop force resultants are the same for both solutions. The deflection $\bar{w}_L$ at the end of a ramp of length $\bar{L}$ is given by the convolution integral,

$$\bar{w}_L = \frac{-\bar{\theta}}{\gamma\bar{L}}\int_0^{\bar{L}} e^{-(\bar{x}-\bar{l})}\cos(\bar{x}-\bar{l})\,d\bar{l} + \frac{\nu\bar{P}}{\gamma} + \frac{\theta}{\gamma}$$

(6)

where

$$\bar{x} \geq \bar{L} \text{ and } \bar{x} = 0 \text{ at the cold end of the ramp.}$$

The peak value of $\bar{N}_\phi$ at $\bar{x} = \bar{L}$ is given by $F\bar{\theta}$, where

$$F = \frac{1}{2\bar{L}}(1 - e^{-\bar{L}}(\cos\bar{L} - \sin\bar{L}))$$

(7)

The yield condition will be closely matched provided that the ramp is short, so that the bending stresses at $\bar{x} = \bar{L}$ are small compared with yield. The attenuation factor F can also be used to remove the pessimism of an assumed step profile if the elastically calculated stresses for a real profile are known.

It is convenient to give separate consideration to long and short traverses, relative to the cylinder geometry parameter $(ah)^{1/2}$. The sections that follow consider the travel of a step change in temperature.

Long Travels

A structure is said to 'shakedown' if the cycle of stresses and strains reduces to one which is entirely elastic. The shakedown limit can be identified by finding the maximum loading case which will result in elastic behaviour in the steady cyclic condition. This steady cycle of stress is obtained by superposing a self-equilibrating set of residual stresses on to the elastically calculated stresses (see Ref. [7]).

This section will consider a long traverse $(\Delta \bar{x} \gg \beta(ah)^{1/2})$ such that most of the vessel swept by the thermal load is unaffected by the local bending stresses left at the ends of the travel. For this case any residual stresses must be invariant with respect to axial distance $\bar{x}$. From considerations of static equilibrium a uniform component of residual stress is not admissible. It follows that for a thin vessel the only allowable residuals must describe constant bending moments $\bar{M}_{xR}, \bar{M}_{\phi R}$. However, the elastic solution predicts a first yield that occurs uniformly through the section at the instantaneous position of the thermal step, $\bar{x} = 0$. Any non-zero residual moment $\bar{M}_{xR}$ would result in yielding at a vessel surface for a less severe combination of the loads $\bar{\theta}, \bar{P}$. The shakedown and first yield load combinations therefore coincide, and are shown for a von Mises material as a locus in $(\bar{\theta}, \bar{P})$ space in Fig. 2. Superposition of a bending residual does afford some advantage for shakedown in the more realistic problem of a temperature profile of finite length; the algebra required here is expanded in Appendix 1.

The ratchet behaviour due to a travelling step $\bar{\theta}$ in excess of shakedown will be estimated by considering the cycle of stresses and strains at the mid-surface at $\bar{x} = 0$. It is assumed that $\bar{N}_{\phi}$ decays rapidly with distance from the discontinuity, so that yielding may be considered to be localised at $\bar{x} = 0$. Since $\bar{M}_x$ is zero at $\bar{x} = 0$, it is then sufficient to consider only the mid-surface (see Fig. 1).

For travel of a hot front when the load combination $(\bar{\theta}, \bar{P})$ is within the first yield locus, the cycle of stresses is given by AC_1AB_1D in Fig.

A. M. Goodman

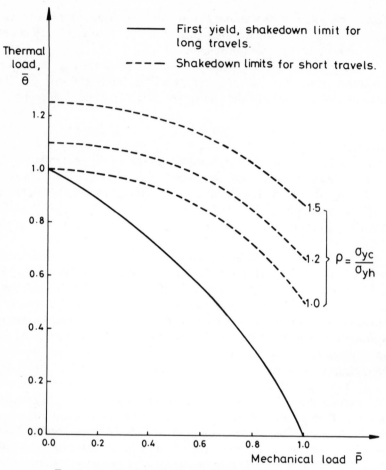

FIG. 2. Load loci for the first yield and shakedown limits.

3, where $AC_1 = AB_1$. As $\bar{\theta}$ is increased, yield occurs first in compression at the hot side of the thermal step when, from eqns (2) and (4), the first yield value of $\bar{\theta}$, $\bar{\theta}_{yh}$, is given by

$$\bar{\theta}_{yh} = -\frac{1}{2}[\bar{P} - (4 - 3\bar{P}^2)^{1/2}] \quad (0 < \bar{P} < 1 \cdot 0) \tag{8}$$

The direction of the plastic strain vector at B_2 in Fig. 3 shows that an inward radial ratchet results for $\bar{\theta}$ slightly in excess of $\bar{\theta}_{yh}$.

The theorem of Frederick and Armstrong [8] states that a repeated

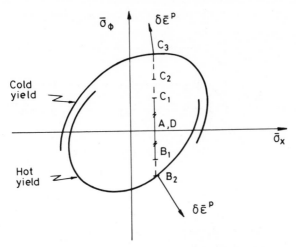

FIG. 3.   von Mises yield loci in terms of hoop and axial stress.

cycle of loading will lead to a steady and unique cycle of deviatoric stresses and strains in regions of creep or plasticity within a structure. This applies to the hydrostatic components of stress if these regions have a free surface, as is the case here at the surfaces of the vessel walls. The stresses in elastic regions which may have yielded during an earlier part of the loading history are not necessarily unique. However, yield occurs simultaneously at all points of a section swept by a step change in temperature. Consequently, if a solution is postulated for the steady cyclic behaviour of the cylinder considered here, which satisfies equilibrium, compatibility and the constitutive laws, then this can be identified as the correct solution.

It is supposed that the solution for travel of a hot front, for all values of $\bar{\theta}$ greater than $\bar{\theta}_{yh}$, describes the cycle of hoop stress AC_2AB_2D in Fig. 3, where $AC_2 = AB_2$. That is, the cylinder retains the first yield stresses and strains given by eqn (1) with $\bar{\theta} = \bar{\theta}_{yh}$, except at the trailing edge of the step, where a plastic strain is introduced. If this plastic strain is just sufficient to balance the increase in thermal strain beyond first yield in the hoop direction, i.e. if

$$\delta\bar{\epsilon}_\phi^p = -2(\bar{\theta} - \bar{\theta}_{yh}) \tag{9}$$

then the deformed profile will stay the same as at first yield (see Fig. 4). Compatibility of strains and deformations is therefore satisfied. The stress distribution, remaining unchanged since first yield, is also

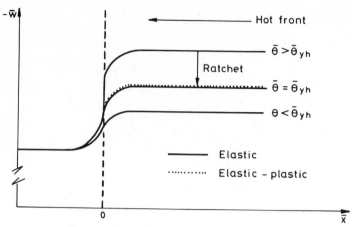

FIG. 4.   Ratchet due to travel of a hot front.

one which satisfies equilibrium. Plastic strain occurs at the one location where the stress is at yield, so satisfying the assumed constitutive law. This solution is therefore correct and the response to a hot front is described by the hysteresis loop $AC_2AB_2B_2'D$ in Fig. 5(a). Examination of this loop shows that the ratchet strain in the hoop direction is given by eqn (9). (N.B. All oblique lines in Fig. 5 have unit slope.)

A long travel of a cold front will result initially in compressive yield ahead of the step with the cycle of stress AB_2DC_2D, with $AB_2 = AC_2$ (see Fig. 3). The plastic strain ahead of the step is in the wrong sense to reduce the thermal deformation behind the step. For a cold front both the yield ahead of the step and the change in thermal strain decrease the circumference, and, hence, the radius, behind the front. It is not now possible to apply the previous argument, and an increasing $\bar{\theta}$ may result in yielding both ahead of the step and at the trailing edge, the cycle of stress being AB_2DC_3D, with $AB_2 \neq AC_3$ (see Fig. 3).

However, an approximate solution can be obtained for long travels of a cold front if it is assumed that local kinematic determinacy is unaffected by moderate yielding (see Refs [9] and [10]). With this assumption, the radial displacement profile retains the antisymmetry about the thermal step which is found for the elastic solution, i.e. $AA^* = DD^*$ in Figs 5(b) and 5(c). For yield ahead of the step only

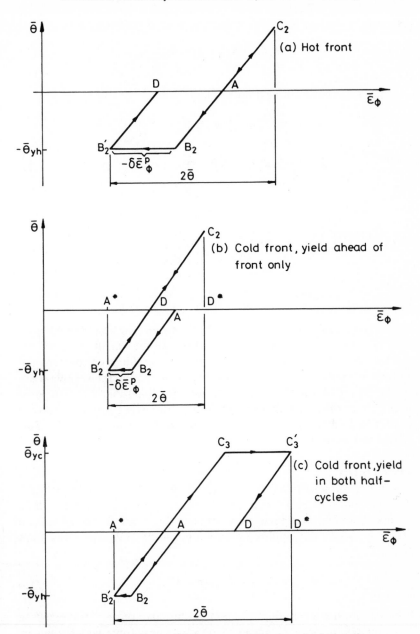

FIG. 5.   Hysteresis loops of hoop stress and hoop mechanical strain.

328   A. M. Goodman

(Fig. 5b),

$$\bar{\theta}_{yh} - \delta\bar{\epsilon}^p_\phi = 2\bar{\theta} - \bar{\theta}_{yh}$$
$$\delta\bar{\epsilon}^p_\phi = -2(\bar{\theta} - \bar{\theta}_{yh}) \tag{10}$$

Yield occurs at both positions when

$$\bar{\theta} = \frac{1}{2}(\bar{\theta}_{yh} + \bar{\theta}_{yc})$$

where

$$\bar{\theta}_{yc} = \frac{1}{2}[\bar{P} + (4\rho^2 - 3\bar{P}^2)^{1/2}] \quad (0 < \bar{P} < 1\cdot0) \tag{11}$$

and

$$\rho = \frac{\sigma_{yc}}{\sigma_{yh}}$$

For all greater $\bar{\theta}$

$$\delta\bar{\epsilon}^p_\phi = C_3C'_3 - B_2B'_2$$

in Fig. 5(c), i.e.

$$\delta\bar{\epsilon}^p_\phi = 2(\bar{\theta} - \bar{\theta}_{yc}) \tag{12}$$

Estimates of incremental hoop plastic strain can be obtained from eqns (9), (10) or (12) as appropriate, the axial increment from eqn (5) and the change in thickness by applying the requirement of constant volume to these plastic strains. The estimates of $\delta\bar{\epsilon}^p_\phi$ for various $\bar{\theta}$ are shown in Fig. 6 for two conditions of axial load and material yield stress. It is observed that, in general, a hot front gives an inward ratchet, a cold front an outward ratchet, and that incremental deformation can take place when $\bar{P} = 0$.

Short Travels

A travel of length $\Delta\bar{x}$ is considered to be short if $\Delta\bar{x} \le \beta(ah)^{1/2}$. Residual hoop stresses can exist over a short section of a long cylinder provided that their decay outside the length traversed is such that they satisfy the equilibrium condition in the hoop force resultant $\bar{N}_{\phi R}$,

$$\int_{-\infty}^{\infty} \bar{N}_{\phi R}\, d\bar{x} = 0 \tag{13}$$

For equilibrium of a ring element within the section traversed, a constant $\bar{N}_{\phi R}$ must be balanced by a radial shear force which changes in a linear manner with axial distance $\bar{x}$, which in turn requires a

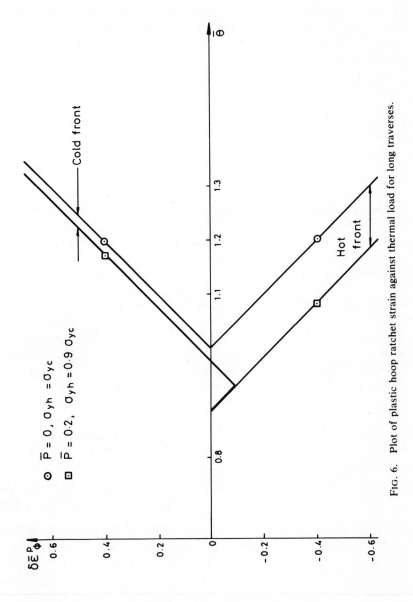

FIG. 6. Plot of plastic hoop ratchet strain against thermal load for long traverses.

parabolic variation in axial bending moment. A residual stress over
the section which maximises the elastic stress range in the hoop
direction (B_2C_3 in Fig. 3) will also optimise the shakedown limit. This
residual is half the difference in the hoop stresses at first yield at the
discontinuity,

$$\bar{\sigma}_{\phi R} = \frac{\bar{\theta}_{yc} - \bar{\theta}_{yh}}{2} \tag{14}$$

and is independent of $\bar{x}$. The shakedown limit (see Fig. 2) is then
given by

$$\bar{\theta} \leqslant \frac{1}{2}(\bar{\theta}_{yc} + \bar{\theta}_{yh}) \tag{15}$$

A further small improvement will result from the axial bending
residual after the manner described in Appendix 1 for long travels.

Continued radial ratcheting over a short axial length of a cylinder
will be restrained by the regions outside the section traversed. The
steady cycle of stresses and strains must trace a hysteresis loop in the
hoop direction, which is closed (see Fig. 7). This is sufficient to define
the plastic strains. Additionally, the hoop residual cannot increase
beyond yield if the material is perfectly plastic. For travel of a hot
front the stress will start at yield at point A in Fig. 7, following ABB'
during the traverse, leaving a residual $\bar{N}_\phi = -\bar{\theta}_{yh}$ behind the front,
point B'. The loop is then closed along B'A'A during the second
half-cycle. This reverse half-cycle amounts to the uniform heating or
cooling of a short length of a long cylinder, when the change in hoop
mechanical strain, due to the restraint of the remainder of the
cylinder, is equal and opposite to the change in thermal strain. This
can only be the case if the first half-cycle starts and finishes as
described at the corner points A and B' of the closed hysteresis loop
(see Fig. 7).

The hoop plastic strain range ϵ_ϕ^p (Fig. 7) is given by

$$\bar{\epsilon}_\phi^p = 2\bar{\theta} - (\bar{\theta}_{yc} + \bar{\theta}_{yh}) \tag{16}$$

While the plastic strains in the hoop direction in the two half-cycles
are balanced, the shape of the yield locus requires that there will be a
ratchet in the axial direction unless $\bar{P} = 0$. This axial strain increment
can be obtained using eqns (16) and (5); the expression obtained is

$$\delta\bar{\epsilon}_x^p = \bar{\epsilon}_\phi^p \left(\frac{2\bar{P} - \bar{\theta}_{yc}}{2\bar{\theta}_{yc} - \bar{P}} + \frac{2\bar{P} + \bar{\theta}_{yh}}{2\bar{\theta}_{yh} + \bar{P}} \right) \tag{17}$$

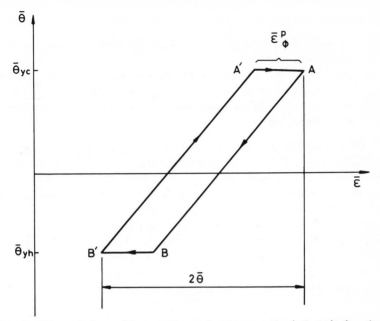

FIG. 7.   Hysteresis loop of hoop stress against hoop mechanical strain for plastic cycling due to short travels.

It should be noticed that this mode of incremental deformation may be insidious in practice. The requirement of constant volume during plastic deformation means that the vessel wall is thinning over the section traversed, but the usual check on radial growth will not detect any incremental deformation.

Intermediate behaviour between that considered for short travels and long travels may be expected for distances of travel of the same order as $\beta(ah)^{1/2}$.

COMPARISON WITH COMPUTATION

Details of Computer Representation

The PATAS thin-shell program, which in its normal form considers monotonic loading only, has been described previously by Goodman [4, 5]. Modifications to accommodate the cyclic loading under consideration here involved a small number of approximations which must be considered when results are compared with those obtained from the analysis:

(1) A convergence tolerance of 3% on the material yield stress is allowed.
(2) A finite length of travel must be considered so that suitable non-cyclic boundary constraints can be applied.
(3) A steep linear thermal ramp is substituted for the step change in temperature.

A very thin cylinder was considered, having a radius-to-thickness ratio greater than 400:1. The gradient of the thermal ramp was set at $d\bar{\theta}/d\bar{x} = 6\cdot79$. The yield stress was assumed either to be constant or to vary linearly with temperature; in Table 1, where the parameters are listed for each analysis, the ratio of Young's modulus to yield stress is quoted for the extremes of temperature. At the completion of any one travel the temperatures are scaled down to zero, when the original profile is regenerated ready for the next traverse. It was noted that plastic yielding occurred during the second stage of this procedure when short travels were being analysed.

Long Travels

A length of travel of $\Delta\bar{x} = 4\cdot67$ was set as a compromise between computing costs and the ideal of indefinite length of travel. The parameters for these analyses are listed in Table 1.

The qualitative description for long travels, given in the Analysis section above, may be compared with the computed hysteresis loops shown in Figs 8–11. Figures 8 and 9 refer to the particular case of $\bar{P} = 0$ and $\sigma_{yh} = \sigma_{yc}$ for a hot front and a cold front, respectively. The

TABLE 1
Parameters for long-travel computations

Analyses No.	Front	$\bar{\theta}$	$\bar{P}$	E/σ_{yh}	E/σ_{yc}
1	Hot	2·376	0	800	800
2	Cold	2·376	0	800	800
3	Hot	2·263	0	1 143	800
4	Cold	2·263	0	1 143	800
5	Hot	1·584	0·1	800	800
6	Cold	1·584	0·1	800	800
7	Hot	1.584	0·25	800	800
8	Cold	1·584	0·25	800	800

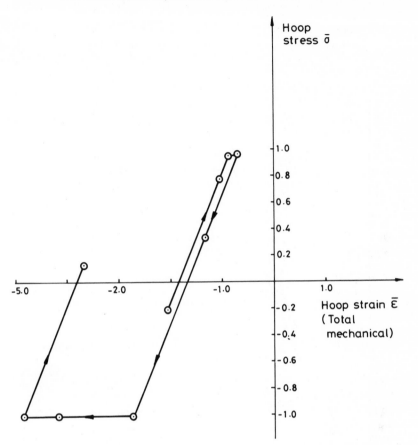

FIG. 8. Computed hysteresis loop, analysis 1 (hot front). ⊙ Computed points.

analytical solutions predict plastic yielding on one side only, the stress in the other half of the loop just reaching yield. The small computed reverse plastic yielding can be attributed to the approximations of a digital computing code, particularly the 3% tolerance noted in (1) above. The other loops confirm the prediction of yield on one side for a hot front and on both sides for a cold front, as exemplified by Figs 10 and 11.

The analytical estimate of ratchet is compared in Table 2 with the computed increment for the second traverse. Since the length of travel is finite, some interaction can be expected between the two extremes of the travel. This is seen as non-zero stress levels at the

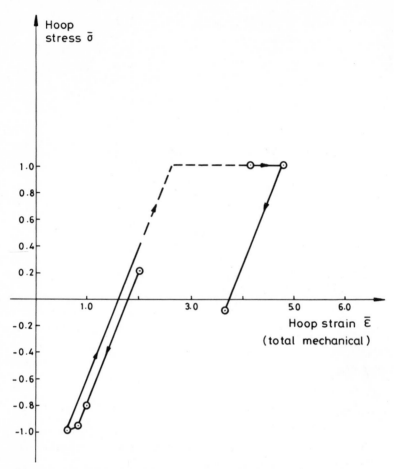

FIG. 9.   Computed hysteresis loop, analysis 2 (cold front). ⊙ Computed points.

ends of the hysteresis loops (see Figs 8–11), as a ratchet which decreases with number of cycles (Table 2) and an increasingly humped deformation profile (Fig. 12). The increment during the second traverse is least affected by the finite length of travel.

The discrepancy between the analytical and computed estimates of ratchet, referred to the computed ratchet, varies for hot fronts from − 2·5% (Analysis 1) to + 16·3% (Analysis 5). The same comparison applied to the cold front calculations is less satisfactory, giving − 25% (Analysis 6) to + 33% (Analysis 4). Since the comparison is favour-

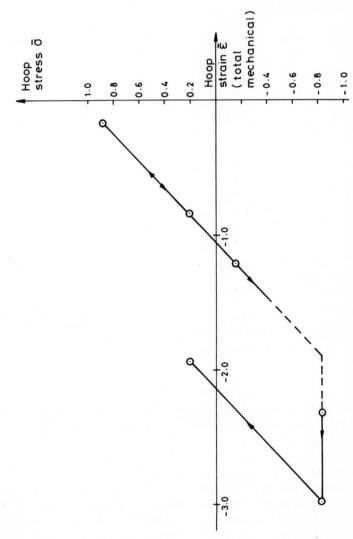

Fig. 10. Computed hysteresis loop, analysis 7 (hot front). ⊙ Computed points.

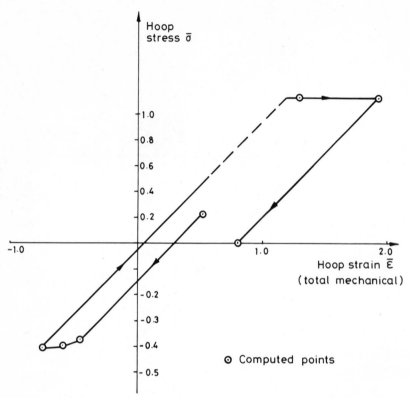

FIG. 11.   Computed hysteresis loop, analysis 8 (cold front). ⊙ Computed points.

able in the case of Analysis 2, it may be deduced that the assumption of antisymmetry about $\bar{x} = 0$ is less reliable in the presence of other asymmetric effects, such as a temperature-dependent yield stress or an axial load.

Short Travels

Computer analyses of travel of a hot front have been completed for travels of $\Delta\bar{x} = 2\cdot34$, $1\cdot17$, $0\cdot583$ and $0\cdot467$. The loads were $\bar{\theta} = 3\cdot39$ and $\bar{P} = 0\cdot129$. The material was assumed to have the temperature-dependent yield shown in Table 1, Analyses 3 and 4.

The computed circumferential hysteresis loop for the fourth cycle of the shortest travel examined is shown in Fig. 13. The loop, which is nearly closed, may be compared with the schematic of Fig. 7. The

TABLE 2
Computed and analytic estimates of ratchet strains

| Analysis No. | Traverse | Plastic ratchet strain, $\delta\bar{\epsilon}^p$ | | | |
| | | Computed | | From analysis | |
		Hoop	Axial	Hoop	Axial
1	$1\to2$	-1.984	0.995	-1.928	0.968
	$2\to3$	-1.892	0.897		
	$3\to4$	-1.868	0.884		
2	$1\to2$	1.960	-0.937	1.928	-0.968
	$2\to3$	1.926	-0.913		
	$3\to4$	1.903	-0.899		
3	$1\to2$	-2.160	1.086	-2.000	1.006
4	$1\to2$	0.857	-0.377	1.143	-0.572
	$2\to3$	0.789	-0.343		
5	$1\to2$	-0.784	0.433	-0.912	0.408
6	$1\to2$	0.667	-0.276	0.496	-0.224
	$2\to3$	0.643	-0.248		
	$3\to4$	0.614	-0.220		
7	$1\to2$	-1.068	0.687	-1.152	0.408
	$2\to3$	-0.937	0.624		
8	$1\to2$	0.509	-0.077	0.456	-0.160

computed hoop plastic strain range of 0.217% may also be compared with the estimate of 0.298% from eqns (11) and (16).

The computed ratchet strains per cycle are given in Table 3. Equations (16) and (17) predict no radial ratchet, 0.047% per cycle axially and -0.047% per cycle through the thickness. It is seen that the steady cyclic computed ratchet strains approach the analytical estimate as the length of travel is reduced. Several cycles are required for a close approach to the steady cyclic condition.

Table 3 also shows the intermediate behaviour. The results for $\Delta\bar{x} = 2.34$ differ little from the 'long' travel calculations. At $\Delta\bar{x} = 1.17$ an intermediate behaviour is seen in which there is a reduced inward radial ratchet well balanced by the axial ratchet, so that there is little change in thickness.

Figure 14 shows the computed distribution of the stress resultants with axial distance $\bar{x}$ at the beginning of the fourth cycle, $\Delta\bar{x} = 0.467$. This is identical in nature with the distribution envisaged for short travels in the Analysis section above.

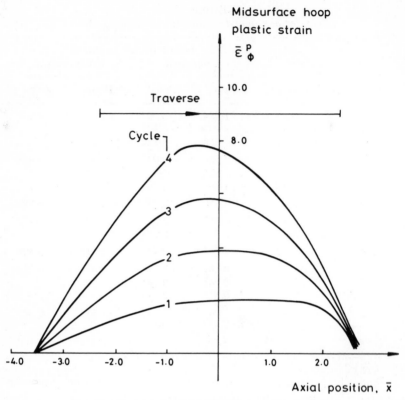

FIG. 12.   Computed radial deformation profiles, analysis 1.

DISCUSSION

Two theoretical analyses of the problem, one by approximate closed form analysis and the other involving detailed computation, have resulted in predictions of component response which are considered to agree with one another. The greatest uncertainty in any practical application will be concerned with the rather simple material behaviour which has been assumed. Many steel alloys, including the austenitic stainless steels, show non-linear but continuous strain-hardening. It can be shown that any component of such a material which is subjected to cyclic loads above shakedown will not ratchet indefinitely. Instead, the strain-hardening will decrease the incremental deformation from cycle to cycle in such a way that a finite

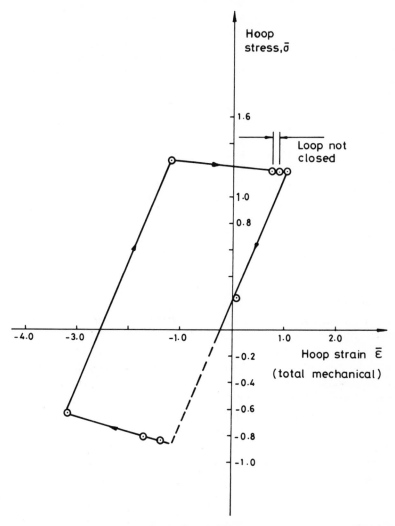

FIG. 13.   Computed hysteresis loop, fourth cycle (travel $\Delta \bar{x} = 0.467$). ⊙ Computed points.

asymptote is approached [11]. However, there is some suggestion that many materials exhibit a partially fading 'memory' of previous events; see for example Ref. [12]. Such an effect can certainly be expected as temperatures approach the creep range, and such factors as creep relaxation, thermal softening and anelastic strains need to be

TABLE 3
Computed ratchet strains, short travel cycles

Travel, $\Delta \bar{x}$	Cycle	Plastic ratchet strain, $\delta \bar{\epsilon}^p$		
		Hoop	Axial	Through-thickness
2·34	$1 \rightarrow 2$	$-3·429$	1·977	1·452
	$2 \rightarrow 3$	$-3·098$	1·875	1·223
1·17	$1 \rightarrow 2$	$-2·469$	1·017	1·566
	$2 \rightarrow 3$	$-0·914$	0·812	0·103
	$3 \rightarrow 4$	$-0·800$	0·743	0·057
	$4 \rightarrow 5$	$-0·720$	0·697	0·023
0·583	$1 \rightarrow 2$	$-0·434$	0·617	$-0·183$
	$2 \rightarrow 3$	$-0·297$	0·549	$-0·251$
0·467	$1 \rightarrow 2$	$-0·320$	0·549	$-0·229$
	$2 \rightarrow 3$	$-0·206$	0·491	$-0·286$
	$3 \rightarrow 4$	$-0·160$	0·457	$-0·297$
	$4 \rightarrow 5$	$-0·114$	0·446	$-0·331$

considered. In these circumstances it might be expected that a small but continuing ratchet per cycle would be necessary to maintain a given condition of hardening and residual stress in the material.

The experimental results of Yamamoto and Kano [13] are useful here. These authors tested the response to the more usual form of cyclic thermal loading (as analysed by Bree) using a sodium loop, and concluded that an acceptable description of component behaviour could be obtained by assuming a perfectly plastic material with a yield stress taken at a high proof strain, possibly in excess of 0·2%. This may be interpreted as meaning that the steeper strain-hardening for proof strains of less than about 0·2% is reliable, but that the same cannot be said of the more gradual hardening at greater strains.

CONCLUSIONS

(1)   An asymmetric cycle of thermal loading may give rise to incremental deformation in the absence of a steady mechanical load. The Bree Diagram cannot be applied and under normal conditions of operation the cyclic thermal component of load cannot be allowed to exceed shakedown.

(2)   A cyclic thermal load problem of this type has been analysed

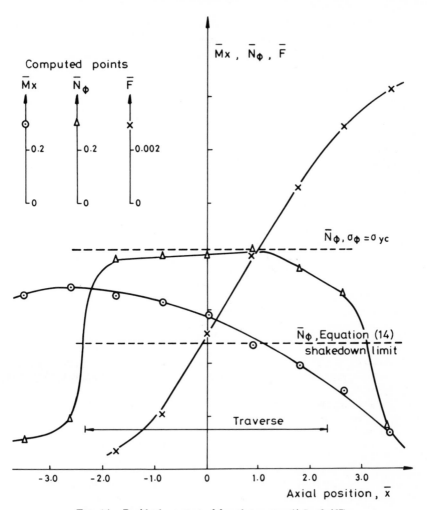

FIG. 14. Residuals at start of fourth traverse ($\Delta \bar{x} = 0.467$).

both approximately in closed form and in more detail using a computer code. The analyses compare favourably with each other.

(3) For this case simple formulae have been obtained which enable the incremental deformation and high strain fatigue damage to be assessed where a small number of severe cycles are expected in the component lifetime.

(4) Independent experiments on plastic thermal ratcheting show that strain-hardening is reliable for strains of less than 0·2%, but that continuing strain-hardening may not halt incremental deformation.

ACKNOWLEDGEMENTS

This paper is published by permission of the Central Electricity Generating Board.

REFERENCES

1. EDMUNDS, H. G. and BEER, F. J., Notes on incremental collapse in pressure vessels. *J. Mech. Eng. Sci.*, **3**, 1961.
2. BREE, J., Elastic behaviour of thin tubes subjected to internal pressure and intermittent high heat fluxes with application to fast nuclear reactor fuel elements, *J. Strain Anal.* **2**, 1967, 226–238.
3. 'ASME. Boiler and Pressure Vessel Code, Code Case 1592', American Society of Mechanical Engineers, 1974.
4. GOODMAN, A. M., 'A User's Guide to PATAS: Part 1, Elastic and Elastic-Plastic Analysis of Shell Structures', CEGB Report RD/B/N1751, 1971.
5. GOODMAN, A. M., A User's Guide to PATAS: Part 2, Analysis of Creep and Large Deformations of Shell Structures', CEGB Report RD/B/N2584, 1973.
6. TIMOSHENKO, S. and WOINOWSKY-KRIEGER, S., *Theory of Plates and Shells*, McGraw-Hill, New York, 1959.
7. MELAN, E., The stress condition of a Mises-Hencky continuum with variable loading, *Sb. Akad. Wiss. Wien*, **145**, 1938, 73.
8. FREDERICK, C. O. and ARMSTRONG, P. J., Convergent internal stresses and steady cyclic states of stress, *J. Strain Anal.* **1**, 1966, 154–159.
9. CALLADINE, C. R., Some calculations to assess the effect of the stress–strain rate relationship on creep in the neighbourhood of an opening in a pressurised thin spherical shell. *J. Mech. Eng. Sci.*, **9**, 1967, 198–210.
10. LECKIE, F. A., *Plasticity Concepts in the Creep Range* (Ed. J. Hult), 2nd I.U.T.A.M. Conference, Gothenburg, 1970.
11. AINSWORTH, R. A., Bounding solutions for creeping structures subjected to load variations above the shakedown limit, *Int. J. Solids Struct.*, **13**, 1977, 971–980.
12. KREMPL, E., *Cyclic Creep—An Interpretative Literature Survey*, Welding Research Council Bulletin 195, 1974.
13. YAMAMOTO, S., KANO, T. and YOSHITOSHI, A., Reprint from the 1976 Elevated Design Symposium by the American Society of Mechanical Engineers, 1976.

APPENDIX 1: SHAKEDOWN LIMIT

A small improvement on the shakedown limits given in the main text can be obtained for non-zero bending residuals $\bar{M}_{xR}$ and $\bar{M}_{\phi R}$, except for the particular case when $\bar{P} = 0$ and the material yield stress is independent of temperature. An axial change of temperature $\bar{\theta}$ takes place along a linear ramp of length $\bar{L}$, which is sufficiently short for first yield to occur at one end of the ramp, either $\bar{x} = 0$ or $\bar{x} = \bar{L}$. From the shape of the yield loci (Fig. 3) it can be seen that yield will occur first in compression at the hot end of the ramp, $\bar{x} = \bar{L}$, at the surface where the bending stresses are also compressive. Consider first the superposition of a set of residual bending stresses to maximise the combination of loads $(\bar{\theta}, \bar{P})$, for which the section at $\bar{x} = \bar{L}$ will remain elastic when the ramp is stationary. These stresses must balance the elastically calculated bending stresses, $\bar{M}_{xE}$ and $\bar{M}_{\phi E}(= \nu\bar{M}_{xE})$, so that

$$\bar{M}_{xR} = -\bar{M}_{xE}, \qquad \bar{M}_{\phi R} = \nu\bar{M}_{xR} \tag{A.1}$$

and yield occurs simultaneously at all points through this section.
The elastic bending moment is given by

$$\bar{M}_{xE} = \gamma \frac{d^2\bar{w}}{d\bar{x}^2}$$

when, from eqn (6), the residual bending moment at $\bar{x} = \bar{L}$ is

$$\bar{M}_{xR} = \frac{-\bar{\theta}}{\bar{L}}[1 - e^{-\bar{L}}(\cos \bar{L} + \sin \bar{L})] \tag{A.2}$$

The surface fibre stresses, $\bar{\sigma}_s$, due to this moment are given by

$$\bar{\sigma}_s = \pm G\bar{\theta}$$

where $\qquad G = \dfrac{3^{1/2}}{2(1-\nu^2)^{1/2}} \dfrac{1}{\bar{L}}[1 - e^{-\bar{L}}(\cos \bar{L} + \sin \bar{L})]$

The uniform components of stress through this section are given by

$$\bar{\sigma}_x = \bar{P}$$
$$\bar{\sigma}_\phi = -F\bar{\theta} \tag{A.3}$$

When the residual stresses are included, the section will remain elastic for any mechanical load $\bar{P}$ and any thermal load less than $\bar{\theta}_{SH}$, where

$$\bar{\theta}_{SH} = \frac{[\bar{P}^2 + 4(1 - \bar{P}^2)]^{1/2} - \bar{P}}{2F} \tag{A.4}$$

Consideration of $\bar{P} = 0$ shows that the positive root should be taken. In the absence of any residual moments, the stresses at the surface are

$$\bar{\sigma}_x = \bar{P} + G\bar{\theta}, \qquad \bar{\sigma}_\phi = -F\bar{\theta} \pm \nu G\bar{\theta}$$

The positive sign gives the more severe shear stresses. The value of $\bar{\theta}$ at the elastic limit, $\bar{\theta}_{yl}$, is then the smaller root of the equation

$$\bar{\sigma}^2[G^2 + (\nu G - F)^2 + GF - \nu G^2] + \bar{\theta}(2G\bar{P} + F\bar{P} - \nu G\bar{P}) + \bar{P}^2 - 1 = 0$$

$$(A.5)$$

For the data $\bar{P} = 0{\cdot}1$, $\bar{L} = 0{\cdot}2334$

$$\bar{\theta}_{SH} = 1{\cdot}070, \qquad \bar{\theta}_{yl} = 0{\cdot}995$$

a difference in the thermal load to first yield of 7%.

If now a travelling ramp change in temperature is considered, it is seen that a uniform bending residual which reduces the bending stresses at the hot end of the ramp will increase the bending stresses at the cold end. Since the cold yield stress of the material usually exceeds the hot yield stress, some advantage will remain, but this will be a small effect. The inclusion of residual bending moments is not considered worthwhile.

17

The 'Reduced-Deflection Method' for Calculating Structural Buckling Loads

D. J. F. Ewing

Central Electricity Research Laboratories

SUMMARY

Non-linear buckling analyses of structures are normally carried out by the direct method, i.e. one attempts to calculate the full non-linear deflection field $\mathbf{u}$ *as the load parameter* λ *increases. The 'reduced-deflection method' is an alternative approach based on filtering out the 'buckling' components of* $\mathbf{u}$ *(which grow fastest) from the linearly elastic 'non-buckling' remainder. One works with the reduced deflection* $\mathbf{w} = \mathbf{u} - \lambda \mathbf{u}_o$*, where* $\mathbf{u}_o$ *is a separately calculated 'primary' displacement field, usually but not always the linearly elastic response of the structure.* $\mathbf{w}$ *is rotation-dominant (i.e. its rotations are much larger than its strains) but* $\mathbf{u}_o$ *is not. Expressing the potential energy in terms of* $\mathbf{w}$ *and* λ *(assuming conservative loading) leads to a reduced potential energy* $V_R(\mathbf{w}, \lambda)$*. The second-degree terms of* V_R *generate the eigenvalue problem of classical buckling theory. If the third- or higher-degree terms are destabilising, large shortfalls below the classical buckling load can occur, which can be calculated by the method. Expressing* $\mathbf{w}$ *and so* V_R *in terms of the* N *lowest eigenmodes* $\mathbf{w}_1, \ldots, \mathbf{w}_N$ *leads to a discretisation of the original non-linear problem into one involving* N *degrees of freedom (the amplitudes* q_i *of the* $\mathbf{w}_i$*). For complicated geometries, the finite-element method can be used to do the numerical work. The method is illustrated by examples, including the buckling of an end-loaded cylinder (following Koiter). Mathematically, the method is a variant of Koiter's general theory of elastic stability. The ideas of the method extend to non-elastic buckling, e.g. creep collapse.*

NOTATION

A	area in reference configuration
C, c	numerical constants ($C = c\mu$ and $c = 3(3)^{1/2}/2$)
D	tensor of linear-elastic moduli (Cartesian formulation) *or* material rigidity modulus (finite-element formulation)
d	small displacement from reference configuration
E	Young's modulus
e_1, e_2, e_3	orthogonal triad of vectors in shell mid-surface (Fig. 2)
F	nominal traction (force per unit area of the *reference* configuration): $F = \lambda F_{oo} = \lambda(F_o + F_s)$
g	nominal body force per unit reference volume
h	shell thickness
K	Koiter coefficient
M	number of buckling modes (all with eigenvalue λ_c)
m	circumferential wavenumber (w is proportional to $\cos m\theta$)
m_o	$= (3a^2/4h^2)^{1/4}$
N	(i) membrane stress resultants due to primary loading (ii) number of degrees of freedom, i.e. eigenmodes considered
n	'secondary' membrane stress resultants set up by the reduced deflection w
P	prestress work
p, p_w, p_m'	pressure, wind pressure-head, dimensionless pressure coefficients (eqn 86)
q	degree of freedom
t	ratio $-z/a$
$U(v)$	linear elastic strain energy calculated for an arbitrary displacement field v
$U_f(v)$	finite-strain elastic energy
$U_3(v), U_4(v)$	third- and fourth-degree terms in the homogeneous expansion $U_f = U + U_3 + U_4$
U_{IJK}, U_{IJKL}	coefficients defined by eqns (40), (41)
u, v, w	displacement components of w: for a shell $w = ue_1 + ve_2 + we_3$ (Fig. 2)
u	non-linear displacement from reference configuration
u_{oo}	linear elastic response of unit load $F_{oo}(u_{oo} = u_o + u_s)$
V	volume in reference configuration *or* total potential energy (according to context)

v	general trial displacement
W	work function of applied loading
$W''(\mathbf{w})$	second-degree terms in expansion of $W(\lambda \mathbf{u}_o + \mathbf{w})$ in powers of $\mathbf{w}$
$\mathbf{w}$	reduced deflection $\mathbf{u} - \lambda \mathbf{u}_o$
$\mathbf{w}_I$	eigenmode of the linearised-buckling equation corresponding to eigenvalue λ_I (if $\lambda_I = \lambda_c$, then $\mathbf{w}_I$ is a buckling mode)
x, y, z	Cartesian coordinates
x_i	Cartesian coordinates (in Cartesian tensor analysis)
$\boldsymbol{\epsilon}(\mathbf{v})$	linear-elastic strain calculated from $\mathbf{v}$
$\boldsymbol{\epsilon}_f(\mathbf{v})$	finite strain calculated from $\mathbf{v}$
η	perturbation parameter
θ, θ_o	polar coordinate angle; parametric angle (eqn 82)
λ, λ_c	load factor; critical load factor at instability
μ	coefficient *or* root of cubic according to context
ν	Poisson's ratio
σ	stress
ϕ	meridional angle (Fig. 2)
ω	rotation (eqn 6)

Subscripts

b	bending
c	critical
f	finite strain
L	linearised
m	membrane
o	primary
oo	unit
s	secondary
R	reduced

Suffixes

i, j, k, l, p	Cartesian (ranging from 1 to 3)
I, J, K, L	finite-degree-of-freedom (ranging from 1 to N)
α, β	shell surface (ranging over θ and ϕ)

Some symbols used only once are defined as they arise.

INTRODUCTION AND REVIEW OF IMPERFECTION-
SENSITIVITY

The possibility of failure by buckling must always be considered in
the stress analysis of any thin-walled structure. Unfortunately, classical linearised buckling theory [1] sometimes grossly overestimates
actual buckling loads [2, 3]. Notorious examples are the cylinder
under axial compression and the pressurised hemisphere. Buckling
loads in the range 15–60% of the theoretical ones are observed [2, 3]
and the experiments show wide scatter. The opposite phenomenon is
observed in the simple buckling of plates: although substantial
deformation is observed at the classical buckling load, the buckled
shape starts to stiffen up as the load is increased beyond the buckling
value.

These structures are all examples of ones which would undergo a
bifurcation buckle (i.e. a sudden change in gradient of the deflection-
load path) if they were 'perfect', i.e. free from all geometrical irregularities and with loadings free from all eccentricities. For such
structures, it had long been suspected that small irregularities in the
nominally perfect geometry and/or loading were responsible for the
shortfall, and finally Koiter [4–6] was able to demonstrate this
theoretically.

Koiter examines slightly imperfect structures where the corresponding 'perfect' structure would undergo a bifurcation in its
load–displacement path (Fig. 1) at a negligibly small elastic displacement. q is a generalised coordinate measuring the deflection. The
assumptions of classical linearised buckling theory are then satisfied
for the 'perfect' structure, so that classical theory predicts the correct
load parameter λ_c (Fig. 1). The assumptions in question are (i)
everything remains elastic until buckling begins, (ii) geometry changes
before buckling are negligible and (iii) stresses before buckling are as
calculated by linear elastic theory. However, the local shape of the
post-buckling path is determined by higher-order terms not considered in the linearised theory. If this post-buckling path turns *down-
wards* (Fig. 1a, b) the equilibrium at the bifurcation point is *unstable*
and the *imperfection-sensitivity* phenomenon arises. The load–
deflection paths of slightly irregular structures are shown as light lines
in Fig. 1; the irregularities have 'rounded the corner' off the sharp
bifurcation in Fig. 1(a) or (b), converting these to a 'snap-through' or
'limit-point' instability in standard terminology [7] with the load

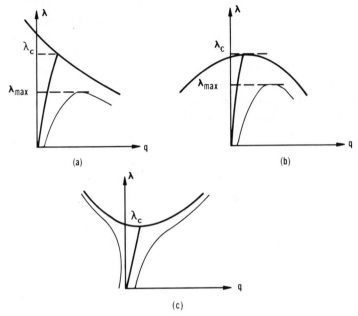

FIG. 1.   Classification of bifurcation points: (a) asymmetric; (b) unstable-symmetric; (c) stable-symmetric. q is a measure of deflection and λ is a load parameter. Light lines are the equilibrium paths of slightly perturbed structures.

passing through a simple maximum. Conversely, if the post-buckling path is concave *upwards* (Fig. 1c) the equilibrium at the bifurcation point is *stable* and the 'imperfect' structure stiffens as it passes the classical buckling load (as for the plate).

Koiter derives a general quantitative theory of this effect. He shows that if the amplitude of the irregularity is proportional to a parameter η, then the load drop is typically proportional to $\eta^{1/2}$ or $\eta^{2/3}$. This fractional power law is responsible both for the observed shortfall in buckling loads and for the wide scatter typical of experiments. Thompson [7], following Sewell [8], rederived Koiter's results for conservative systems whose potential energy $V(q_1, q_2, \ldots, q_n, \lambda)$ was assumed known in terms of a finite number of degrees of freedom. Roorda [9, 10] confirmed the theory by experiments on rigid frames. It is not possible to mention here the many other contributors to this field, but a summary of the state of the art up to 1970 (when the present method was being developed) is given in a review by Hutchinson and Koiter [11].

More recently, developments in the finite-element method apparently allow one to overcome the inadequacies of classical theory, by calculating the non-linear response of the 'imperfect' structure step by step [12]. However, the disadvantages of this method are (i) its inherent cost and complexity, (ii) the results are sensitive to the amplitude of the initial irregularity assumed, (iii) it is possible to miss out a bifurcation altogether, and (iv) the load drop is also sensitive to the shape of the initial irregularity. In the classic problem of the axially loaded cylinder, which is analysed as a worked example later in this paper, it is possible for an irregularity in the root mean square radius of only 1% of the shell thickness to produce drops in buckling load of up to 25%, depending on the irregularity's shape.

The idea of the reduced-deflection method is to filter out the fast-growing 'buckling' constituents of the non-linear deformation from the slow-growing 'non-buckling' ones. It corresponds mathematically to a variant of Koiter's method. It is not restricted to the analysis of geometrically near-perfect shells which would undergo a bifurcation (if geometrically perfect), but it then becomes equivalent to Koiter's method. It provides a rapid and rational way of setting up the equations of classical linearised buckling theory and then of estimating the effects of the higher-order terms neglected by this theory. It has also been helpful in setting up the equations of creep collapse of a finite-length cylindrical tube under pressure [13].

By working in terms of a reduced deflection $\mathbf{w}$, one may form a reduced potential energy $V_R(\mathbf{w}, \lambda)$ entirely equivalent to, but more convenient than, the original potential energy $V(\mathbf{u}, \lambda)$ (where $\mathbf{u}$ is the actual deflection from the reference configuration and λ is a load parameter). The second-degree terms of V_R correspond to those of classical linearised buckling theory. The higher-degree terms can be analysed asymptotically by Koiter's method (of replacing the forcing term generating the buckle by η times itself and considering an asymptotic expansion in powers of η) to recover Koiter's results. Alternatively, V_R can be expanded as a sum of N eigenmodes $\mathbf{w}_1, \mathbf{w}_2, \ldots, \mathbf{w}_N$ (calculated from classical theory) to convert the system into one involving λ and N degrees of freedom $q_1, q_2, \ldots, q_N$ (the amplitudes of the eigenmodes). This method requires the solution of one linear elastic problem (to generate the initial stress distribution), one linearised buckling problem, and the evaluation of a suitable number of surface or volume integrals to give the required coefficients. The finite-element method could be used to do the numerical work. $V_R(\mathbf{w}, \lambda)$ may also form an efficient starting point for a direct analysis by conventional incremental means. It may be possible to use fewer degrees of freedom because the slow-growing part of $\mathbf{u}$ no longer has to be allowed for. However, this last approach has still to be tested.

MATHEMATICAL PRELIMINARIES

Conventions

Symbols are defined in the Notation section, except for those used in a special local sense, which are defined as they arise. Vector, matrix and Cartesian tensor notation (suffix notation) are all used as appropriate. Lower case Roman letters (e.g. $\mathbf{u}$) are used for vectors. Greek letters are second-order tensors (e.g. $\boldsymbol{\sigma}$ stands for stress). Upper case Roman letters stand for fourth-order tensors. Subscripted vectors and tensors have their subscripts raised to superscripts when using Cartesian suffix notation (see e.g. eqn 1).

The notation can be interpreted also in finite-element terms; e.g. $\mathbf{D}$ stands both for the tensor of elastic moduli (in a Cartesian coordinate context) and the material rigidity matrix in a finite-element context [12, 14]. In the finite-element $\boldsymbol{\epsilon}$ is interpreted as $(\epsilon_x \epsilon_y \epsilon_z \gamma_{xy} \gamma_{yz} \gamma_{zx})$, where γ_{xy} are engineering shear strains (twice the tensor components ϵ_{xy}).

Problem Formulation

Any general theory capable of describing large elastic displacements and/or buckling must include somewhere the idea of *finite strain* (see, e.g., Ref. [15] or [16]). Classical linear elastic theory cannot predict bifurcation or loss of stability; by Kirchhoff's uniqueness theorem all solutions are unique and any equilibrium position is stable. Let $\mathbf{u}$ denote departure from the reference configuration. The finite strain ϵ_f is defined in Cartesian coordinates Ox_i $(i = 1, 2, 3)$ by

$$\epsilon_{ij}^f = \tfrac{1}{2}(u_{i,j} + u_{j,i} + u_{p,i}u_{p,j}) \tag{1}$$

in contrast to the linear elastic strain

$$\epsilon_{ij} = \tfrac{1}{2}(u_{i,j} + u_{j,i}) \tag{2}$$

The physical interpretation of ϵ_f is that if $\mathbf{x}^* = \mathbf{x} = \mathbf{u}$, then $|\mathbf{dx}^*| \equiv |\mathbf{dx}|$ if and only if ϵ_f vanishes. This is the 'Lagrangian' definition of finite strain (i.e. **u** and ϵ_f for a material point are expressed in terms of the point's original coordinates **x** rather than its current coordinates **x*** as in the 'Eulerian' description). Lagrangian coordinates are used throughout the analysis.

Let A_u be the fixed part of the body's undeformed surface A. Let **F** be the *nominal tractions* acting on the remainder A_F of A (i.e. **F** is force per unit undeformed area [16]). Let **g** be the nominal body force per unit undeformed volume. Suppose that **F** and **g** are *conservative*, i.e. can be derived from a work function $W(\mathbf{u})$ by an identity

$$dW \equiv \int_A \mathbf{F}\, d\mathbf{u}\, dA + \int_V \mathbf{g}\, d\mathbf{u}\, dV \qquad (3)$$

for all d**u**. In particular, for dead loading, defined here as when **F** and **g** are independent of **u**, i.e. are constant in magnitude *and* direction as the surface deforms, then

$$W(u) = \int_A \mathbf{Fu}\, dV + \int_V \mathbf{gu}\, dV \qquad (4)$$

Non-conservative loadings can occur (e.g. 'flutter' of an aircraft's wing is caused by energy being drawn from the passing windstream in a closed displacement cycle), for which the left hand side of eqn (3) is not a perfect differential. Such cases are excluded from the present analysis.

Although large deflections are allowed for, the strains are assumed small (otherwise plastic flow will occur) and so the elastic strain energy is assumed to be given by

$$U_f(\mathbf{u}) = \tfrac{1}{2}\int_V \epsilon_{ij}^f D_{ijkl}\epsilon_{kl}^f\, dV \equiv \tfrac{1}{2}\int_V \epsilon_f^T \mathbf{D}\epsilon_f\, dV \qquad (5)$$

assuming that the reference configuration is unstressed. Replacing ϵ_f by ϵ gives the ordinary linear elastic energy $U(\mathbf{u})$. Large displacements and small strains can only coexist if the rotations ω_{ij} dominate ϵ_{ij}, where

$$\omega_{ij} = \tfrac{1}{2}(u_{i,j} - u_{j,i}) \qquad (6)$$

Thus, in a buckle the term $\tfrac{1}{2}u_{p,i}u_{p,j}$ in eqn (1) is expected to be of the same order as ϵ_{ij} or even to be larger.

For simplicity **g** will henceforward be neglected and **F** will be assumed to be $\lambda \mathbf{F}_{oo}$, where λ is a load parameter. Thus, $W(u) = \lambda W_{oo}(\mathbf{u})$ and the body has a total potential energy

$$V(\mathbf{u}, \lambda) = U_f(\mathbf{u}) - \lambda W_{oo}(\mathbf{u}) \qquad (7)$$

The problem is to calculate **u** in terms of λ as λ increases from zero and, in particular, to predict any bifurcations, 'snap-throughs' or losses of stability.

THE REDUCED-DEFLECTION METHOD

The Reduced Potential Energy
The idea of the method is to filter out the rapidly growing 'buckling' constituents of **u** from the slowly growing ones. As already noted, the rotations must dominate the strains in any buckling mode. Accordingly, divide the applied load $\mathbf{F}_{oo}$ into two parts, viz. a primary part $\mathbf{F}_o$ whose corresponding *linear elastic* response $\mathbf{u}_o$ is *not* rotation-dominant, and a secondary part $\mathbf{F}_s$ whose linear elastic response $\mathbf{u}_s$ *is* rotation-dominant. Explicitly, the rotations $\boldsymbol{\omega}(\mathbf{u}_o)$ formed from $\mathbf{u}_o$ must be of the same order as, or smaller than, the linear elastic strains $\boldsymbol{\epsilon}(\mathbf{u}_o)$. When $\mathbf{u}_o = \mathbf{u}_{oo}$ (i.e. when $\mathbf{F}_s$ vanishes), we recover the case considered by Koiter. Now define the reduced deflection

$$\mathbf{w} = \mathbf{u} - \lambda \mathbf{u}_o \tag{8}$$

and express V in terms of **w** and λ. If U_f is first expanded in powers of **u** as

$$U_f(\mathbf{u}) = U(\mathbf{u}) + U_3(\mathbf{u}) + U_4(\mathbf{u})$$

then, in Cartesian coordinates,

$$U_3(\mathbf{u}) = \tfrac{1}{2} \int_V u_{p,i} u_{p,j} D_{ijkl} \epsilon_{kl}(\mathbf{u}) \, \mathrm{d}V \tag{9}$$

$$U_4(\mathbf{u}) = \tfrac{1}{8} \int_V u_{p,i} u_{p,j} D_{ijkl} u_{q,k} u_{q,l} \, \mathrm{d}V \tag{10}$$

It follows that

$$U_3(\lambda \mathbf{u}_o + \mathbf{w}) = U_3(\mathbf{w}) + \lambda P_0(\mathbf{w}) + \text{small terms} \tag{11}$$

$$U_4(\lambda \mathbf{u}_o + \mathbf{w}) = U_4(\mathbf{w}) \qquad\qquad + \text{small terms} \tag{12}$$

where $P_0(\mathbf{w})$ is the 'prestress work' of the 'unit loading' $\mathbf{F}_o$, and represents the second-order work done by the 'pre-existing' stresses $\boldsymbol{\sigma}_0$ created by $\mathbf{F}_o$. In Cartesian coordinates,

$$P_0(\mathbf{w}) = \int_V \sigma_{ij}^0 w_{p,i} w_{p,j} \, \mathrm{d}V \tag{13}$$

P_o is called an 'initial stress' term by Zienkiewicz [12]. The 'small terms' in eqns (11) and (12) are of order $\lambda \sigma_c / \lambda_c E$ compared with those retained, i.e. are negligible, because $\mathbf{u}_o$ is not rotation-dominant. σ_c is the buckling stress calculated by linearised buckling theory in the next subsection, and λ_c is the corresponding buckling load parameter. If the loading is dead, then

$$V(\mathbf{u}, \lambda) = V_R(\mathbf{w}, \lambda) + \text{terms independent of } \mathbf{w}$$

where $V_R(\mathbf{w}, \lambda)$ is the 'reduced potential energy', given by

$$V_R(\mathbf{w}, \lambda) = U(\mathbf{w}) + \lambda P_o(\mathbf{w}) + U_3(\mathbf{w}) + U_4(\mathbf{w}) - \lambda \int \mathbf{F}_s \mathbf{w} \, dA \qquad (14)$$

$$\equiv U_f(\mathbf{w}) + P_o(\mathbf{w}) - \lambda \int \mathbf{F}_s \mathbf{w} \, dA \qquad (15)$$

The $\mathbf{F}_o\mathbf{w}$ term cancels out the $U\langle \mathbf{u}_o, \mathbf{w} \rangle$ cross-product term, by linear equilibrium. (Here and later the notation $Q\langle \mathbf{u}, \mathbf{v} \rangle$ is used to denote the cross-product term in the expansion of a general homogeneous quadratic form $Q(\mathbf{u} + \mathbf{v})$; formally,

$$Q\langle \mathbf{u}, \mathbf{v} \rangle = Q(\mathbf{u} + \mathbf{v}) - Q(\mathbf{u}) - Q(\mathbf{v}) \qquad (16)$$

so that $Q\langle \mathbf{u}, \mathbf{u} \rangle = 2Q(\mathbf{u})$.) Working with $V_R(\mathbf{w}, \lambda)$ is equivalent to working with $V(\mathbf{u}, \lambda)$, but more convenient.

Non-dead loads create complications, because $W(\lambda \mathbf{u}_o + \mathbf{w})$ must be expanded in terms of $\mathbf{w}$ by a Taylor series. The higher-order second, third, ... terms $\lambda W_2(\mathbf{w}), \lambda W_3(\mathbf{w}), \ldots$ coming from the expansion are sometimes important. The classic example is the long shell under lateral pressure p and free to undergo a pure bending buckle with wavenumber $m = 2$ (i.e. deformation proportional to $\cos 2\theta$); the buckling pressure is 25% lower than that calculated by assuming that p remains dead [3, 17]. However, for shell buckling under pressure loading at *high wavenumber m* (discussed below) $W_2, W_3, \ldots$ are of order $1/m^2$ compared with the $U(\mathbf{w}), U_3(\mathbf{w}), \ldots$ terms [6], i.e. are negligible. If greater accuracy is required, a term $\lambda W_2(\mathbf{w})$ can always be added on to eqn (16). Non-dead loading will be ignored in the rest of the theoretical development.

Recovery of Linearised Buckling Theory

Equations (14) and (15) are valid for all dead loadings $\mathbf{F}_s$: they depend only on $\mathbf{u}_o$ having non-dominant rotations. But it is natural to examine what happens for 'small' $\mathbf{F}_s$. Neglecting the third- and fourth-degree $\mathbf{w}$ terms, V_R reduces to V_{RL}, where

$$V_{RL} = U(\mathbf{w}) + \lambda P_o(\mathbf{w}) - \lambda \int \mathbf{F}_s \mathbf{w} \, dA \qquad (17)$$

This potential V_{RL} corresponds to that of classical linearised theory [16–18]. Thus, the equilibrium equations for $\mathbf{w}$ may be set up from the variational principle $\delta V_{RL} \equiv 0$, i.e. from the variational identity

$$U\langle \mathbf{w}, \mathbf{v} \rangle + \lambda P_o \langle \mathbf{w}, \mathbf{v} \rangle = \lambda \int \mathbf{F}_s \mathbf{v} \, dA \qquad (18)$$

for all $\mathbf{v}$ satisfying the boundary conditions. (The bracket notation is defined in eqn 16.) Removal of derivatives of $\mathbf{v}$ by integration by parts generates the differential equations satisfied by $\mathbf{w}$. Alternatively, eqn (18) provides the direct starting point for a numerical finite-element solution. The eigenvalues $\lambda_1 \leqslant \lambda_2 \leqslant \ldots, \lambda_J \leqslant \ldots$ and eigenfields $\mathbf{w}, \mathbf{w}_2, \ldots, \mathbf{w}_J, \ldots$ of linearised buckling theory satisfy the identity in $\mathbf{v}$

$$U\langle \mathbf{w}_J, \mathbf{v} \rangle + \lambda_J P_o \langle \mathbf{w}_J, \mathbf{v} \rangle \equiv 0 \qquad (19)$$

Thus, λ_1 equals λ_c and $\mathbf{w}_1$ equals w_c (i.e. the buckling mode). The term 'buckling mode' is reserved for eigenfields whose eigenvalue equals λ_c (there may be more than one).

The modes are orthogonal:

$$U\langle \mathbf{w}_I, \mathbf{w}_J \rangle = 0 \text{ for } I \neq J \qquad (20)$$

This follows if $\lambda_I \neq \lambda_J$, while if $\lambda_I = \lambda_J$, w_I and w_J may be chosen orthogonal by well-known techniques. Also,

$$U\langle \mathbf{w}_J, \mathbf{v} \rangle = 0(\lambda_J \sigma_c / \lambda_c E) \qquad (21)$$

and so is negligible (unless $\lambda_N \gg \lambda_c$) for all $\mathbf{v}$ that are *not* rotation-dominant, from eqn (19).

It is known from simple examples [1] that if the load that would lead to bifurcation buckling of a 'perfect' structure is instead applied slightly eccentrically, then the deflection as calculated by classical linearised theory normally grows like $\lambda/(\lambda_c - \lambda)$. This is confirmed by the present theory as follows. Try the general solution

$$\mathbf{w} = \sum_{J=1}^{\infty} q_J \mathbf{w}_J + \mathbf{w}^* \qquad (22)$$

where w^* is orthogonal to all eigenfields and is (presumably) not rotation-dominant. The convergence and completeness of this expansion is not considered here. The orthogonality equations (eqns 20 and 21) show that $q_J = U\langle \mathbf{w}, \mathbf{w}_J \rangle / U\langle \mathbf{w}_J, \mathbf{w}_J \rangle$. Eliminating $U\langle \mathbf{w}, \mathbf{w}_J \rangle$, using eqns (18) and (19), gives finally

$$q_J = \lambda \lambda_J \mu_J / (\lambda_J - \lambda) \text{ for } \mu_J = \int_A \mathbf{F}_s \mathbf{w}_J \, \mathrm{d}A / U\langle \mathbf{w}_J, \mathbf{w}_J \rangle \qquad (23\mathrm{a}, \mathrm{b})$$

For small λ, eqns (23) agree with the linear elastic solution, but the $\mathbf{w}_1$ component dominates (provided $\mu_1 \neq 0$) as $\lambda \to \lambda_1$: the standard result.

In particular, the case $\mu_1 = 0$ (and *a fortiori* $\mathbf{F}_s = 0$) corresponds to a bifurcation. In this connection, it is interesting to note that, since $\mathbf{u}_0$ is the linearly elastic response to $\mathbf{F}_0$,

$$U\langle \mathbf{u}_0, \mathbf{w} \rangle = \int_A \mathbf{F}_0 \mathbf{w} \, \mathrm{d}A$$

for all $\mathbf{w}$, in particular $\mathbf{w}_J$. Equation (21) therefore shows that

$$\int_A \mathbf{F}_0 \mathbf{w}_J \, \mathrm{d}A = \text{negligible} \qquad (24)$$

at least, for the lower λ_J. The point of this observation is that a rigorous analysis of the non-linear equations at a bifurcation point shows that, if $\mathbf{F}^*$ denotes applied load per unit deformed area, then

$$\int_S \mathbf{F}^* \mathbf{v} \, \mathrm{d}S = 0 \qquad (25)$$

exactly, $\mathbf{v}$ being a field proportional to the jump in gradient $\mathrm{d}\mathbf{u}/\mathrm{d}\lambda$ and S being the deformed body surface. (Hill [19] observes that this is a necessary condition for bifurcation of general non-elastic structures.) Physically, eqn (25) simply expresses the fact that the rate of work of the applied load is momentarily the same along the old and the new equilibrium paths.

Application to Shells, Plates and Struts

The theory translates directly for shells, plates and struts. For example, for a straight strut of length L under a thrust T, with $\mathbf{w}$ equal to $(v(z), w(z))$ along and normal to its length,

$$P(\mathbf{w}) = -\tfrac{1}{2}T \int_0^L \{(\mathrm{d}w/\mathrm{d}z)^2 + (\mathrm{d}v/\mathrm{d}z)^2\} \, \mathrm{d}z \qquad (26)$$

z being distance along the strut. For a shell of revolution (Fig. 2)

$$P(\mathbf{w}) = \tfrac{1}{2} \int_A N_{\alpha\beta}(\partial\mathbf{w}/\partial s_\alpha)(\partial\mathbf{w}/\partial s_\beta)\,\mathrm{d}A \qquad (27)$$

where $N_{\alpha\beta}$ are the membrane (in-plane) stress resultants (positive in tension). The Greek suffixes α and β range over θ and ϕ and when repeated as here are summed over θ and ϕ. Contributions from the bending moments are normally negligible. If in the notation of Fig. 2

$$\mathbf{w} = u\mathbf{e}_1 + v\mathbf{e}_2 + w\mathbf{e}_3 \qquad (28)$$

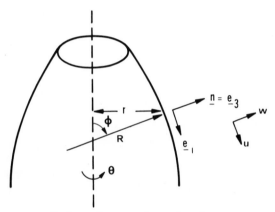

FIG. 2.   Shell notation. e_2 and v are into the paper.

then

$$\partial\mathbf{w}/\partial s_\theta = [(\partial u/\partial\theta - v\cos\phi)\mathbf{e}_1 + (\partial w/\partial\theta - v\sin\phi)\mathbf{e}_3]/r \qquad (29)$$

and

$$\partial\mathbf{w}/\partial s_\phi = [\mathbf{e}_2\partial v/\partial\phi + (\partial w/\partial\phi - u)\mathbf{e}_3]/R \qquad (30)$$

neglecting pure strain terms. The corresponding strain energy for the strut is

$$U(\mathbf{v}) = \tfrac{1}{2}EA \int_0^L (\mathrm{d}v/\mathrm{d}z)^2\,\mathrm{d}z + \tfrac{1}{2}EI \int_0^L (\mathrm{d}^2w/\mathrm{d}z^2)^2\,\mathrm{d}z \qquad (31)$$

(where here A is the cross-sectional area) and for the shell is the usual sum of membrane (stretching) or bending energies [14, 20].

From eqns (26) and (31),

$$V_{RL}(\mathbf{w}) = \tfrac{1}{2}(\bar{E}A - T) \int_0^L (dv/dz)^2 \, dz$$

$$+ \tfrac{1}{2} \int_0^L \{EI(d^2w/dz^2)^2 - T(dw/dz)^2\} \, dz \qquad (32)$$

The boundary conditions at $z = 0$ are $v = w = dw/dz = 0$. The critical T for buckling is found such that V_{RL} first vanishes for cases in which v and w are not both zero; evidently this corresponds to $dv/dz = 0$ (since the dv/dz term only increases V_{RL}) and the standard energy expression of Timoshenko [1] is obtained, with $v = 0$, i.e. an inextensional buckle. The analytical solution therefore continues along classical lines. However, the physical origins of the prestress work terms are different for the two methods. Here, because 'Lagrangian' coordinates are used, $P(\mathbf{w})$ arises from the non-linear terms in the expression for finite strain. In Timoshenko's analysis finite strain theory is not used and $P(\mathbf{w})$ appears instead as the second-order work done by the applied dead load as the strut buckles. The difference arises because Timoshenko is really using an embedded coordinate system, deforming with the strut. For simple structures this approach works very well, but it becomes increasingly hard to apply to complex structures.

The higher-degree terms U_3 and U_4 are evaluated later. Formulae for plates follow directly from the corresponding (more complex) ones for shells.

Simplifications at High Wavenumber

Fixing the base of a shell of revolution prevents it from undergoing pure bending deformation. Shells of moderate aspect ratio normally buckle at *high wavenumber* m. For a shell of revolution, with w involving one or more Fourier components proportional to $\cos m\theta$ or $\sin m\theta$, the average m is taken. More generally, m represents the ratio L/l, where L is a typical shell dimension and l is a buckling wavelength. If m is moderately large, the expression for $P(\mathbf{w})$ simplifies to

$$P(\mathbf{w}) = \tfrac{1}{2} \int_A N_{\alpha\beta}(\partial w/\partial s_\alpha)(\partial w/\partial s_\beta) \, dA \qquad (33)$$

with error order $1/m^2$, because u, v are small compared with w (order

$1/m$). Similarly, the bending energy U_b, namely [20]

$$U_b(\mathbf{w}) = \frac{Eh^3}{24(1 - \nu^2)} \int_A [(\kappa_\theta + \kappa_\phi)^2 - 2(1 - \nu)(\kappa_\theta \kappa_\phi - \kappa_{\theta\phi}^2)] \, dA \qquad (34)$$

simplifies by putting

$$\kappa_\theta = \partial^2 w / \partial s_\theta^2, \qquad \kappa_\phi = \partial^2 w / \partial s_\phi^2, \qquad \kappa_{\theta\phi} = \partial^2 w / \partial s_\theta \partial s_\phi \qquad (35)$$

Further, changes of the radii of curvature R, r (Fig. 2) can be ignored in forming these derivatives. The error is again of the order of $1/m^2$. Similarly,

$$U_3(\mathbf{w}) = \tfrac{1}{2} \int n_{\alpha\beta}(\mathbf{w}) \frac{\partial w}{\partial s_\alpha} \frac{\partial w}{\partial s_\beta} \, dA \qquad (36)$$

$$U_4(\mathbf{w}) = \frac{Eh}{8(1 - \nu^2)} \int \left(\frac{\partial w}{\partial s_\alpha} \frac{\partial w}{\partial s_\alpha} \right)^2 dA \qquad (37)$$

where $n_{\alpha\beta}$ are the linear elastic membrane stress resultants corresponding to $\mathbf{w}$.

In the high wavenumber limit u and v only appear in the expression for the membrane strain energy U_m. Hence, they can be calculated in terms of w so as to minimise this. A calculation in powers of $1/m$ is sufficient. This method was adopted in the writer's analysis of hyperboloidal cooling-tower buckling mentioned later when applications of the method are given.

Higher-order Analysis

Returning to eqn (14), account is now taken of the higher-order terms. Koiter's method, of replacing $\mathbf{F}_s$ by $\eta \mathbf{F}_s$ and seeking a power series solution for η, is used. The substitution

$$\mathbf{w} = \sum q_J(\lambda) \mathbf{w}_J + \mathbf{w}^* \qquad (38)$$

is made, where $\mathbf{w}^*$ is not rotation-dominant. It is normally possible to take only a finite number N of eigenfields, because as λ_J/λ_1 gets large, the $\mathbf{w}_J$ get increasingly less rotation-dominant. The contribution to V_R from $\mathbf{w}^*$ is bounded above by $0(\eta^2)$ and so is neglected. Substituting in eqn (14) gives a potential energy involving only a *finite* number of degrees of freedom:

$$V_R(q_1, q_2, \ldots, q_N) = \sum_{J=1}^{N} (1 - \lambda/\lambda_J) q_J^2 U(\mathbf{w}_J) + U_{IJK} q_I q_J q_K$$
$$+ U_{IJKL} q_I q_J q_K q_L - \epsilon F_J q_J \qquad (39)$$

where the summation convention is used for repeated capital letters. The coefficients U_{IJK} and U_{IJKL} are generated from the identities in q

$$U_3(q_J \mathbf{w}_J) \equiv U_{IJK} q_I q_J q_K \qquad (40)$$

$$U_4(q_J \mathbf{w}_J) \equiv U_{IJKL} q_I q_J q_K q_L \qquad (41)$$

again using the summation convention. Thus, $U_{111} = U_3(\mathbf{w}_1)$ and $U_{1111} = U_4(\mathbf{w}_1)$. Equilibrium equations may now be generated by forming $\partial V_R / \partial q_I$ $(I = 1, 2 \ldots, N)$ which must all vanish. Koiter [4, 6] takes the theory further by solving for the $(N - M)$ higher coordinates $q_{M+1}, \ldots, q_M$ in terms of the M buckling mode coordinates $q_1, \ldots, q_M$ (here M (≥ 1) is the number of independent buckling modes with $\lambda_1 = \lambda_2 = \cdots = \lambda_N = \lambda_c$). There are special difficulties in doing this for the case of unequal but closely spaced lower eigenvalues.

The analysis can take account of geometrical imperfections. With $\mathbf{F}_s = 0$ for simplicity, let the shell mid-surface be initially a small displacement $\eta \mathbf{d}$ from its 'perfect' reference configuration. $\mathbf{u}$ is again measured relative to this configuration. To first order in η, V_R becomes

$$V_R(\mathbf{w}, \lambda) = U(\mathbf{w}) + \lambda P_o(\mathbf{w}) + U_3(\mathbf{w}) + U_4(\mathbf{w}) - U\langle \mathbf{w}, \eta \mathbf{d} \rangle \qquad (42)$$

since it can be proved that

$$U_f = \tfrac{1}{2} \int_v \{\epsilon_f(u) - \epsilon_f(\eta \mathbf{d})\}^T \mathbf{D} \{\epsilon_f(u) - \epsilon_f(\eta \mathbf{d})\} \, \mathrm{d}V \qquad (43)$$

if the imperfect configuration is stress-free. It follows that

$$\mathbf{w} = \eta \sum_{J=1}^{N} \frac{\lambda_J}{\lambda_J - \lambda} \frac{U\langle \mathbf{w}_J, \mathbf{d} \rangle}{U\langle \mathbf{w}_J, \mathbf{w}_J \rangle} \mathbf{w}_J + \mathbf{w}^* \qquad (44)$$

$\mathbf{w}^*$ being orthogonal to all eigenfields $\mathbf{w}_J$. In particular, $\mathbf{w} \to \eta \mathbf{d}$ (as required) as $\lambda \to 0$.

Finally, one may recover Koiter's general asymptotic results [4–6] in the small-η limit mentioned in the Introduction. The simplest example is when $\lambda_2 > \lambda_1$ and is not close to it while $U_3(\mathbf{w}_1) \neq 0$. V_R

then reduces to

$$V_R(q_1) = (1 - \lambda/\lambda_c)q_1^2 U_1 + U_{111}q_1^3 - \eta F_1 q_1 \qquad (45)$$

with error $0(\eta^2)$, U_1 being $U(\mathbf{w}_c)$ and U_{111} being $U_3(\mathbf{w}_c)$. This is a one-degree-of-freedom system. Straightforward algebra shows that the maximum load point (given by $dV_R/dq_1 = d^2 V_R/dq_1^2 = 0$) occurs when

$$q_1 = \tfrac{1}{3}(\eta F_1/U_{111})^{1/2} + 0(\eta) \qquad (46)$$

(assuming ηF_1 and U_{111} have the same sign) and that then

$$\lambda_{\max}/\lambda = 1 - \{U_{111}\eta F_1/U_1^2\}^{1/2} + 0(\eta) \qquad (47)$$

Thus, the load drop is proportional to $\eta^{1/2}$, Koiter's result. More generally, if $U_{111} = 0$, the reduced potential energy becomes

$$V_R(q_1) = (1 - \lambda/\lambda_c)q_1^2 U_1 + K U_1 q_1^4 - \eta F_1 q_1 \qquad (48)$$

Stable symmetric bifurcation corresponds to $K > 0$, $\eta = 0$ and unstable symmetric bifurcation corresponds to $K < 0$, $\eta = 0$ (Figs 1 and 3). K is the *Koiter coefficient* and is found by an auxiliary calculation (see below). When $K < 0$, the maximum load point of the imperfect structure corresponds to

$$q_1 = (-\eta F_1/8 K U_1)^{1/3} \qquad (49)$$

and

$$1 - \lambda_{\max}/\lambda_c = 3(\eta^2 F_1^2 |K|/8 U_1^2)^{1/3} \qquad (50)$$

i.e. Koiter's 2/3 power law. If $U_{111} = 0$, i.e. $U_3(\mathbf{w}_c) = 0$, then at first sight unstable symmetric buckling ($K < 0$) seems to be excluded, since $U_4(\mathbf{v}_c)$ is positive definite (at least, for dead loading) and so $U_{1111} > 0$. However, the most dangerous buckling combination is of the form $q\mathbf{w}_c + q^2\mathbf{v}$, where $\mathbf{v}$ is some combination of higher eigenfields [4–6] and at $\lambda = \lambda_c$ the term $U_3(q\mathbf{w}_c + q^2\mathbf{v})$, which is of order η^4, may be negative and outweigh the $U_4(q\mathbf{w}_c)$ term (which is also of order η^4). The coefficient K is defined from

$$K = \min [U_3(q\mathbf{w}_c + q^2\mathbf{v}) + U_4(q\mathbf{w}_c)]/U_1 q^4 \qquad (51)$$

and is in principle obtainable knowing $\mathbf{w}_c$ by a calculus-of-variations argument. In practice, looking directly for a combination of higher buckling modes is normally a simpler process.

Repeated or closely spaced eigenvalues require special treatment, but it is normally possible to work with a linear combination of buckling modes (see Applications, below).

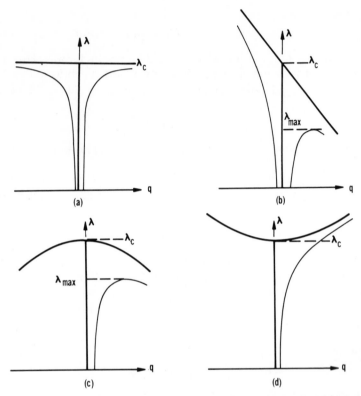

FIG. 3. Buckling paths for perfect and imperfect structures (heavy and light lines) for (a) linearised buckling theory; (b) asymmetric; (c) unstable-symmetric; (d) stable-symmetric. q represents amplitude of reduced deflection.

For structures which do not fail by bifurcation buckling, the criterion 'F_s small' used hitherto can be expressed more precisely as

$$\int_A \mathbf{F}_s \mathbf{u}_s \, dA \ll \int_A \mathbf{F}_o \mathbf{u}_o \, dA \qquad (52)$$

Comparison with Koiter's Theory

The reduced-deflection method is very similar to Koiter's general theory [4, 5]. Koiter expands the total potential energy in terms of an arbitrary departure ($\mathbf{v}$ say) from the current non-linear equilibrium configuration $\mathbf{u}$. The structure is then assumed to lose stability at a bifurcation, normally with negligible displacement (so $\mathbf{u} = \lambda \mathbf{u}_o$) and $\mathbf{v}$

then refers to a neighbouring equilibrium point, so that $\mathbf{v} \simeq \mathbf{w}$.

The reduced-deflection method is not restricted to loadings which cause bifurcation buckling. However, so far the method has only been applied to problems whose secondary loading $\mathbf{F_s}$ is 'small' in the sense of eqn (52). This case can be treated in Koiter's analysis by taking $\mathbf{F_s}$ as a perturbation.

The use of the reduced deflection and reduced potential energy as defined here (together with the systematic use of the rotation-dominance idea when expanding U_f) seems to result in a useful simplification at the cost of a small loss of generality. Koiter's theory can sometimes be applied where the reduced-deflection method fails, namely for structures which undergo significant non-linear deflections before undergoing a bifurcation. An example is a pin-ended circular arch under a central point load (Fig. 4a). The initial deflection is symmetrical but the arch suddenly fails by an S-shaped antisymmetric buckle. (This can be demonstrated with a postcard.) If the deflected shape can be calculated just before this bifurcation, Koiter's analysis will then give the slope of the post-buckling path. The problem was analysed in this way by Walker [21], who used the finite-degree-of-

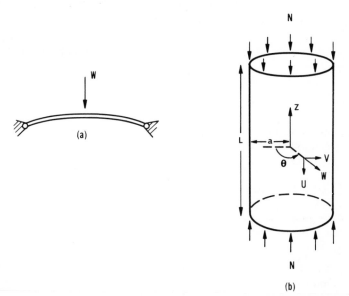

FIG. 4. (a) Pin-ended circular arch under central point load; (b) shell under axial membrane thrust N.

freedom analogue of Koiter's theory. The arch resists the applied load predominantly by bending forces and so the initial symmetric linear elastic deflection is almost entirely pure bending i.e. is rotation-dominant. The entire loading is therefore secondary, u_o is null, and the reduced-deflection method gives no information.

APPLICATIONS

The End-Loaded Cylinder

This is the classic example of imperfection-sensitivity. It shows very clearly the dangerous effect of a combination of buckling modes, each of which would be harmless by itself. It is one of the few shell post-buckling problems which can be solved algebraically in closed form, and so provides a good worked example, Koiter considered the effect of axisymmetric imperfections [5, 22]; the present analysis considers a slightly more dangerous combination.

A shell of radius a, thickness h is under a uniform axial membrane thrust N (Fig. 4b). The perfect shell would undergo a bifurcation buckle and so Koiter's method and the reduced-deflection method are equivalent. The 'primary' deflection u_o is just the axial compression Nh/E produced per unit length in a perfect shell; it is negligible as usual. It is required to determine the post-buckling path and the effect of imperfections.

The high wavenumber approximation applies and, to keep the algebra as simple as possible, ν will be assumed to vanish. (Allowing for it merely results in small correction terms like $(1 - \nu^2)^{1/2}$ at the end of the analysis.) Put $z = -at$, so that t denotes dimensionless distance down the shell (Fig. 4(a)). The strains are

$$\epsilon_t = \partial u / a \partial t, \; \epsilon_{t\theta} = (\partial v / \partial t + \partial u / \partial \theta)/2a$$

$$\epsilon_\theta = (w + \partial v / \partial \theta)/a \tag{53}$$

Assume a linearised buckling mode of form

$$w = h \cos kt \cos l\theta \tag{54a}$$

and determine u and v to minimise the membrane energy U_m of the mode. Omitting the details,

$$u = hkl^2 \sin kt \cos l\theta / a(k^2 + l^2)^2 \tag{54b}$$

$$v = -hl(2k^2 + l^2) \cos kt \sin l\theta / a(k^2 + l^2)^2 \tag{54c}$$

and so

$$\epsilon_\theta = [hk^4/a(k^2 + l^2)^2] \cos kt \cos l\theta \qquad (55a)$$

$$\epsilon_t = l^2 \epsilon_\theta / k^2 \qquad (55b)$$

$$\epsilon_{t\theta} = [hk^3 l/a(k^2 + l^2)^2] \sin kt \sin l\theta \qquad (55c)$$

The U_m so minimised is (for $l \neq 0$)

$$U_m(\mathbf{w}) = \tfrac{1}{2} U_o k^4/(k^2 + l^2)^2 \qquad (56)$$

$$U_o = \pi E h^3 L/2a \qquad (57)$$

and is twice this expression when $l = 0$. The high wavenumber approximation (see Simplifications at High Wavenumber, above) gives, for $l \neq 0$,

$$U_b = \tfrac{1}{2} U_o (k^2 + l^2)^2 h^2/12a^2 \qquad (58)$$

$$P = -(\pi L h^2/4a)Nk^2 \qquad (59)$$

and so

$$V_{RL}(\mathbf{w}) = \tfrac{1}{2} U_o [k^4/(k^2 + l^2)^2 + (k^2 + l^2)^2 h^2/12a^2 - Nk^2/Eh] \qquad (60)$$

When $l = 0$, the expressions for U_b, U_m and P are all doubled, and so is V_{RL}. The critical N for buckling, namely N_c, is therefore

$$N_c = Eh \min [k^2/(k^2 + l^2)^2 + (k^2 + l^2)^2 h^2/12a^2 k^2] \qquad (61)$$

whether or not $l = 0$, and so

$$N_c = Eh^2/a(3)^{1/2} \qquad (62)$$

which is attained for

$$k^2 + l^2 = 2m_o k \qquad (63)$$

$$m_o = (3a^2/4h^2)^{1/4} \qquad (64)$$

l (but not necessarily k) must be an integer. In particular, when $l = 0$, the classical axisymmetric solution [1] is recovered.

It will be seen that the solution consists of a large range of distinct buckling modes, all with the same N_c.

Now put $\lambda = N/N_c$ (so that $\lambda_c = 1$). Equation (60) gives, after substituting and rearranging,

$$V_{RL}(\mathbf{w}) = (U_o k^2/4m_o^2)(1 - \lambda) \qquad (65)$$

for $l \neq 0$, and is twice this when $l = 0$.

Now consider higher-order terms and the effect of imperfections. Consider the initial geometric irregularity w_o such that

$$w_o = h\{\eta_1 \cos 2mt + \eta_2 \cos m\theta \cos (mt - t_o)\} \tag{66}$$

where t_o is a phase angle. Let it excite the buckling mode

$$w = h[q_1 \cos 2mt + 2q_2 \cos m\theta \cos (mt - t_o)] \tag{67}$$

(The factor 2 in the amplitude $2q_2$ appears for algebraic convenience.) m is the integer nearest to m_o. If m_o is an integer, then each term of eqn (67) is a buckling mode, with $k = 2m$, $l = 0$ and with $k = l = m$, respectively. If m_o is not an integer, then, strictly speaking, eqn (67) is not a buckling mode, but the error is of relative order $1/m^2$, i.e. is of the same order as the terms already neglected in the high wavenumber expansion. It follows from eqn (65) that

$$V_{RL}(\mathbf{w}) = U_o(1 - \lambda)(2q_1^2 + q_2^2) + P\langle w, w_o\rangle \tag{68}$$

where

$$P\langle w, w_o\rangle = - N \int_0^{2\pi} \int_0^{L/a} (\partial w/\partial t)(\partial w_o/\partial t) \, d\theta \, dt \tag{69}$$

$$= - U_o\lambda(4\eta_1 q_1 + \eta_2 q_2) \tag{70}$$

using the result (from eqn 62)

$$N = \lambda Eh/2m^2 \tag{71}$$

Now consider the higher-order terms. The calculation of $U_3(\mathbf{w})$ from eqns (36) and (67) is long, but straightforward. Since $\nu = 0$, $n_{\theta\theta}$ equals $Eh\epsilon_\theta(\mathbf{w})$ etc. and $\epsilon_\theta(\mathbf{w})$ etc. are given by eqns (55). On writing

$$U_3 = U_3^{(\theta)} + U_3^{(\theta t)} + U_3^{(t)} \tag{72}$$

where

$$U_3^{(\theta)} = \tfrac{1}{2}Eh \int \epsilon_\theta(w)(\partial w/\partial\theta)^2 \, d\theta \, dt \tag{73a}$$

$$U_3^{(\theta t)} = \tfrac{1}{2}Eh \int 2\epsilon_{\theta t}(w)(\partial w/\partial\theta)(\partial w/\partial t) \, d\theta \, dt \tag{73b}$$

$$U_3^{(t)} = \tfrac{1}{2}Eh \int \epsilon_t(w)(\partial w/\partial t)^2 \, d\theta \, dt \tag{73c}$$

(the integrations being from $\theta = 0$ to 2π and from $t = 0$ to L/a), it is found eventually that

$$U_3^{(\theta)} = U_3^{(\theta t)} = U_3^{(t)} = (3^{1/2}/2)U_o q_1 q_2^2 \cos 2t_o \tag{74}$$

$$U_3 = (3(3)^{1/2}/2)U_o q_1 q_2^2 \cos 2t_o \tag{75}$$

It follows that the final reduced potential energy, assuming no other modes are excited, becomes

$$V_R(w) = U_o\{(1-\lambda)(2q_1^2 + q_2^2) - \lambda(4\eta_1 q_1 + \eta_2 q_2)$$
$$+ (3(3)^{1/2}/2)q_1 q_2^2 \cos 2t_o\} + \cdots \tag{76}$$

The neglected terms are order q^4, η^2 or ηq^2. The third-order terms are most destabilising when $\cos 2t_o = -1$, $t_o = \pi/2$ and this is hereafter assumed. The equilibrium path of the imperfect structure is given by $\partial V_R/\partial q_I = 0$ for $I = 1, 2$ and so are (writing $c = 3(3)^{1/2}/2$)

$$4(1-\lambda)q_1 - cq_2^2 = 4\lambda\eta_1 \tag{77a}$$

$$2(1-\lambda)q_2 - 2cq_1 q_2 = \lambda\eta_2 \tag{77b}$$

In particular, the post-buckling path of the perfect structure ($\eta_1 = \eta_2 = 0$) corresponds to $q_2 = \pm 2q_1$. Eliminating q_1 between eqns (77) gives

$$4(1-\lambda)^2 q_2 - cq_2(4\lambda\eta_1 + cq_2^2) = 2\lambda(1-\lambda)\eta_2 \tag{78}$$

Differentiating to find λ_{max} gives

$$4(1-\lambda_{max})^2 - c(4\lambda_{max}\eta_1 + cq_2^2) = 2c^2 q_2 \tag{79}$$

i.e.

$$c^2 q_2^3 = \lambda_{max}(1-\lambda_{max})\eta_2 \tag{80}$$

This equation can now be used to express λ_{max} in terms of η_1 and η_2. Finally, let

$$\eta_1 = \eta \cos \theta_o \tag{81a}$$

$$\eta_2 = \eta(2)^{1/2} \sin \theta_o \tag{81b}$$

$$\bar{w}_{rms} = \tfrac{1}{2}h\eta \tag{81c}$$

where $\bar{w}_{rms}$ is the root mean square deflection of the assumed imperfection w_o, from eqn (66). Then, after further algebra, it can be shown that

$$(1-\lambda_{max})^2 = C\lambda_{max}\eta \tag{82}$$

for

$$C = 3(3)^{1/2}\mu/2 \tag{83}$$

where μ is the positive root of the cubic

$$32(\mu - \cos \theta_0)^3 = 27\mu \sin^2 \theta_0 \qquad (84)$$

In particular, for axisymmetric imperfections $\theta_0 = 0$ and $\mu = 1$, so that $C = 2\cdot60$, which is in numerical agreement with Koiter's analysis [4, 5, 22]. μ attains a maximum of $1\cdot587$ for $\theta_0 = \cos^{-1} 0\cdot7622 = 40\cdot3°$, for which $C = 4\cdot12$. Thus, a variation in radius whose r.m.s. value is only 1% of the shell thickness but which has the form of this 'most dangerous' imperfection, leads to a 25% drop in buckling load, according to eqn (82) (with $\eta = 0\cdot02$ from eqn 81c).

A more general analysis would Fourier analyse the initial irregularity into a general combination of modes. For fixed values of the Fourier coefficient ratios, then, if η measures the r.m.s. deflection in radius by eqn (81c), eqn (82) again results in general, but with a smaller coefficient C. (See also Ref. [11].)

The Wind-loaded Hyperboloid (Cooling Tower)

This was the original practical problem that led to the present method. A hyperboloid of revolution is subjected to a pressure distribution

$$p = p_w \sum p'_m \cos m\theta \qquad (85)$$

where p_w is the wind pressure-head and $p'_0, p'_1, p'_2, \ldots$ are dimensionless pressure coefficients. (The writer analysed this problem, using linearised buckling theory and treating the pressure loading as dead, in 1971; the work is not available in the open literature but is reviewed in Ref. [23].) For uniform all-round pressure buckling, the experimental buckling pressure was about 85% of its calculated value. But for Der and Fidler's wind-tunnel tests [24], the experimental wind pressure-heads were, in general, only some 30–50% of the calculated values.

For hyperboloidal shells of the geometry used in the wind-tunnel tests [24] the pressure coefficients in eqn (85) are, from Ref. [25],

$$p'_0 = 0\cdot23, \quad p'_1 = 0\cdot28, \quad p'_2 = 0\cdot60, \quad p'_3 = 0\cdot47$$

$$p'_4 = 0\cdot06, \quad p'_5 = -0\cdot12, \quad p'_6, p'_7, \ldots \text{negligible} \ldots \qquad (86)$$

To apply the reduced-deflection method, one assumes that the lowest four harmonics, p'_0 to p'_3 inclusive, correspond to the primary load $\mathbf{F}_0$ and that the others $p'_4, p'_5, \ldots$, etc., generate the rotation-dominant remainder. Membrane theory is reliable up to about the third harmonic for the calculation of linear elastic stresses for a typical tower [25]. Expanding **w** as a sum $\Sigma\ q_I \mathbf{w}_I$, where I ranges from 4 to 10 and where $\mathbf{w}_I$ are buckling modes proportional to $\cos I\theta$ and calculated for *uniform* pressure buckling (with circumferential wavenumber I), a formula similar to eqn (39) is obtained:

$$V_R(q_1, q_2, \ldots, q_N) = (U_{IJ} - \lambda P_{IJ})q_I q_J - F_J q_J + U_{IJK} q_I q_J q_K + \cdots \quad (87)$$

where U_{IJ} is a diagonal matrix (as in eqn 39) but P_{IJ} is not, because of the interactions between the p_0', p_1', p_2' and p_3' harmonics. The summation convention is again used. The $F_J q_J$ terms are formed from the $p_4', p_5', p_6', p_7', \ldots$ terms. (Although $p_6', p_7', \ldots$ are negligible, it costs little to retain them temporarily in the analysis).

The quadratic terms give rise to a linearised buckling problem, which was solved by matrix methods. An account of the similar but simpler procedure for cylinder buckling appears in Ref. [26]. The third-order and higher-order terms control the stability at the bifurcation point. The fourth, fifth, ... loading harmonics 'round off' the corner of this bifurcation point to give the kind of load–displacement curve sketched in Fig. 3.

For uniform pressure loading, the third-order terms defined by eqn (36) vanish when $\mathbf{w} = \mathbf{w}_c$, the critical buckling mode. The buckling load is fairly well defined, and so the bifurcation point must be either stable symmetric or unstable symmetric. A search for the minimising K in eqn (51) was accordingly made. K is bounded above by $U_4(\mathbf{w}_c)/u_1$ and, in fact, no significant decrease in K could be found by combining $q\mathbf{w}_c$ with some combination $q^2\mathbf{v}$. It was concluded, therefore, that the uniform pressure loading was indeed *stable symmetric* (as it is for the laterally pressurised cylinder). The shortfall between theory and experiment can be attributed to movement of the nominally rigid base support. But base movement cannot account for the much larger discrepancy between experiment and theory in the wind-tunnel tests. Croll [27] had studied the deformation of cooling-tower-type model shells under a single lateral point load, and had demonstrated that a snap-through instability arose. So the higher-degree terms appear to have been destabilising in his experiments (but large deflections would have occurred before buckling, complicating the situation). From eqn (36) it is clear that non-zero third-order terms can arise from a link between, for example, the $\cos 4\theta$, $\cos 7\theta$ and

$\cos 11\theta$ harmonics (the uniform-pressure buckling wavenumber was 7). Accordingly, it was concluded in the original work (1971) that the higher-order terms were *destabilising*. This conclusion is probably incorrect. Later calculations suggest that the third-order terms were very small and would be dominated by the fourth-order terms. This agrees with the conclusions of Cole, Abel and Billington [23].

In fact, there is a very simple explanation for the observed shortfall [26]. The pressure loading in eqn (85) is assumed to be dead. It is certainly legitimate to ignore the change in *direction* of the wind pressure as the cooling-tower surface deforms; high wavenumber theory suggests these contribute a decrease of perhaps 2%. But it is not legitimate to neglect the change in *magnitude* of the pressure coefficients p_0' to p_3'. These will almost certainly be significant, especially near the separation points, and will therefore affect the balance of second-order terms that make up the linearised buckling calculation.

The moral of this is that in the search for a relatively elaborate explanation (destabilising higher-order terms) one should not miss the simple and physically obvious explanation.

CONCLUSIONS

(1)   The reduced-deflection method provides a rapid and efficient method of setting up the equations of classical linearised buckling theory, and of considering the higher-degree terms neglected in this theory. Mathematically, it is a variant of Koiter's general theory.

(2)   The higher-degree terms control the post-buckling behaviour of the structure. If they are destabilising, large drops below the predictions of classical buckling theory can occur. The method can calculate these drops, and so assess the value of classical theory.

(3)   Once the N lowest eigenmodes $\mathbf{w}_1, \mathbf{w}_2, \ldots, \mathbf{w}_N$ of classical buckling theory have been calculated, the original non-linear problem can be discretised into one involving N degrees of freedom (the amplitudes q_I of the $\mathbf{w}_I$). This forms a possible starting point for a full non-linear analysis.

(4)   The reduced-deflection method is illustrated by considering the buckling of an end-loaded circular cylinder (in detail) and the

wind buckling of a hyperboloidal cooling-tower (in outline only).

(5) The method divides up the unit applied loading F_{oo} into two parts, F_o (primary) and F_s (secondary). The corresponding linear elastic responses are u_o and u_s. u_s is rotation-dominant (i.e. its rotations dominate its strains) but u_o is not. The method is not useful for structures which resist F_{oo} mainly by bending, because F_o is then small or null. The method is most useful when F_s is small (in the sense that the work done by F_s to produce u_s is small compared with the work done by F_o to produce u_o).

(6) The method has also proved helpful in setting up the creep buckling equations of a cylindrical tube allowing for end effects and axial temperature variations (described in Ref. [13]).

ACKNOWLEDGEMENT

This work was carried out at the Central Electricity Research Laboratories and is published by permission of the Central Electricity Generating Board.

REFERENCES

1. TIMOSHENKO, S. P., *Theory of Elastic Stability*, McGraw-Hill, 1936.
2. FLUGGE, W., *Stresses in Shells*, Springer-Verlag, 1960.
3. GERARD, G., *Introduction to Structural Stability Theory*, McGraw-Hill, 1962.
4. KOITER, W. T., 'On the Stability of Elastic Equilibrium', English translation: NASA TT F10,833, 1967.
5. KOITER, W. T., Elastic stability and post-buckling behaviour, *Proc. Symp. Non-linear Problems* (Ed. R. E. Langer), Wisconsin University Press, 1963.
6. KOITER, W. T., General equations of elastic stability for thin shells, *Proc. Symp. in Honour of L. H. Donnell* (Ed. D. Muster), University of Houston, Texas, 1967.
7. THOMPSON, J. M. T., Discrete branching points in the general theory of elastic stability, *J. Mech. Phys. Solids*, 13, 1965, 295–310.
8. SEWELL, M. J., The static perturbation technique in buckling problems, *J. Mech. Phys. Solids*, 13, 1965, 247–265.
9. ROORDA, J., Stability of structures with small imperfections, *ASCE, J. Eng. Mech. Div.*, 91, 1965, EM1, Proc. Paper 4230.

10. ROORDA, J., The buckling behaviour of imperfect structural systems, *J. Mech. Phys. Solids*, **13**, 1965, 267–280.
11. HUTCHINSON, J. W. and KOITER, W. T., Post-buckling theory, *Applied Mechanics Reviews*, 1970. (*Trans. ASME*, **23**, 1353–1366).
12. ZIENKIEWICZ, O. C., *The Finite Element Method in Engineering Science*, McGraw-Hill, 1971.
13. EWING, D. J. F., The creep collapse of pressurised non-uniformly-heated tubes, This volume, pp. 137–166.
14. BREBBIA, C. A. and CONNOR, J. J., *Fundamentals of Finite Element Techniques for Structural Engineers'*, Butterworths, 1973.
15. HUNTER, S. C., *Mechanics of Continuous Media*, Ellis Horwood (John Wiley), 1977.
16. HILL, R., On uniqueness and stability in the theory of finite elastic strain, *J. Mech. Phys. Solids*, **5**, 1957, 229–241.
17. PEARSON, C. E., General theory of elastic stability, *Q. Math.*, **14**, 1956, 133–144.
18. TREFFTZ, E., Theory of stability of the elastic equilibrium (in German), *Z. Angew. Math. Mech.*, **13**, 1933, 160.
19. HILL, R., Eigenmodal deformations in elastic–plastic continua, *J. Mech. Phys. Solids*, **15**, 1967, 371–386.
20. NOVOSHILOV, V. V., *Theory of Thin Shells*, Noordhof, 1959.
21. WALKER, A. C., A nonlinear finite-element analysis of shallow circular arches, *Int. J. Solids Struct.*, **5**, 1969, 97–107.
22. KOITER, W. T., The effect of axisymmetric imperfections on the buckling of cylindrical shells under axial compression, *Proc. Kon. Nederl. Akad. Wetenschap. B*, **66**, 1963, 265–279.
23. COLE, P. P., ABEL, A. F. and BILLINGTON, D. P., Buckling of cooling-tower shells: state of the art, *J. Struct. Div. ASCE*, **101**, 1975, ST6, Proc. Paper 11364, 1185–1203.
24. DER, T. J. and FIDLER, R., A model study of the buckling behaviour of hyperbolic shells, *Proc. Inst. Civil Eng.* **41**, 1968, 105–118.
25. ALBASINY, E. L. and MARTIN, D. W. Bending and membrane equilibrium in cooling towers, *J. Eng. Mech. Div. ASCE*, **95**, 1967, EM3, Proc. Paper 5256.
26. EWING, D. J. F., Discussion of paper 'Buckling of cylindrical shells by wind pressure' (by Wang, Y. S. and Billington, D. P., *J. Mech. Engr. Div. ASCE*, **100**, 1974, Proc. Paper 10874), *J. Mech. Eng. Div. ASCE*, **101**, 1975, EM5, Proc. Paper 11598 (part), 713–715.
27. CROLL, J. G. A., The buckling of hyperboloidal cooling towers under lateral loading, *Proc. Int. Colloq. Prog. Shell Structures*, IASS Vol. VII, 1969.

18

Non-Linear Analysis of Shells and Beams Using Degenerate Isoparametric Elements

B. Kråkeland and O. Mo

AS Computas, Høvik, Norway

SUMMARY

The paper is concerned with the non-linear analysis of shells and beams considering the combined effect of large displacements and material non-linearities. An updated Lagrangian description of motion is adopted for the geometrically non-linear problem.

The theory is formulated within the framework of the finite-element method, adopting doubly curved shell and beam elements. Both elements are derived by degeneration of three-dimensional isoparametric elements. Inelastic material behaviour is modelled with the flow theory of plasticity, adopting the von Mises yield criterion. Two hardening rules are adopted; the isotropic hardening rule and the so-called overlay model.

The applicability of the method presented is illustrated by two numerical examples.

INTRODUCTION

Stiffened shells and beams are being increasingly used in aerospace structures, marine structures, nuclear power plant and other structures of high complexity. Owing to the complex loading and severe environmental conditions, a high degree of precision is needed in determining the safety of such structures.

The requirements of weight optimisation and material economy have led to utilisation of improved materials and resulted in 'thinner' structures for which geometric non-linearities are important. There has also been a general trend towards the use of a greater part of the load-carrying capacity of structures. A consequence of this is that material non-linearities steadily become more important in design work and the determination of the non-linear behaviour prior to collapse is often needed in order to assess the safety of a structure. This is especially the case for shell structures, where overloading into the buckling range may be very dangerous.

Various finite-element models have been suggested in the literature for non-linear shell and beam problems. Some authors have used flat triangular or quadrilateral shell elements and straight beam elements, and updated the geometry of the structure during deformation. In the work by Murray and Wilson [1] only the linear element stiffness matrices were assembled in the deformed configuration.

The shell and beam can also be considered as three-dimensional continua for the analysis of which the definition of stress and strain at finite deformations may be used. Obviously, for most practical applications this method may be both complicated and costly. The method adopted in the present work entails simplifications of the three-dimensional theory in accordance with the basic assumptions for shells and beams. The formulation is related to doubly curved shell and beam elements. Both elements are obtained by degeneration of three-dimensional isoparametric elements. Stress and strain are referred to local 'tangential' axes and simplifications are introduced which restrict the theory to cases of small strain and large rotations [2].

THEORETICAL DEVELOPMENT

Basic Equations

In an analysis dealing with both geometric and material non-linearities an incremental solution scheme is the most suitable approach. An incremental form of the principle of virtual work based on an updated Lagrangian description of motion is considered in this paper. The basic quantities are the incremental displacements referred to the coordinate axes in the updated configuration at the beginning of the increment. Denoting this configuration by C_1, the following incremental virtual work equation is used to determine a new configuration C_2 after the load increment [3]:

$$\int_V C_{ijkl}\Delta E_{kl}\delta\Delta E_{ij}\,\mathrm{d}V + \int_V {}^1\sigma_{ij}\delta\Delta\eta_{ij}\,\mathrm{d}V$$
$$= \int_A T_i\delta\Delta u_i\,\mathrm{d}A + \int_V F_i\delta\Delta u_i\,\mathrm{d}V - \int_V {}^1\sigma_{ij}\delta\Delta\epsilon_{ij}\,\mathrm{d}V \quad (1)$$

In this equation,

Δu_i are the incremental displacements between C_1 and C_2.

ΔE_{ij} $(=\Delta\epsilon_{ij}+\Delta\eta_{ij})$ are the components of Green's strain tensor corresponding to Δu_i. ($\Delta\epsilon_{ij}$ and $\Delta\eta_{ij}$ are the linear and quadratic parts of ΔE_{ij}.)

C_{ijkl} are the components of the incremental constitutive tensor.

${}^1\sigma_{ij}$ are true (Cauchy) stresses in C_1.

T_i are prescribed surface tractions in C_2.

F_i are body forces in C_2.

In eqn (1) the following relationships have been used [3]:

$$\Delta S_{ij} = C_{ijkl}\Delta E_{kl} \tag{2}$$

and

$$S_{ij} = {}^1\sigma_{ij} + \Delta S_{ij} \tag{3}$$

where S_{ij} is the Kirchhoff stress tensor.

It should be noted that residual forces in C_1 are accounted for by the last term of eqn (1).

Characteristics of the Isoparametric Shell Element

The degenerate isoparametric element [4, 5] is shown in Fig. 1. The element is of the parabolic type with eight nodal points.

Geometry

The global Cartesian coordinates X, Y, Z for a point in the element are given by the element curvilinear coordinates ξ, η, ζ from the relation [4]

$$\begin{bmatrix} X \\ Y \\ Z \end{bmatrix} = \sum_{i=1}^{8} N_i(\xi, \eta) \begin{bmatrix} X_i \\ Y_i \\ Z_i \end{bmatrix} + \frac{1}{2}\sum_{i=1}^{8} N_i(\xi, \eta)\zeta t_i v_{3i} \tag{4}$$

where $N_i(\xi, \eta)$ is the shape function for the middle surface corresponding to node i, t_i is the shell thickness and v_{3i} is a vector of unit length in the thickness direction at node i (see Fig. 1).

Displacement field

The displacement field is defined in terms of the nodal displacements u_i, v_i, w_i in the directions of the global axes, and two rotations of the vector v_{3i} about orthogonal directions normal to v_{3i}. Two such directions are given by the vectors v_{1i} and v_{2i} (of unit length) (see Fig. 1).

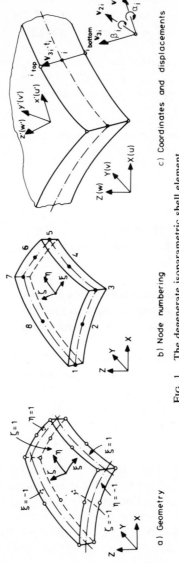

a) Geometry   b) Node numbering   c) Coordinates and displacements

FIG. 1.   The degenerate isoparametric shell element.

The displacements in the directions of the global axes are given by [4]

$$\begin{bmatrix} u \\ v \\ w \end{bmatrix} = \sum_{i=1}^{8} N_i(\xi, \eta) \begin{bmatrix} u_i \\ v_i \\ w_i \end{bmatrix} + \frac{1}{2} \sum_{i=1}^{8} N_i(\xi, \eta) \zeta t_i \phi_i \begin{bmatrix} \alpha_i \\ \beta_i \end{bmatrix} \tag{5}$$

where

$$\phi_i = [-v_{2i}, v_{1i}] \tag{6}$$

The components of strains and stresses are referred to local Cartesian coordinates x', y', z'. The local derivatives of the local displacements are given by

$$[\mathbf{e}_u, \mathbf{e}_v, \mathbf{e}_w] = \begin{bmatrix} \dfrac{\partial u'}{\partial x'} & \dfrac{\partial v'}{\partial x'} & \dfrac{\partial w'}{\partial x'} \\[2mm] \dfrac{\partial u'}{\partial y'} & \dfrac{\partial v'}{\partial y'} & \dfrac{\partial w'}{\partial y'} \\[2mm] \dfrac{\partial u'}{\partial z'} & \dfrac{\partial v'}{\partial z'} & \dfrac{\partial w'}{\partial z'} \end{bmatrix} = \theta^{T} \mathbf{J}^{-1} \begin{bmatrix} \dfrac{\partial u}{\partial \xi} & \dfrac{\partial v}{\partial \xi} & \dfrac{\partial w}{\partial \xi} \\[2mm] \dfrac{\partial u}{\partial \eta} & \dfrac{\partial v}{\partial \eta} & \dfrac{\partial w}{\partial \eta} \\[2mm] \dfrac{\partial u}{\partial \zeta} & \dfrac{\partial v}{\partial \zeta} & \dfrac{\partial w}{\partial \zeta} \end{bmatrix} \theta \tag{7}$$

Here θ is the orthogonal transformation matrix between local and global axes and $\mathbf{J}$ is the Jacobian matrix.

In the local coordinate system the x' axis is chosen tangential to the curvilinear ξ axis and the z' axis is normal to the surface $\zeta = $ constant (see Fig. 1). The coordinate system is right-handed.

Assuming ζ to be parallel to z', the following simplified form is achieved for $\theta^{T} \mathbf{J}^{-1}$ (see also Ref. [5]):

$$\mathbf{A} = \theta^{T} \mathbf{J}^{-1} = \begin{bmatrix} A_{11} & 0 & 0 \\ A_{21} & A_{22} & 0 \\ 0 & 0 & A_{33} \end{bmatrix} \tag{8}$$

The derivatives of the displacements with respect to the curvilinear coordinates are computed from the expressions of eqn (5). Substitution of these terms and eqn (8) into eqn (7) yields the following:

$$\begin{bmatrix} \mathbf{e}_u \\ \mathbf{e}_v \\ \mathbf{e}_w \end{bmatrix} = \sum_{i=1}^{8} [\bar{\mathbf{B}}_i \theta^{T}, \frac{1}{2} t_i(\zeta \bar{\mathbf{B}}_i + \bar{\mathbf{C}}_i) \theta^{T} \phi_i] \begin{bmatrix} u_i \\ v_i \\ w_i \\ \alpha_i \\ \beta_i \end{bmatrix} \tag{9}$$

378   B. Kråkeland and O. Mo

where

$$\bar{B}_i = \begin{bmatrix} B_1 & 0 & 0 \\ B_2 & 0 & 0 \\ 0 & 0 & 0 \\ 0 & B_1 & 0 \\ 0 & B_2 & 0 \\ 0 & 0 & 0 \\ 0 & 0 & B_1 \\ 0 & 0 & B_2 \\ 0 & 0 & 0 \end{bmatrix}_i \qquad \bar{C}_i = \begin{bmatrix} 0 & 0 & 0 \\ 0 & 0 & 0 \\ C_1 & 0 & 0 \\ 0 & 0 & 0 \\ 0 & 0 & 0 \\ 0 & C_1 & 0 \\ 0 & 0 & 0 \\ 0 & 0 & 0 \\ 0 & 0 & C_1 \end{bmatrix}_i \qquad (10)$$

with

$$B_{1i} = A_{11}N_{i,\xi}$$
$$B_{2i} = A_{21}N_{i,\xi} + A_{22}N_{i,\eta} \qquad (11)$$
$$C_{1i} = A_{33}N_i$$

Element matrices

The small-displacement stiffness matrix and the geometric stiffness matrix are derived from the first and second terms of eqn (1) after deleting the quadratic strain expressions from the first term of the equation. Substitution of the displacement derivatives of eqn (9) into eqn (1) yields the following expression for the nodal submatrix k_{Iij} of the incremental stiffness matrix:

$$k_{Iij} = \int_{-1}^{1}\int_{-1}^{1}\int_{-1}^{1} \begin{bmatrix} \theta & 0 \\ 0 & \frac{1}{2}t_i\phi_i^T\theta \end{bmatrix} \bar{k}_{Iij} \begin{bmatrix} \theta^T & 0 \\ 0 & \frac{1}{2}t_j\theta^T\phi_j \end{bmatrix} \det(J)\, d\xi\, d\eta\, d\zeta$$

where   (12)

$$\bar{k}_{Iij} = \bar{k}_{Lij} + \bar{k}_{Gij} \qquad (13)$$

i.e. k_{Iij} contains contributions from the small-displacement and geometric stiffness matrices. The small-displacement stiffness matrix is shown in Ref. [5] and will not be given here.

The contribution from the geometric stiffness matrix is given by

$$\bar{k}_{Gij} = \begin{bmatrix} \bar{B}_i^T \\ \zeta\bar{B}_i^T + \bar{C}_i^T \end{bmatrix} \begin{bmatrix} \bar{\sigma} & 0 & 0 \\ 0 & \bar{\sigma} & 0 \\ 0 & 0 & \bar{\sigma} \end{bmatrix} [\bar{B}_j, \zeta\bar{B}_j + \bar{C}_j] \qquad (14)$$

where $\bar{\sigma}$ is the matrix of Cauchy stresses given by

$$\bar{\sigma} = \begin{bmatrix} \sigma_{x'} & \tau_{x'y'} & \tau_{x'z'} \\ & \sigma_{y'} & \tau_{y'z'} \\ \text{SYM.} & & 0 \end{bmatrix} \tag{15}$$

The nodal load vector is derived from the first and second terms on the right-hand side of eqn (1). By recalculating the load vector in the currently updated reference configuration, account may be taken of displacement-dependent loading, e.g. hydrostatic pressure.

The nodal forces in equilibrium with the internal stress field are calculated from the last term of eqn (1). Virtual displacements at node i yield the following expression for the nodal forces:

$$\mathbf{Q}_i = \begin{bmatrix} Q_x \\ Q_y \\ Q_z \\ M_1 \\ M_2 \end{bmatrix}_i = \int_{-1}^{1} \int_{-1}^{1} \int_{-1}^{1} \begin{bmatrix} \theta & 0 \\ 0 & \frac{1}{2}t_i\phi_i^{\mathrm{T}}\theta \end{bmatrix} \begin{bmatrix} \mathbf{B}_i^{\mathrm{T}} \\ \zeta\mathbf{B}_i^{\mathrm{T}} + \mathbf{C}_i^{\mathrm{T}} \end{bmatrix} \begin{bmatrix} \sigma_{x'} \\ \sigma_{y'} \\ \tau_{x'y'} \\ \tau_{x'z'} \\ \tau_{y'z'} \end{bmatrix} \det(\mathbf{J}) \, d\xi \, d\eta \, d\zeta \tag{16}$$

Here Q_{xi}, Q_{yi}, Q_{zi} are forces referred to global axes and M_{1i}, M_{2i} are moments about the local nodal axes in the directions of v_{1i} and v_{2i}. $\mathbf{B}_i$ and $\mathbf{C}_i$ are 5×3 matrices which contain the terms B_{1i}, B_{2i} and C_{1i} similar to eqn (10).

Depending on the strategy for solving the non-linear problem, the incremental stiffness matrix need not be updated for every equilibrium iteration cycle (see section on Solution of Non-linear Equations). In this case no updating of the reference configuration is required and the nodal forces in equilibrium with the internal stress field are calculated from the following virtual work expression:

$$\mathbf{Q}^{\mathrm{T}}\delta\Delta\mathbf{v} = \int_{V} \mathbf{S}^{\mathrm{T}}\delta\Delta\mathbf{E} \, dV \tag{17}$$

Here $\Delta\mathbf{v}$ is the vector of incremental nodal displacements between the reference configuration and the current configuration, $\Delta\mathbf{E}$ is the vector of the corresponding Green strains and $\mathbf{S}$ is the vector of Kirchhoff stresses. The nodal force vector may be obtained from eqn (17) by substitution of the displacement derivatives of eqn (9).

All element matrices are evaluated using numerical integration and improved accuracy is obtained with the 'reduced integration technique', in which 2×2 gaussian integration points are used in the plane of the element (see Ref. [5]). In the elastic case either analytical or two-point gaussian integration is used in the thickness direction. However, for elasto-plastic materials a higher-order integration scheme is required through the thickness of the shell. In the present study five integration points through the thickness are found to give sufficient accuracy after yielding (see Ref. [2]).

Characteristics of the Isoparametric Beam Element

The three-node degenerate isoparametric beam element is shown in Fig. 2. The element is eccentric with rectangular cross-section. The derivation of the element is similar to that of the shell element described in the previous section.

Geometry

The global Cartesian coordinates for a point in the element are given by the relation

$$
\begin{bmatrix} X \\ Y \\ Z \end{bmatrix} = \sum_{i=1}^{3} N_i(\xi) \begin{bmatrix} X_i \\ Y_i \\ Z_i \end{bmatrix} + \sum_{i=1}^{3} N_i(\xi) \left(\frac{t_{1i}}{2} \eta + e_{1i} \right) \mathbf{v}_{2i} + \sum_{i=1}^{3} N_i(\xi)
$$
$$
\times \left(\frac{t_{2i}}{2} \zeta + e_{2i} \right) \mathbf{v}_{3i} \tag{18}
$$

Here the shape function $N_i(\xi)$ depends only on the curvilinear coordinate along the beam. The beam thicknesses (t_{1i} and t_{2i}) and eccentricities (e_{1i} and e_{2i}) are shown in Fig. 2(b). The vectors $\mathbf{v}_{2i}$ and $\mathbf{v}_{3i}$ are of unit length and parallel to the sides of the rectangular cross-section $\zeta = \pm 1$ and $\eta = \pm 1$, respectively.

Displacement field

The displacement field is defined in terms of the nodal displacements u_i, v_i, w_i in the directions of the global axes, and the rotations α, β and γ about the vectors $\mathbf{v}_{1i}$, $\mathbf{v}_{2i}$ and $\mathbf{v}_{3i}$ as shown in Fig. 2(c). The displacement field is given by

$$
\begin{bmatrix} u \\ v \\ w \end{bmatrix} = \sum_{i=1}^{3} N_i(\xi) \begin{bmatrix} u_i \\ v_i \\ w_i \end{bmatrix} + \sum_{i=1}^{3} N_i(\xi) \left(\frac{t_{1i}}{2} \eta + e_{1i} \right) \phi_{3i} \begin{bmatrix} \alpha_i \\ \gamma_i \end{bmatrix}
$$
$$
+ \sum_{i=1}^{3} N_i(\xi) \left(\frac{t_{2i}}{2} \zeta + e_{2i} \right) \phi_{2i} \begin{bmatrix} \beta_i \\ \alpha_i \end{bmatrix} \tag{19}
$$

where

$$
\phi_{3i} = [\mathbf{v}_{3i}, -\mathbf{v}_{1i}]
$$
$$
\phi_{2i} = [\mathbf{v}_{1i}, -\mathbf{v}_{2i}] \tag{20}
$$

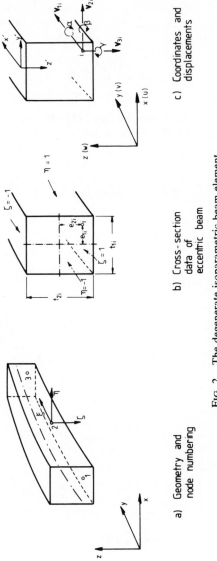

FIG. 2. The degenerate isoparametric beam element.

The components of strains and stresses are referred to the local Cartesian coordinates. In the local coordinate system the y' axis is chosen parallel to the surfaces $\zeta = \pm 1$. The x' axis is chosen tangential to the ξ axis (see Fig. 2), and, hence, the following form results for $\theta^T J^{-1}$ in eqn (7):

$$A = \theta^T J^{-1} = \begin{bmatrix} A_{11} & 0 & 0 \\ A_{21} & A_{22} & A_{23} \\ A_{31} & A_{32} & A_{33} \end{bmatrix} \tag{21}$$

By substitution of eqn (19) and eqn (21) into eqn (7) the following expression is obtained for the derivatives of the displacements with respect to the local axes:

$$\begin{bmatrix} e_u \\ e_v \\ e_w \end{bmatrix} = \sum_{i=1}^{3} \left[\bar{B}_i \theta^T, \left\{ \left(\left(\frac{t_{1i}}{2} \eta + e_{1i} \right) \bar{B}_i + \frac{t_{1i}}{2} \bar{C}_i \right) \theta^T \phi'_{3i} \right. \right.$$

$$\left. \left. + \left(\left(\frac{t_{2i}}{2} \zeta + e_{2i} \right) \bar{B}_i + \frac{t_{2i}}{2} \bar{D}_i \right) \theta^T \phi'_{2i} \right\} \right] \begin{bmatrix} u_i \\ v_i \\ w_i \\ \alpha_i \\ \beta_i \\ \gamma_i \end{bmatrix} \tag{22}$$

In eqn (22)

$$\bar{B}_i = \begin{bmatrix} B_1 & 0 & 0 \\ B_2 & 0 & 0 \\ B_3 & 0 & 0 \\ 0 & B_1 & 0 \\ 0 & B_2 & 0 \\ 0 & B_3 & 0 \\ 0 & 0 & B_1 \\ 0 & 0 & B_2 \\ 0 & 0 & B_3 \end{bmatrix}_i, \quad \bar{C}_i = \begin{bmatrix} 0 & 0 & 0 \\ C_1 & 0 & 0 \\ C_2 & 0 & 0 \\ 0 & 0 & 0 \\ 0 & C_1 & 0 \\ 0 & C_2 & 0 \\ 0 & 0 & 0 \\ 0 & 0 & C_1 \\ 0 & 0 & C_2 \end{bmatrix}_i, \quad \bar{D}_i = \begin{bmatrix} 0 & 0 & 0 \\ D_1 & 0 & 0 \\ D_2 & 0 & 0 \\ 0 & 0 & 0 \\ 0 & D_1 & 0 \\ 0 & D_2 & 0 \\ 0 & 0 & 0 \\ 0 & 0 & D_1 \\ 0 & 0 & D_2 \end{bmatrix}_i \tag{23}$$

where the quantities involved are given by

$$\begin{array}{ll} B_{1i} = A_{11} N_{i,\xi}; & C_1 = A_{22} N_i \\ B_{2i} = A_{21} N_{i,\xi}; & C_2 = A_{32} N_i \\ B_{3i} = A_{31} N_{i,\xi}; & D_1 = A_{23} N_i \\ & D_2 = A_{33} N_i \end{array} \tag{24}$$

The following symbols have also been used in eqn (22):

$$\phi'_{3i} = [v_{3i}, 0, -v_{1i}]$$

and

$$\phi'_{2i} = [-v_{2i}, v_{1i}, 0]$$

Element matrices

The nodal submatrix $\mathbf{k}_{Iij}$ of the incremental stiffness matrix is given by

$$\mathbf{k}_{Iij} = \int_{-1}^{1}\int_{-1}^{1}\int_{-1}^{1} \begin{bmatrix} \theta & 0 & 0 \\ 0 & \phi_{3i}'^{\mathrm{T}}\theta & \phi_{2i}'^{\mathrm{T}}\theta \end{bmatrix} \bar{\mathbf{k}}_{Iij} \begin{bmatrix} \theta^{\mathrm{T}} & 0 \\ 0 & \theta^{\mathrm{T}}\phi'_{3j} \\ 0 & \theta^{\mathrm{T}}\phi'_{2j} \end{bmatrix} \det(\mathbf{J})\, d\xi\, d\eta\, d\zeta \quad (25)$$

where

$$\bar{\mathbf{k}}_{Iij} = \bar{\mathbf{k}}_{Lij} + \bar{\mathbf{k}}_{Gij} \quad (26)$$

i.e. $\mathbf{k}_{Iij}$ contains contributions from the small-displacement and geometric stiffness matrices. The matrices $\bar{\mathbf{k}}_{Lij}$ and $\bar{\mathbf{k}}_{Gij}$ are given by

$$\bar{\mathbf{k}}_{Lij} = \begin{bmatrix} \mathbf{B}_i^{\mathrm{T}} \\ \left(\dfrac{t_{1i}}{2}\eta + e_{1i}\right)\mathbf{B}_i^{\mathrm{T}} + \dfrac{t_{1i}}{2}\mathbf{C}_i^{\mathrm{T}} \\ \left(\dfrac{t_{2i}}{2}\zeta + e_{2i}\right)\mathbf{B}_i^{\mathrm{T}} + \dfrac{t_{2i}}{2}\mathbf{D}_i^{\mathrm{T}} \end{bmatrix} \cdot \mathbf{D} \cdot \quad (27)$$

$$\left[\mathbf{B}_j, \left(\dfrac{t_{1j}}{2}\eta + e_{1j}\right)\mathbf{B}_j + \dfrac{t_{1j}}{2}\mathbf{C}_j, \left(\dfrac{t_{2j}}{2}\zeta + e_{2j}\right)\mathbf{B}_j + \dfrac{t_{2j}}{2}\mathbf{D}_j\right]$$

and

$$\bar{\mathbf{k}}_{Gij} = \begin{bmatrix} \bar{\mathbf{B}}_i^{\mathrm{T}} \\ \left(\dfrac{t_{1i}}{2}\eta + e_{1i}\right)\bar{\mathbf{B}}_i^{\mathrm{T}} + \dfrac{t_{1i}}{2}\bar{\mathbf{C}}_i^{\mathrm{T}} \\ \left(\dfrac{t_{2i}}{2}\zeta + e_{2i}\right)\bar{\mathbf{B}}_i^{\mathrm{T}} + \dfrac{t_{2i}}{2}\bar{\mathbf{D}}_i^{\mathrm{T}} \end{bmatrix} \cdot \begin{bmatrix} \bar{\sigma} & 0 & 0 \\ 0 & \bar{\sigma} & 0 \\ 0 & 0 & \bar{\sigma} \end{bmatrix} \cdot \quad (28)$$

$$\left[\bar{\mathbf{B}}_j, \left(\dfrac{t_{1j}}{2}\eta + e_{1j}\right)\bar{\mathbf{B}}_j + \dfrac{t_{1j}}{2}\bar{\mathbf{C}}_j, \left(\dfrac{t_{2j}}{2}\zeta + e_{2j}\right)\bar{\mathbf{B}}_j + \dfrac{t_{2j}}{2}\bar{\mathbf{D}}_j\right]$$

Here $\mathbf{D}$ is the elastic or elasto-plastic constitutive matrix and $\bar{\sigma}$ is the matrix of Cauchy stresses:

$$\bar{\sigma} = \begin{bmatrix} \sigma_{x'} & t_{x'y'} & t_{x'z'} \\ & 0 & 0 \\ \text{SYM.} & & 0 \end{bmatrix} \tag{29}$$

The matrices $\mathbf{B}_i$, $\mathbf{C}_i$, and $\mathbf{D}_i$ are given by

$$\mathbf{B}_i = \begin{bmatrix} B_1 & 0 & 0 \\ B_2 & B_1 & 0 \\ B_3 & 0 & B_2 \end{bmatrix}_i, \quad \mathbf{C}_i = \begin{bmatrix} 0 & 0 & 0 \\ C_1 & 0 & 0 \\ C_2 & 0 & 0 \end{bmatrix}_i \tag{30}$$

and

$$\mathbf{D}_i = \begin{bmatrix} 0 & 0 & 0 \\ D_1 & 0 & 0 \\ D_2 & 0 & 0 \end{bmatrix}_i$$

Virtual displacements at node i yield the following expression for the nodal forces in equilibrium with the internal stress field:

$$\mathbf{Q}_i = \begin{bmatrix} Q_x \\ Q_y \\ Q_z \\ M_1 \\ M_2 \\ M_3 \end{bmatrix}_i = \int_{-1}^{1} \int_{-1}^{1} \int_{-1}^{1} \begin{bmatrix} \theta & 0 & 0 \\ 0 & \phi_{3i}^{\prime\mathrm{T}}\theta & \phi_{2i}^{\prime\mathrm{T}}\theta \end{bmatrix} \cdot \begin{bmatrix} \mathbf{B}_i^{\mathrm{T}} \\ \left(\dfrac{t_{1i}}{2}\,\eta + e_{1i} \right) \mathbf{B}_i^{\mathrm{T}} + \dfrac{t_{1i}}{2}\,\mathbf{C}_i^{\mathrm{T}} \\ \left(\dfrac{t_{2i}}{2}\,\zeta + e_{2i} \right) \mathbf{B}_i^{\mathrm{T}} + \dfrac{t_{2i}}{2}\,\mathbf{D}_i^{\mathrm{T}} \end{bmatrix} \cdot$$

$$\begin{bmatrix} \sigma_{x'} \\ \tau_{x'y'} \\ \tau_{x'z'} \end{bmatrix} \det(\mathbf{J}) \, d\xi \, d\eta \, d\zeta \tag{31}$$

where Q_{xi}, Q_{yi} and Q_{zi} are forces referred to global axes and M_{1i}, M_{2i} and M_{3i} are moments about the local nodal axes in the directions of v_{1i}, v_{2i} and v_{3i}.

The element matrices are evaluated by numerical integration with two gaussian integration points along the beam axis. In the beam cross-section 2×2 gaussian integration points are used in the elastic case. For elasto-plastic cross-sections higher-order integration schemes are required.

Treatment of Large Displacements and Finite Rotations

Basic quantities for evaluation of the element matrices are the global nodal coordinates and the directions of the local nodal axes in the reference configuration.

The nodal coordinates for configuration m are readily obtained from the relation

$$\begin{bmatrix} X_i \\ Y_i \\ Z_i \end{bmatrix}_{(m)} = \begin{bmatrix} X_i \\ Y_i \\ Z_i \end{bmatrix}_{(m-1)} + \begin{bmatrix} \Delta u_i \\ \Delta v_i \\ \Delta w_i \end{bmatrix} \tag{32}$$

where i is the node number and Δu_i, Δv_i, Δw_i are the incremental nodal displacements between the two consecutive reference configurations.

Similarly, the vector v_{3i} in configuration m is obtained from the relations:

$$\mathbf{v}^*_{3i(m)} = \mathbf{v}_{3i(m-1)} + \tan(\Delta\beta_i)\mathbf{v}_{1i(m-1)} - \tan(\Delta\alpha_i)\mathbf{v}_{2i(m-1)} \tag{33}$$

$$\mathbf{v}_{3i(m)} = \mathbf{v}^*_{3i(m)}/|\mathbf{v}^*_{3i(m)}| \tag{34}$$

The procedure in eqn (33) for updating the normal vector is illustrated in Fig. 3.

It should be noted that the simple addition process assumed in eqn (33) is valid only for small values of the incremental nodal rotations. However, the error which is introduced due to finite incremental rotations may be controlled by reducing the size of the load steps. For beam elements, rotation about v_{3i} is included in the formulation. However, this rotation component ($\Delta\gamma_i$ in Fig. 3) gives only a small

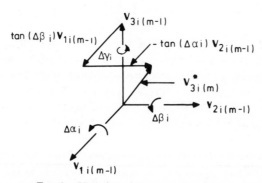

FIG. 3.   Updating of the normal vector.

(second-order) contribution to $v^*_{3i(m)}$, and the term is neglected when writing eqn (33).

Having computed $v_{3i(m)}$, the vector $v_{2i(m)}$ is formed normal to the plane through $v_{3i(m)}$ and the tangent vector to the ξ-axis of the element with the lowest element number meeting at the node. $v_{1i(m)}$ is formed normal to $v_{2i(m)}$ and $v_{3i(m)}$.

Restricting the formulation to small strains and with the previously defined local coordinate systems, the local Cartesian axes at the integration points can be considered as tangential to the same material line elements throughout deformation. In this case it can be shown [2] that the Kirchhoff stresses computed from eqn (3) correspond to true (Cauchy) stresses in the updated configuration after deformation. These stresses form the input to eqn (1) for the next load step.

Solution of the Non-Linear Equations

To ensure that the solution satisfies equilibrium throughout deformation, equilibrium iterations may be needed at each level of loading in addition to the residual force correction in the incremental formulation of eqn (1).

The Newton–Raphson iteration corresponds to updating the tangent stiffness matrix at each iteration cycle. To save computer time, modified Newton–Raphson iteration may be applied by updating the tangent stiffness at some iteration cycles and keeping it constant during subsequent iterations.

In order to characterise the non-linear behaviour of a multi-degree-of-freedom system, the following scalar quantity may be computed at load level i (see Ref. [6]):

$$S_p = \frac{\dot{r}^{1T} R^1_{ref}}{\dot{r}^{iT} R^i_{ref}} \tag{35}$$

S_p is referred to as the 'current stiffness parameter'. R^1_{ref} and R^i_{ref} are the initial and current reference load vectors which by multiplication of the load factors ρ_1 and ρ_i give the loading after the first and current load step. A dot denotes differentiation with respect to the load factor, and r^1 and r^i are the displacement vectors obtained after the first and current load step. Superscript T denotes transpose of the vectors.

It is readily seen from the expression above that the initial value of the current stiffness parameter is equal to unity. It is less than unity for systems which become 'softer' than the initial system and greater than unity for systems which become 'stiffer'. The unstable behaviour is characterised by a value of S_p less than zero, and the maximum or buckling load corresponds to $S_p = 0$. The current stiffness parameter has been used to guide the present numerical solution algorithm.

More detailed discussions of the current stiffness parameter may be found in Ref. [6].

Material Modelling

The coupled effect of large displacements and non-linear material behaviour is often of fundamental importance in the ultimate strength analysis of shells. In the present formulation the flow theory of plasticity (Prandtl–Reuss) and the von Mises yield criterion are adopted. Two hardening rules have been considered, namely the isotropic hardening rule and the 'overlay model' (see Ref. [7]).

NUMERICAL EXAMPLES

The theory outlined above has been implemented in a computer program. To illustrate the applicability of the method, some numerical examples will be considered.

Cantilever Beam with End Moment

An example which is much used for checking the accuracy of non-linear programs is a cantilever beam loaded by a moment M at the free end. The beam was divided into four equal elements as shown in Fig. 4.

Deformed configurations obtained at three levels of loading are shown in Fig. 4, in which the exact solutions are represented by circular arcs. Computed values are given in Table 1, in which M_{min} and M_{max} are the minimum and maximum moments computed from the stress states at the integration points.

The example confirms the accuracy of the present formulation during large displacements and rotations.

Cylindrical Shell with Axial Load and Hydrostatic Pressure

The accuracy of the present non-linear formulation for shells is studied in Ref. [2]. The example given here is of practical interest and concerns a non-linear analysis of an outer cell wall of a concrete offshore gravity platform (Type Condeep Statfjord A). A linear buck-

388

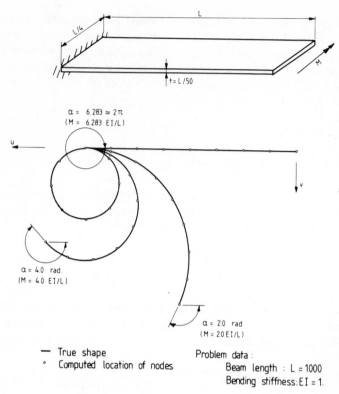

FIG. 4. Cantilever beam with end moment.

TABLE 1
Computed results for cantilever beam with end moment ($L = 1000$, $EI = 1$)

	$\alpha(\text{rad})^a$	2·0	4·0	2π
Exact solution	u	545·35	1189·20	1000
	v	708·07	413·41	0
Computed values	$\alpha(\text{rad})$	1·982	3·943	6·206
	u	539·37	1184·14	1014·21
	v	707·83	430·60	1·85
	$M_{min} \times 10^3$	1·9974	3·9506	6·1561
	$M_{max} \times 10^3$	1·9999	4·0062	6·3964

^{a}With the values indicated in Fig. 4, $\alpha = M \times 10^3$.

ling analysis in Ref. [8] showed that the critical buckling mode of the wall is symmetrical, corresponding to three half-waves in the circumferential direction. For this reason only one half of the wall was considered in the model (see Fig. 5). The idealised boundary conditions are also shown in Fig. 5, together with the actual water pressure heads which were chosen for the reference loads. The specific weight

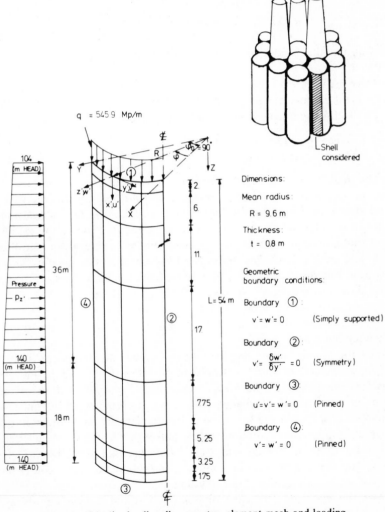

FIG. 5. Idealised cell wall, geometry, element mesh and loading.

of sea-water was set equal to $10\cdot3\,\text{kN/m}^3$. In addition, the specific weight of the cell wall and the spherical shell at the top of the cylinder was accounted for.

The non-linear behaviour of reinforced concrete was idealised with an elastic–ideally plastic material model with uniaxial yield strength $\sigma_y = 30\,\text{MN/m}^2$. In addition, initially imperfect geometry (out-of-roundness) of the form

$$\Delta R = -\Delta e_{max} \sin\left(3\phi - \frac{\pi}{2}\right) \sin\frac{\pi Z}{L} \tag{36}$$

was assumed. The out-of-roundness amplitude Δe_{max} was set equal to $0\cdot030$ m.

Two analyses with purely elastic material were also carried out, one with ideal geometry and the other with the imperfection of eqn (36).

The following elastic material data were used:

Young's modulus: $E = 27\,\text{GN/m}^2$
Poisson's ratio: $\nu = 0\cdot15$

The load factor is plotted against the current stiffness parameter in Fig. 6. The actual loading is obtained by multiplying the reference load shown on Fig. 5 by the load factor. In all cases the calculations were terminated when the external loading got so close to the critical load that convergence was not obtained during the iteration (see Ref. [2]). The critical load was determined by extrapolation of the curves for S_p in Fig. 6 to zero value. The linear buckling load from Ref. [8] is also shown in the figure. It is seen that this load exceeds that of the present analysis by about $23\cdot5\%$. This confirms what is generally recognised—that the linear buckling analysis tends to overestimate the buckling load.

For the elastic cases the load versus radial displacement at two nodes is plotted in Fig. 7. It is seen that the displacement pattern changes drastically as the load is increased. For small values of the load inward radial displacements were observed at all nodes. However, for increasing loads the radial displacements changed into a buckling mode corresponding to one half-wave in the vertical direction and three half-waves in the circumferential direction of the complete shell. The buckling mode corresponds to outward radial displacements along the symmetry line.

For the elasto-plastic case diagrams of radial displacements are shown along two vertical lines in Fig. 8. It is interesting to note that the plastic yielding results in a radial displacement field corresponding

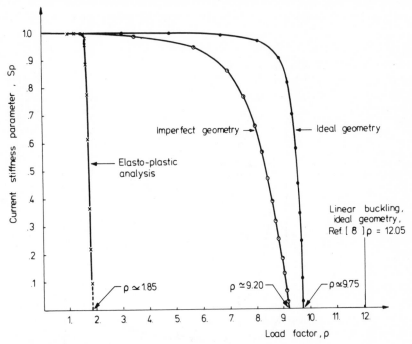

FIG. 6. Non-linear behaviour measured by the current stiffness parameter.

to three half-waves along the symmetry line (see Fig. 8). Near the upper and lower boundaries inward radial displacements were calculated, whereas the displacement field in the mean vertical region corresponds to three half-waves in the circumferential direction of the complete shell. In Fig. 9 the computed circumferential force per unit vertical length has been compared with the values $N_\phi = pR$. The value corresponding to the uniaxial yield stress is also shown in the figure. The reason why the computed values are larger may be explained from the biaxial stress state with vertical compression stresses. The development of plastic zones at the two surfaces of the shell is illustrated in Fig. 10. The plastic yielding is essentially governed by the circumferential compression stresses. At the highest load level yielding occurs at almost all integration points in a cross-section near the bottom of the shell. It may be concluded from this analysis that the shell collapses owing to material failure; the geometrically non-linear effects are believed to have only a small influence on the load-carrying capacity of the structure.

The present material model is not intended to give a complete

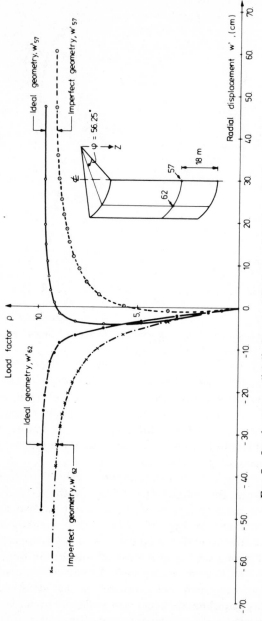

FIG. 7. Load versus radial displacement for two nodal points, elastic analysis.

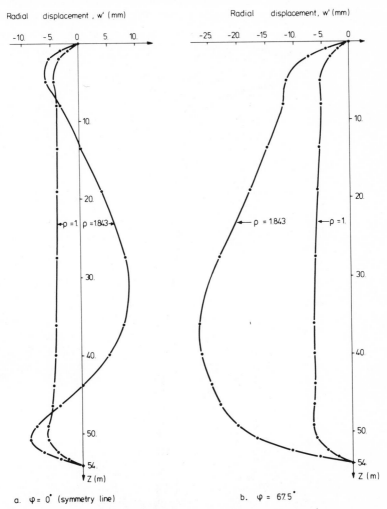

a. $\varphi = 0°$ (symmetry line)   b. $\varphi = 67.5°$

Fig. 8. Radial displacements for elasto-plastic analysis.

account of the many non-linear effects present in reinforced concrete. One effect which is not accounted for is the cracking of concrete caused by tensile and shear stresses. An examination of the results of the analysis reveals that only two stress points exhibit tensile stresses and the error introduced is believed to be small.

The total computer time for the elasto-plastic analysis was 716 s Central Arithmetic Units (CAU) on a UNIVAC 1110 computer. The

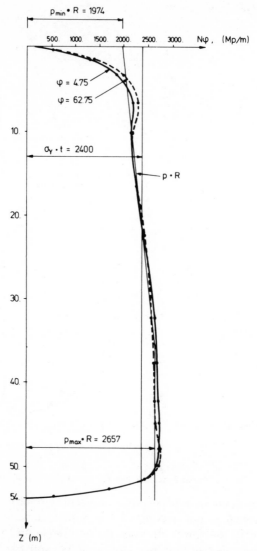

FIG. 9.   Circumferential force N_ϕ for elasto-plastic analysis ($\rho = 1\cdot843$).

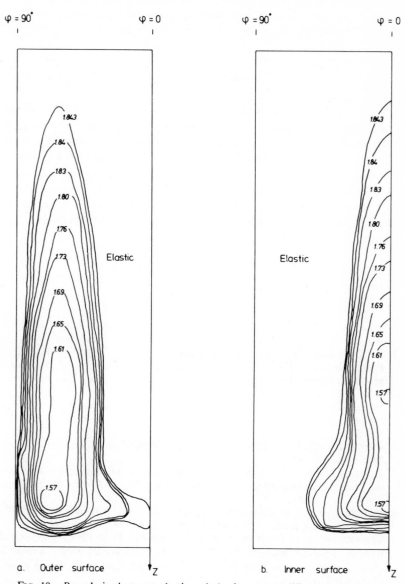

FIG. 10. Boundaries between elastic and plastic zones at different load levels.

solution procedure used 12 load increments. In addition, a total of 22 iteration cycles was required with updating of the tangent stiffness matrix for the first iteration cycle within each increment.

CONCLUSION

A formulation for the non-linear analysis of shells and beams by degenerate isoparametric elements has been presented. The effects of both large-displacement and elasto-plastic material behaviour have been studied and a method for updating the geometry of the elements during deformation has been discussed. Some numerical examples have been presented to demonstrate the capability of the method.

It may be concluded that the present degenerate isoparametric elements are both accurate and computationally efficient when applied to non-linear shell and beam problems.

REFERENCES

1. MURRAY, D. W. and WILSON, E. L., Finite element large deflection analysis of plates, *J. Eng. Mech. Div.*, *ASCE*, **95**, 1969, 143–165.
2. KRÅKELAND, B., 'Large Displacement Analysis of Shells Considering Elasto-plastic and Elasto-viscoplastic Materials', Division of Structural Mechanics, The Norwegian Institute of Technology, University of Trondheim, Norway, 1977.
3. BATHE, K. J., RAMM, E. and WILSON, E. L., Finite element formulations for large deformation dynamic analysis, *Int. J. Num. Meth. Eng.*, **9**, 1975, 353–386.
4. ZIENKIEWICZ, O. C., *The Finite Element Method in Engineering Science*, McGraw-Hill, 1971.
5. ZIENKIEWICZ, O. C., TAYLOR, R. L. and TOO, J. M., Reduced integration technique in general analysis of plates and shells, *Int. J. Num. Meth. Eng.*, **3**, 1971, 275–290.
6. BERGAN, P. G., HORRIGMOE, G., KRÅKELAND, B. and SØREIDE, T., Solution of non-linear finite element problems (to be published).
7. ZIENKIEWICZ, O. C., NAYAK, G. C. and OWEN, D. R. J., 'Composite and overlay models in numerical analysis of elasto-plastic continua', Proceedings of International Symposium on Foundations of Plasticity, Warsaw, September 1972.
8. FURNES, O., 'Condeep Statfjord A Platform. Structural Stability of Cell Walls Implosion Analysis', Report No. 75-96-C, Det norske Veritas, Oslo, Norway, 1975.

19

Stability of Shells of Revolution Using the Finite-Element Method

M. K. Exeter

University of Nottingham

and

R. D. Henshell

PAFEC Ltd

SUMMARY

This paper outlines an eigenvalue problem approach to predicting the axisymmetric pressure load which causes shells of revolution to buckle non-axisymmetrically. The eigenvalue problem is derived using the finite-element technique. Some difficulties associated with this method of solution are described. The results from analysing torispherical pressure vessel ends are also given.

INTRODUCTION

The method of solution presented in this paper is based on a general shell approach rather than the usual thin-shell one. When this general theory is used in combination with the finite-element technique, the stability equations are far less complex, relative to those derived using a thin-shell theory. However, this does mean that the numerical toil is greater than would normally occur with the thin-shell approach. Nevertheless, by virtue of the computer's immunity to drudgery, the simplicity of the equations far outweighs the increased numerical effort.

Both the theory and the results described in this paper are discussed more fully in Ref. [1].

STRAIN–DISPLACEMENT AND STRESS–STRAIN
RELATIONSHIPS

Non-linear strain–displacement relations for a general three-dimensional material continuum have been derived by Novozhilov [2, 3]. If the strains are small, these equations can be written in the form

$$\epsilon_{xx} = \frac{\partial u_x}{\partial x} + \frac{1}{2}\left[\left(\frac{\partial u_y}{\partial x}\right)^2 + \left(\frac{\partial u_z}{\partial x}\right)^2\right] \tag{1}$$

$$\epsilon_{xy} = \frac{\partial u_y}{\partial x} + \frac{\partial u_x}{\partial y} + \left(\frac{\partial u_z}{\partial x}\frac{\partial u_z}{\partial y}\right) \tag{2}$$

where u_x, etc., are displacement components, ϵ_{xx} is a direct strain component and ϵ_{xy} is a shear strain component. The subscripts indicate the direction of the strain and displacement components in a three-dimensional Cartesian coordinate system. Other strain terms such as ϵ_{yy}, ϵ_{yz}, etc., can be obtained by permutation of the x, y, z letters. These strain–displacement relations can be expressed in cylindrical polar coordinates (x, r, θ) by a transformation procedure (see Appendix 1).

By invoking Hooke's law, it is possible to obtain a stress–strain relationship which can be written in matrix form as

$$\{\sigma\} = [D]\{\epsilon\} \tag{3}$$

where $\{\sigma\}$ and $\{\epsilon\}$ are column vectors containing the stress and corresponding strain components, and $[D]$ is a stress–strain matrix containing the Poisson's ratio and Young's modulus values.

DISPLACEMENT FUNCTIONS

The usual isoparametric displacement functions for an axisymmetric deformation mode are

$$u_x = [N]\{\delta_{e_x}\}_a \tag{4}$$

$$u_r = [N]\{\delta_{e_r}\}_a \tag{5}$$

where u_x and u_r are the axial and radial displacements respectively, $[N]$ is a shape function, and $\{\delta_{e_x}\}_a$ and $\{\delta_{e_r}\}_a$ are the nodal values of the axial and radial displacement components. The subscript a indicates a purely axisymmetric displacement.

For linear behaviour it can be shown that any non-axisymmetric buckling motion is harmonic and the displacements may be written as

$$u_x = [N]\{\delta_{e_x}\}_{na} \cos n\theta \tag{6}$$

$$u_r = [N]\{\delta_{e_r}\}_{na} \cos n\theta \tag{7}$$

$$u_\theta = [N]\{\delta_{e_\theta}\}_{na} \sin n\theta \tag{8}$$

where u_θ is the circumferential displacement. n is the number of circumferential waves. The subscript na signifies a non-axisymmetric displacement ($n \neq 0$ since this case is given by eqns 4 and 5).

The well-known isoparametric procedure is adopted in which $[N]$ is written in terms of curvilinear coordinates (ξ, η) which are related to cylindrical-polar coordinates (x, r) (see Refs. [4–10]).

BASIC CONCEPTS—STRAIN AND DISPLACEMENT COMPONENTS

In order to investigate the stability of a shell of revolution buckling non-axisymmetrically, it is necessary to view the deformation of the buckled shell as being composed of two component parts: an axisymmetric and a non-axisymmetric deformation component. Therefore, the displacement matrix $\{\delta_e\}$, which represents the total nodal displacements for an axisymmetric finite element modelling the shell wall, can be considered to be composed of the axisymmetric nodal displacement components $\{\delta_e\}_a$ and the non-axisymmetrical nodal displacement components $\{\delta_e\}_{na}$ such that

$$\{\delta_e\} = \{\delta_e\}_a + \{\delta_e\}_{na} \tag{9}$$

Similarly, the strain vector $\{\epsilon\}$ can be considered to be composed of those strains caused by the purely axisymmetric deformation $\{\epsilon\}_a$ and those strains caused by the non-axisymmetric deformation $\{\epsilon\}_{na}$. Both of these strain components can be subdivided into linear first-order effects, $\{\epsilon_L\}$, and second-order non-linear effects, $\{\epsilon_{NL}\}$. Therefore,

$$\{\epsilon\} = \{\epsilon\}_a + \{\epsilon\}_{na} \tag{10}$$

where

$$\{\epsilon\}_a = \{\epsilon_L\}_a + \{\epsilon_{NL}\}_a \tag{11}$$

$$\{\epsilon\}_{na} = \{\epsilon_L\}_{na} + \{\epsilon_{NL}\}_{na} \tag{12}$$

By using the appropriate finite-element displacement expression, each of these strain effects can be expressed in terms of the corresponding nodal displacements. For instance, using the linear parts of the strain expressions given in Appendix 1, we have

$$
\{\epsilon_L\}_{na} =
\begin{Bmatrix}
\epsilon_{xx} \\
\epsilon_{rr} \\
\epsilon_{\theta\theta} \\
\epsilon_{xr} \\
\epsilon_{r\theta} \\
\epsilon_{\theta x}
\end{Bmatrix}_{L_{na}}
=
\begin{Bmatrix}
\dfrac{\partial u_x}{\partial x} \\[2mm]
\dfrac{\partial u_r}{\partial r} \\[2mm]
\dfrac{u_r}{r} + \dfrac{1}{r}\dfrac{\partial u_\theta}{\partial \theta} \\[2mm]
\dfrac{\partial u_r}{\partial x} + \dfrac{\partial u_x}{\partial r} \\[2mm]
\dfrac{\partial u_\theta}{\partial r} + \dfrac{1}{r}\dfrac{\partial u_r}{\partial \theta} - \dfrac{u_\theta}{r} \\[2mm]
\dfrac{1}{r}\dfrac{\partial u_x}{\partial \theta} + \dfrac{\partial u_\theta}{\partial x}
\end{Bmatrix}_{na}
\tag{13}
$$

By substitution from eqns (6)–(8), it follows that

$$
\{\epsilon_L\}_{na} =
\begin{bmatrix}
\dfrac{\partial [N]}{\partial x}\cos n\theta & 0 & 0 \\[2mm]
0 & \dfrac{\partial [N]}{\partial r}\cos n\theta & 0 \\[2mm]
0 & \dfrac{[N]}{r}\cos n\theta & \dfrac{n[N]}{r}\cos n\theta \\[2mm]
\dfrac{\partial [N]}{\partial r}\cos n\theta & \dfrac{\partial [N]}{\partial x}\cos n\theta & 0 \\[2mm]
0 & \dfrac{-n[N]}{r}\sin n\theta & \left(\dfrac{\partial [N]}{\partial r} - \dfrac{[N]}{r}\right)\sin n\theta \\[2mm]
\dfrac{-n[N]}{r}\sin n\theta & 0 & \dfrac{\partial [N]}{\partial x}\sin n\theta
\end{bmatrix}
\begin{Bmatrix}
\{\delta_{e_x}\}_{na} \\
\{\delta_{e_r}\}_{na} \\
\{\delta_{e_\theta}\}_{na}
\end{Bmatrix}
\tag{14}
$$

This can be written as

$$
\{\epsilon_L\}_{na} = [B_L]_{na}\{\delta_e\}_{na} \tag{15}
$$

Taking differentials, it follows that

$$
d\{\epsilon_L\}_{na} = [B_L]_{na}\,d\{\delta_e\}_{na} \tag{16}
$$

To obtain the corresponding linear axisymmetric strain expressions,

u_θ should be set to zero and terms containing variations with respect to θ should be deleted. The following expressions are obtained:

$$\{\epsilon_L\}_a = [B_L]_a \{\delta_e\}_a \tag{17}$$

and

$$d\{\epsilon_L\}_a = [B_L]_a \, d\{\delta_e\}_a \tag{18}$$

The non-linear, non-axisymmetric terms can be expressed as

$$\{\epsilon_{NL}\}_{na} = \begin{Bmatrix} \epsilon_{xx} \\ \epsilon_{rr} \\ \epsilon_{\theta\theta} \\ \epsilon_{xr} \\ \epsilon_{r\theta} \\ \epsilon_{\theta x} \end{Bmatrix}_{NL_{na}} = \begin{Bmatrix} \frac{1}{2}\left[\left(\frac{\partial u_r}{\partial x}\right)^2 + \left(\frac{\partial u_\theta}{\partial x}\right)^2\right] \\ \frac{1}{2}\left[\left(\frac{\partial u_x}{\partial r}\right)^2 + \left(\frac{\partial u_\theta}{\partial r}\right)^2\right] \\ \frac{1}{2}\left[\left(\frac{1}{r}\frac{\partial u_x}{\partial \theta}\right)^2 + \left(\frac{1}{r}\frac{\partial u_r}{\partial \theta} - \frac{u_\theta}{r}\right)^2\right] \\ \frac{\partial u_\theta}{\partial x}\frac{\partial u_\theta}{\partial r} \\ \frac{\partial u_x}{\partial r}\frac{1}{r}\frac{\partial u_x}{\partial \theta} \\ \left(\frac{1}{r}\frac{\partial u_r}{\partial \theta} - \frac{u_\theta}{r}\right)\frac{\partial u_r}{\partial x} \end{Bmatrix}_{na} \tag{19}$$

Again, $\{\epsilon_{NL}\}_a$ can be obtained from eqn (19) by deleting the appropriate terms.

WORK DONE EQUATIONS

By considering a small surface area of the shell subjected to the pressure load, it is possible to determine the first- and second-order work done effects by the pressure load on a finite element when it deforms in a general non-axisymmetric manner. In terms of the ξ, η, θ coordinate system, the total work done is given by

$$\begin{aligned} WD = &\int_{-1}^{+1}\int_0^{2\pi} P r\, u_r \frac{\partial x}{\partial \xi} \, d\theta \, d\xi - \int_{-1}^{+1}\int_0^{2\pi} P r\, u_x \frac{\partial r}{\partial \xi} \, d\theta \, d\xi \\ &+ \frac{1}{2}\int_{-1}^{+1}\int_0^{2\pi} P r \frac{\partial u_x}{\partial \xi} u_r \, d\theta \, d\xi - \frac{1}{2}\int_{-1}^{+1}\int_0^{2\pi} P r \frac{\partial u_r}{\partial \xi} u_x \, d\theta \, d\xi \\ &- \frac{1}{2}\int_{-1}^{+1}\int_0^{2\pi} P \frac{\partial u_r}{\partial \theta} u_\theta \frac{\partial x}{\partial \xi} \, d\theta \, d\xi + \frac{1}{2}\int_{-1}^{+1}\int_0^{2\pi} P \frac{\partial u_x}{\partial \theta} u_\theta \frac{\partial r}{\partial \theta} \, d\theta \, d\xi \end{aligned} \tag{20}$$

where P is the uniform pressure acting on the element.

The first two integrals in eqn (20) represent the first-order pressure work done terms associated with a general displacement of a surface element in the x and r coordinate directions. The second pair of integrals in the equation represent the second-order pressure work done effects for the surface element undergoing motion in the same coordinate directions. These second-order terms arise mainly as a consequence of the surface element rotating as it displaces. The first four integrals in eqn (20) are required for (i) axisymmetric and (ii) non-axisymmetric displacements of the shell structure.

Finally, the last two integrals in eqn (20) represent the work done by the pressure force as the surface element displaces in the circumferential direction. They are second-order terms and only occur when the shell deforms non-axisymmetrically.

Pressure Work Done when Element Deforms Axisymmetrically

Substituting the axisymmetric nodal displacements (eqns 4 and 5) into the first two terms of eqn (20) and integrating in the θ direction

$$W_{e_a} = \{F_{e_r}\}_a^T\{\delta_{e_r}\}_a - \{F_{e_x}\}_a^T\{\delta_{e_x}\}_a = \{F_e\}_a^T\{\delta_e\}_a \qquad (21)$$

where

$$\{F_{e_x}\}_a = 2\pi P \int_{-1}^{+1} r\,[N]\frac{\partial r}{\partial \xi}\,d\xi \qquad (22)$$

$$\{F_{e_r}\}_a = 2\pi P \int_{-1}^{+1} r\,[N]\frac{\partial x}{\partial \xi}\,d\xi \qquad (23)$$

$$\{F_e\}_a^T = (\{F_{e_x}\}_a^T, \{F_{e_r}\}_a^T, \{0\}^T) \qquad (24)$$

Pressure Work Done when Element Deforms Purely Non-Axisymmetrically

Substituting the non-axisymmetric nodal displacements (eqns 6–8) into the last four integrals of eqn (20) and integrating in the θ direction gives

$$W_{e_{na}} = \{\delta_{e_x}\}_{na}^T[K_{p,\phi}]\{\delta_{e_r}\}_{na} - \{\delta_{e_r}\}_{na}^T[K_{p,\phi}]\{\delta_{e_x}\}_{na}$$
$$+ \{\delta_{e_r}\}_{na}^T[K_{ppx}]\{\delta_{e_\theta}\}_{na} - \{\delta_{e_x}\}_{na}^T[K_{ppr}]\{\delta_{e_\theta}\}_{na} \qquad (25)$$

where

$$[K_{p,\varphi}] = \frac{P\pi}{2} \int_{-1}^{+1} \frac{\partial [N]^T}{\partial \xi} r[N] \, d\xi \qquad (26)$$

$$[K_{ppx}] = \frac{P\pi n}{2} \int_{-1}^{+1} [N]^T [N] \frac{\partial x}{\partial \xi} \, d\xi \qquad (27)$$

$$[K_{ppr}] = \frac{P\pi n}{2} \int_{-1}^{+1} [N]^T [N] \frac{\partial r}{\partial \xi} \, d\xi \qquad (28)$$

Hence,

$$W_{e_{na}} = \{\delta_e\}_{na}^T \begin{bmatrix} [0] & [K_{p,\varphi}] & -[K_{ppr}] \\ -[K_{p,\varphi}] & [0] & [K_{ppx}] \\ [0] & [0] & [0] \end{bmatrix} \{\delta_e\}_{na} \qquad (29)$$

$$= \{\delta_e\}_{na}^T [F_e]_{na} \{\delta_e\}_{na} \qquad (30)$$

N.B.: In deriving eqns (20), (21) and (30) the second-order work done effects caused by large strains are considered negligible in comparison with rotation effects.

POTENTIAL ENERGY AND STABILITY EQUATIONS

The total work done, W_e, in a general non-axisymmetric displacement can be expressed in the form

$$W_e = \{F_e\}_a^T \{\delta_e\}_a + \{\delta_e\}_{na}^T [F_e]_{na} \{\delta_e\}_{na} \qquad (31)$$

When a finite-element undergoes a general non-axisymmetric deformation, its strain energy U_e can be expressed as

$$U_e = \frac{1}{2} \int (\{\epsilon\}_a + \{\epsilon\}_{na})^T [D](\{\epsilon\}_a + \{\epsilon\}_{na}) \, dV \qquad (32)$$

The integral is taken over the volume of the element. Combining the work done and the strain energy contributions from each element, one obtains the total potential energy of the system V as

$$V = \sum_{\substack{\text{all} \\ \text{elements}}} V_e = \sum_{\substack{\text{all} \\ \text{elements}}} (U_e - W_e) \qquad (33)$$

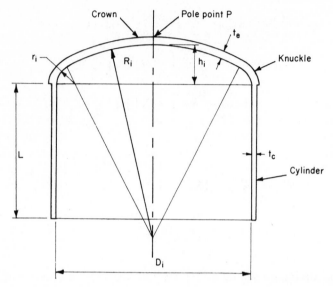

FIG. 1.   Geometry of a torispherical pressure vessel end.

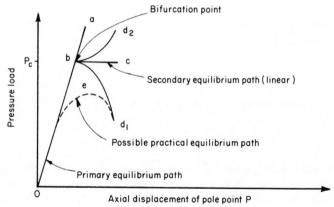

FIG. 2.   Equilibrium paths for a torispherical pressure vessel end.

If the axial displacement of the pole point P of the torispherical pressure vessel end shown in Fig. 1 is considered, then for only axisymmetric deformation of the shell, the primary equilibrium path Oa in Fig. 2 would be obtained. When the pressure reaches the value P_c, the shell can deform non-axisymmetrically (i.e. the shell can buckle) and the pole point P would then follow the secondary equilibrium path bc instead of the primary one. (The line bc is based on a linear non-axisymmetric deformation theory.)

However, as soon as the non-axisymmetric displacements increase, there are geometrically non-linear effects which cause the secondary equilibrium path to deviate from bc to either bd_1 or bd_2, depending upon the shell geometry. The purpose of the present analysis is to find only the position of the point b and not the post-buckling lines bd_1 or bd_2.

Initially, as the pressure is increased from zero, the shell deforms only axisymmetrically and remains in a state of stable equilibrium. The variation of the potential energy with respect to each and all of the axisymmetric nodal displacement components in the shell system is a minimum. Carrying out the usual minimisation procedure of the total potential energy with respect to the axisymmetric nodal displacements gives

$$[S_L]_a\{\delta\}_a = \{F\}_a \tag{34}$$

where

$[S_L]_a$ is the system linear stiffness matrix associated with the axisymmetric nodal displacement components;

$\{\delta\}_a$ is the system axisymmetric nodal displacement matrix; and

$\{F\}_a$ is the system nodal force matrix obtained from the first-order work done terms of each element.

The terms in the matrices in eqn (34) are obtained from the element derivative expression

$$\frac{\partial V_{e_a}}{\partial\{\delta_e\}_a} = \int_{V'_e} [B_L]_a^T[D][B_L]_a|[J]|\mathrm{d}V'_e\{\delta_e\}_a - \{F_e\}_a$$

$$= [S_{L_e}]_a\{\delta_e\}_a - \{F_e\}_a \tag{35}$$

where V'_e is the volume of the finite element in the ξ, η, θ coordinate system. Thus, $[S_L]_a$ is formed by merging in $[S_{L_e}]_a$ matrix contributions. (N.B.: $[J]$ is the Jacobian of the transformation.)

When the pressure reaches the value P_c, the shell is in a state of neutral equilibrium and the variation of the potential energy V, with respect to the system non-axisymmetric nodal displacement components, is stationary. Using this fact, one can obtain the eigenvalue problem

$$([S_L]_{na} + \lambda[S_{GP}])\{\delta\}_{na} = \{\theta\} \tag{36}$$

where

$[S_L]_{na}$ is the system linear stiffness matrix associated with the non-axisymmetric nodal displacements;

$[S_{GP}]$ is the system geometric/pressure stiffness matrix; and

$\{\delta\}_{na}$ is the system non-axisymmetric nodal displacement matrix.

Equation (36) is formed from the element expression

$$\frac{\partial V_e}{\partial\{\partial_e\}_{na}} = ([S_{L_e}]_{na} + [S_{G_e}] + [S_{P_e}]_{na})\{\delta_e\}_{na}$$
$$= ([S_{L_e}]_{na} + [S_{GP_e}])\{\delta_e\}_{na} \tag{37}$$

where

$$[S_{L_e}]_{na} = \int_{V_e'} [B_L]_{na}^T[D][B_L]_{na}|[J]|\,dV_e' \tag{38}$$

$$[S_{G_e}] = \int_{V_e'} [G]^T[M_L]_a[G]|[J]|\,dV_e' \tag{39}$$

and

$$[S_{P_e}]_{na} = \begin{bmatrix} 0 & [K_{P,e^P}]^T - [K_{P,e^P}] & [K_{ppr}] \\ [K_{P,e^P}] - [K_{P,e^P}]^T & 0 & -[K_{ppx}] \\ [K_{ppr}]^T & -[K_{ppx}]^T & 0 \end{bmatrix} \tag{40}$$

In eqn (39)

$$[M_L]_a = \begin{bmatrix} \sigma_{rr} & 0 & 0 & 0 & 0 & 0 \\ 0 & \sigma_{rr} & 0 & \sigma_{xr} & 0 & 0 \\ 0 & 0 & \sigma_{xx} & 0 & 0 & 0 \\ 0 & \sigma_{xr} & 0 & \sigma_{xx} & 0 & 0 \\ 0 & 0 & 0 & 0 & \sigma_{\theta\theta} & 0 \\ 0 & 0 & 0 & 0 & 0 & \sigma_{\theta\theta} \end{bmatrix} \tag{41}$$

where the stress components are associated with the linear axisymmetric deformation, and

$$[G] = \begin{bmatrix} \dfrac{\partial[N]}{\partial r}\cos n\theta & 0 & 0 \\[2ex] 0 & 0 & \dfrac{\partial[N]}{\partial r}\sin n\theta \\[2ex] 0 & \dfrac{\partial[N]}{\partial x}\cos n\theta & 0 \\[2ex] 0 & 0 & \dfrac{\partial[N]}{\partial x}\sin n\theta \\[2ex] \dfrac{-n[N]}{r}\sin n\theta & 0 & 0 \\[2ex] 0 & \dfrac{-n[N]}{r}\sin n\theta & \dfrac{-[N]}{r}\sin n\theta \end{bmatrix} \tag{42}$$

$[S_{GP}]$ is formed from element terms arising from the second-order work done effects $[S_{P_e}]_{na}$ and also the 'geometric stiffness' effects $[S_{G_e}]$.

$[S_{P_e}]_{na}$ is directly dependent on the pressure acting on the shell wall, whereas $[S_{G_e}]$ is directly dependent on the stress acting in the element (and, hence, indirectly dependent on the pressure load).

Equation (34) is used to obtain the axisymmetric nodal displacement components at a given pressure P_1, say. Knowing these displacements, the stress components at any point in an element can be found using the equation

$$\{\sigma_L\}_a = [D][B_L]_a\{\delta_e\}_a \tag{43}$$

The eigenvalue problem (eqn 36) can then be used to predict the pressure load λP_1 at which the shell buckles.

PROGRAM TESTS ON CIRCULAR RING STRUCTURES

Some subroutines were added to the PAFEC scheme [18] in order to employ the above linear-elastic theory. A circular ring structure subjected to an external axisymmetric pressure load was first used as a test for the program. An analytical linear-elastic buckling solution exists for this structure. Initially 8-noded axisymmetric isoparametric elements were used to model the ring structure. It was found that the error between the result given by the program and the corresponding analytical value was often very high owing to round-off effects. In fact, as the ratio of mean wall radius to wall thickness increased, this error became ridiculously large. The causes of the round-off error were:

(1) When the wall thickness is decreased, nodes through the thickness become closer and corresponding terms in the stiffness matrix become large, although the stiffness of the shell decreases. Thus, decreasing differences between increasing numbers are required in the solution process. This problem can be alleviated by using a 6-noded element (see Fig. 3).

(2) The error in the eigenvalue extraction procedure depends upon the largest eigenvalue rather than the smallest. Therefore, when there is a large ratio between the largest and smallest eigenvalues, the accuracy of the smallest eigenvalue will be poor. As it happens, the ring structures used to test the program were rather more critical in this respect than the actual pressure vessels.

ANALYSIS OF TORISPHERICAL PRESSURE VESSEL ENDS

The computer program was used to carry out a linear-elastic eigenvalue analysis of a torispherical pressure vessel end in order to see how the internal buckling pressure value predicted by the program compared with an experimental result [11]. Table 1 gives the relevant nominal dimensions and important data for the vessel (vessel No. 9). (Dimension symbols are defined in Fig. 1.) Six-noded axisymmetric

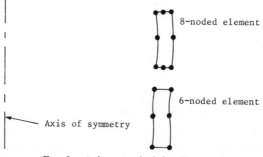

FIG. 3.   Axisymmetric finite elements.

TABLE 1
Pressure vessel data

Vessel No. 9				
D_i(mm)	r_i(mm)	R_i(mm)	h_i(mm)	t_e(mm)
2743	203·2	2743	487·9	3·251
t_c(mm)	L(mm)	t_e/D_i	E^a(GN m^{-2})	ν^a
2·642	1219	0·001 19	183·2	0·31

Vessel No. 4				
D_i(mm)	r_i(mm)	R_i(mm)	h_i(mm)	t_e(mm)
1371	134·6	1403	260·5	3·378
t_c(mm)	L(mm)	t_e/D_i	E^a(GN m^{-2})	ν^a
2·743	609·6	0·002 46	183·2	0·31

[a] N.B. E is Young's modulus; ν is Poisson's ratio.

isoparametric elements were used to model the shell, with a single element width to model the shell thickness.

The buckling pressures predicted by the program, as the number of circumferential waves n was varied, are given in Fig. 4. As can be seen, the lowest buckling pressure of 12·42 bar occurred for 88 circumferential waves around the vessel. The maximum amplitude of the waves occurred in the knuckle region. Unfortunately, this lowest buckling pressure value is a factor of 2·905 greater than the experimental result for this end. This factor is large, although the theoretical buckling pressure compares favourably with a value of 11·03 bar predicted by Thurston and Holston [12] for a shell with very similar dimensions.

The difference between the theoretical and experimental result was due, in the main, to the omission in the program of:

(1) material and geometric non-linear effects,
(2) work-hardening, creep and residual stress effects,
(3) manufacturing imperfections and, in particular, the effects of welds.

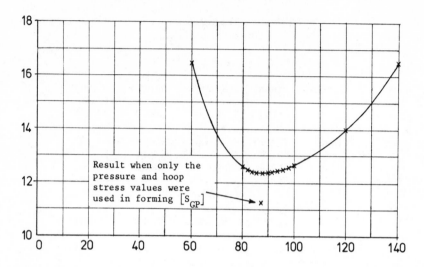

FIG. 4.   Results from linear-elastic program. Ordinate: buckling pressure (bar). Abscissa: number of circumferential waves (n).

Some of the possible causes of the differences between the theoretical and experimental results were examined. In order to show the effects of neglecting certain stresses, a case was computed in which only the pressure and hoop stress values (i.e. not the meridional and normal stress values) were used in the matrix $[S_{GP}]$. A value of $n = 88$ was used. The result is given in Fig. 4; the buckling pressure was reduced by only 11·0%. Further computations on other similar-shaped vessels supported the conclusion that the buckling pressure normally reduces by this magnitude when the meridional and normal stress components are neglected. From this result it was concluded that the hoop stress seems to be the main stress causing the vessel to buckle.

An attempt to ascertain the effect of material non-linearity was made. The program was modified so that the value of Young's modulus used at the gauss points in determining each of the element integral values was based on the hoop stress value in the vessel at the pressure at which the vessel first buckles non-axisymmetrically. A one-dimensional tensile stress–strain diagram for the vessel material was used to find the tangent modulus E at a particular hoop stress value (Fig. 5, curve C1).

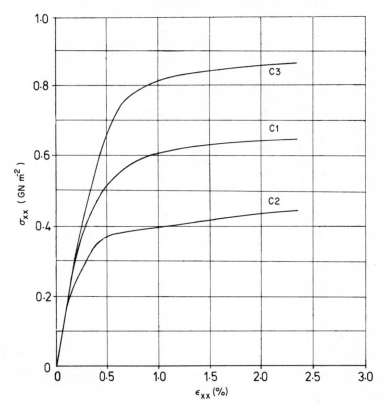

FIG. 5.   Simple, one-dimensional, tensile, stress–strain data.

Justification for using this crude non-linear material approach founded on the hoop stress value was based on the previous observation that this stress is the dominant stress component. An iteration procedure between eqns (34) and (36) has to be used since adjustment of the value of E changes the stress distribution and buckling pressure value (see Fig. 6).

Use of this non-linear stress–strain rule introduced a further problem with eqn (36). Just before the shell buckles, the hoop stress at a point in the shell may have a value σ_{crit} (Fig. 7). But as non-axisymmetric deformation occurs, the value of Young's modulus in the circumferential direction will vary from the value of E_P to E_L. This occurs because the hoop stress value can now either continue

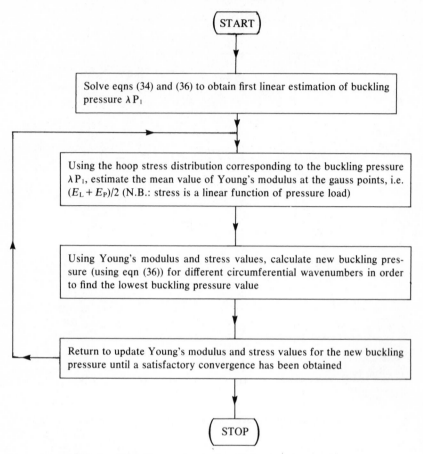

FIG. 6.   Flow diagram for the non-linear material algorithm.

along curve D_1 or move down the linear elastic curve D_2, depending on the way the shell is deforming non-axisymmetrically at that circumferential position (see Fig. 8). As the axisymmetric finite elements used to model the non-axisymmetric deformation component are not capable of taking this into account, a mean value of E, i.e. $(E_P + E_L)/2$, had to be used in eqn (36).

The result from this crude non-linear program showed that the lowest buckling pressure value was 5·637 bar, again with 88 circumferential waves. This buckling pressure was a factor of only 1·319 greater than the experimental value and it was concluded that part of

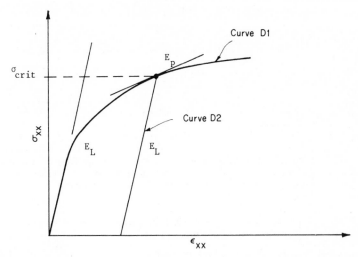

FIG. 7. Variation in Young's modulus in circumferential direction at stress σ_{crit}.

the large difference between the previous theoretical linear-elastic buckling pressure and the experimental value was due to material non-linearity.

Geometric non-linear effects were next included, along with the non-linear material behaviour. Figure 9 shows the algorithm used in the attempt to include both of these effects.

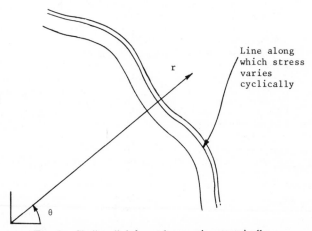

FIG. 8. Shell wall deformed non-axisymmetrically.

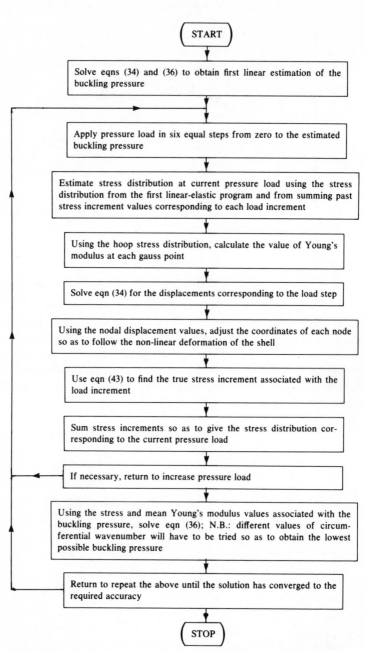

FIG. 9. Flow diagram for the non-linear material/geometric algorithm.

One slight problem was found to arise from using this technique. As the coordinates of the element nodal positions were adjusted at the end of each load increment, the 'mid-side nodes', particularly in the vicinity of the knuckle region, might no longer be mid-way between the adjacent corner nodes. This caused considerable error to arise in the analysis (see Ref. [13]). This problem was overcome by adjusting all mid-side nodes to a mid-way position after changing the coordinate positions.

The results from this non-linear material/geometric algorithm showed that the buckling pressure had increased to 6·550 bar. This was not surprising, since non-linear geometric effects tend to increase the buckling resistance of the shell.

The estimation of buckling pressures is critically dependent upon knowing accurately the tangent modulus of the material and, although best estimates of this quantity were used, they may have been substantially in error, owing to the uncertainty in the curves of Fig. 5. Moreover, the material non-linearity was based totally on curve C_1 of Fig. 5; this curve was not representative of the material in the whole vessel but only of that in the vicinity of the knuckle region, where the amplitudes of the buckling waves were the greatest. The knuckle region material is most heavily work-hardened as a result of manufacture and material outside this region would be more aptly

TABLE 2
Summary of results

Type of algorithm	Lowest buckling pressure(bar)	Number of circumferential waves, n	Factor above experimental buckling value
Vessel No. 9			
Linear elastic	12·48	88	2·919
Non-lin. matl.	5·637	88	1·319
Non-lin. matl./geom.	6·550	76	1·532
Vessel No. 4			
Linear elastic	55·44	59	4·062
Non-lin. matl.	19·15	59	1·403
Non-lin. matl./geom.	22·10	49	1·619

represented by a curve with a lower ultimate tensile strength value (e.g. curve C_2 in Fig. 5). The lower UTS value for this material would most certainly tend to lower the buckling pressure value.

A second vessel (vessel No. 4—see Table 1) was also analysed by the program. Whereas vessel No. 9 showed smallish geometric non-linear behaviour in the experimental investigation, vessel No. 4 displayed quite large non-linear behaviour before the first buckle occurred. The experimental buckling pressure was 13·65 bar. Results obtained from the various programmed algorithms are given in Table 2. Curve C_3 in Fig. 5 was used in representing the non-linear material effects.

CONCLUSIONS

The importance of the material and geometric non-linear effects has been demonstrated. However, after taking these effects into account, there still exists a large relative difference between the theoretical and experimental critical pressure values. Some explanations for this may be found in the crudity of the non-linear material and geometric model, but, undoubtedly, a considerable part of the remaining difference is due to work-hardening, creep and residual stress effects which were known to exist in the vessels under consideration. Manufacturing imperfections and the effects of welds are also certainly significant. All of these latter effects give rise to difficult hurdles in both theoretical and computer programming territories. For design purposes it is possible to allow for some of these unknown effects by using a suitable 'factor of safety'.

It is well known [14–17] that if a secondary equilibrium path involves a negative stiffness such as that shown by line bd_1 in Fig. 2, then imperfections in the shell will cause the practical equilibrium path to follow the dotted line Oed_1. This causes calculations for perfect vessels to overestimate buckling pressures, which is probably the case here.

In conclusion, the power of this general shell approach has been examined in a small way. Problems arising from its use are seen to be tractable. Realisation of its full potential is not achieved in the examples considered, since such a general finite-element program could be applied to axisymmetric problems where thickish and awkward wall shapes would probably cause grave problems for many present-day thin-shell or thin-plate buckling programs.

REFERENCES

1. EXETER, M. K., 'Stability of Shells', B.Sc. Thesis, University of Nottingham, 1976.
2. NOVOZHILOV, V. V., *Foundations of the Non-linear Theory of Elasticity*, Graylock Press, New York, 1953.
3. NOVOZHILOV, V. V., *Theory of Elasticity*, Pergamon Press, 1961.
4. HUEBNER, K. H., *The Finite Element Method for Engineers*, Wiley, 1975.
5. ZIENKIEWICZ, O. C., *The Finite Element Method in Engineering Science*, McGraw-Hill, 1975.
6. BREBBIA, C. A. and CONNOR, J. J., *Fundamentals of Finite Element Techniques*, Butterworths, 1973.
7. DESAI, C. S. and ABEL, J. F., *Introduction to the Finite Element Method*, Van Nostrand Reinhold, 1972.
8. WHITEMAN, J. R., *The Mathematics of Finite Elements and Applications—Parts I and II*, Academic Press, 1973–1977.
9. NORRIE, D. H. and DE VRIES, G., *The Finite Element Method—Fundamentals and Applications*, Academic Press, 1973.
10. ODEN, J. T., *Finite Elements of Non-linear Continua*, McGraw Hill, 1972.
11. CAMPBELL, T. D., 'Strain, Deformations and Buckling in Very Thin Torispherical Pressure Vessel Ends', Ph.D. Thesis, University of Nottingham, 1975.
12. THURSTON, G. A. and HOLSTON, A. A., 'Buckling of Cylindrical Shell End Closures by Internal Pressure', NASA Report CR-540, 1966.
13. HENSHELL, R. D. and SHAW, K. G., Crack tip finite elements are unnecessary, *Int. J. Num. Meth. Eng.*, **9**, 1975, 495–507.
14. THOMPSON, J. M. T. and HUNT, G. W., *A General Theory of Elastic Stability*, Wiley, 1973.
15. BRUSH, D. O. and ALMROTH, B. O., *Buckling of Bars, Plates and Shells*, McGraw-Hill, 1975.
16. CHAJES, A., *Principles of Structural Stability Theory*, Prentice-Hall, 1974.
17. TAUCHERT, T. R., *Energy Principles in Structural Mechanics*, McGraw-Hill, 1974.
18. '*PAFEC 70+ Users Manual*' (Ed. R. D. Henshell), Nottingham University, 1972.

APPENDIX 1: STRAIN–DISPLACEMENT RELATIONS IN CYLINDRICAL POLAR COORDINATES

$$\epsilon_{xx} = \frac{\partial u_x}{\partial x} + \frac{1}{2}\left[\left(\frac{\partial u_r}{\partial x}\right)^2 + \left(\frac{\partial u_\theta}{\partial x}\right)^2\right]$$

$$\epsilon_{rr} = \frac{\partial u_r}{\partial r} + \frac{1}{2}\left[\left(\frac{\partial u_x}{\partial r}\right)^2 + \left(\frac{\partial u_\theta}{\partial r}\right)^2\right]$$

$$\epsilon_{\theta\theta} = \frac{u_r}{r} + \frac{1}{r}\frac{\partial u_\theta}{\partial \theta} + \frac{1}{2}\left[\left(\frac{1}{r}\frac{\partial u_x}{\partial \theta}\right)^2 + \left(\frac{1}{r}\frac{\partial u_r}{\partial \theta} - \frac{u_\theta}{r}\right)^2\right]$$

$$\epsilon_{xr} = \frac{\partial u_r}{\partial x} + \frac{\partial u_x}{\partial r} + \left(\frac{\partial u_\theta}{\partial x}\frac{\partial u_\theta}{\partial r}\right)$$

$$\epsilon_{r\theta} = \frac{\partial u_\theta}{\partial r} + \frac{1}{r}\frac{\partial u_r}{\partial \theta} - \frac{u_\theta}{r} + \left(\frac{\partial u_x}{\partial r}\frac{1}{r}\frac{\partial u_x}{\partial \theta}\right)$$

$$\epsilon_{\theta x} = \frac{1}{r}\frac{\partial u_x}{\partial \theta} + \frac{\partial u_\theta}{\partial x} + \left[\left(\frac{1}{r}\frac{\partial u_r}{\partial \theta} - \frac{u_\theta}{r}\right)\frac{\partial u_r}{\partial x}\right]$$

20

Calculation of Stresses in Austenitic Welds

W. S. Blackburn and A. D. Jackson

C. A. Parsons & Co. Ltd

AND

T. K. Hellen

Berkeley Nuclear Laboratories

SUMMARY

A shallow (24 mm) axisymmetric groove in a long austenitic steel bar of diameter 152 mm has been filled by a single-pass austenitic weld.

The stress–strain–temperature yield surface has been determined from load displacement records obtained on a Gleeble machine from specimens of the parent heat-affected zone and weld metals. The main inconvenience was that (except at room temperature) there was an axial temperature variation along each specimen. The data were then used to adequately predict the deformation of a specimen in which the temperature varied during the loading.

Following a three-dimensional transient non-linear thermal analysis, an axisymmetric thermoelastic–plastic stress analysis was satisfactorily undertaken using Phase 3 of BERSAFE, which allows for loading and unloading. Comparisons are made with measurements using a hole-drilling technique in the weld; poor correlation was found between the experimental and calculated results.

INTRODUCTION

Hibbitt and Marçal [1] attempted an elastic–plastic calculation of the residual stresses arising from the welding of a steel plate. Their results, however, did not agree with experimental values, possibly because of phase transformation in the steel and the use of a two-dimensional program for the stress analysis. The present in-

419

vestigation has therefore been restricted to a steel of type 316, which does not undergo a phase transformation (other than melting and resolidification), and an axisymmetric geometry (three-dimensional elastic–plastic analysis being very time-consuming). The geometry investigated was a 609·6 mm long bar of 152·4 mm diameter, containing an axisymmetric groove at mid-length. Weld metal of the same composition was to be added by a submerged arc process, at such a rate that the temperature analysis could also be treated as axisymmetric. To obtain the shape of the yield surface in a stress–strain-temperature space, data were to be obtained from BSC Gleeble machines in which circular-section cylindrical specimens from the weld, parent and simulated heat-affected zone metal were uniformly strained. The CEGB program BERSAFE was used to carry out elastic–plastic calculations for thermal loading and unloading.

MATERIAL PROPERTIES

The BSC Gleeble machine was used to obtain stress–strain curves for temperatures from room temperature to about 1300 K. At 1300 K the material properties appeared to be stable (see Appendix 1) and were therefore used up to 1700 K. Between 1700 and 1750 K, the steel melts and therefore the yield stress was assumed to fall linearly to a low value (arbitrarily taken as 690 kN m^{-2}) over this range.

Other values used in the calculations (except as otherwise noted) were:

Young's modulus	186 GN m^{-2}
Poisson's ratio	0·3
coefficient of thermal expansion	20×10^{-6} K^{-1}
density	8·055 tonne m^{-3}

The thermal conductivity (K) and specific heat (C) were linearly interpolated from the values given in Table 1.

TABLE 1

T (K)	300	473	673	873	1 073	1 700	1 701	1 750
K kgms^{-3}K^{-1}	71·2	87·5	103·8	120	132·2	132·2	132·2	132·2
C m^{2}s^{-2}K^{-1}	458	535	590	603	603	606	10 968	645

(Melting starts at 1 700 K. The greatly increased specific heats at higher temperatures are due to the latent heat of fusion.)

MANUFACTURE OF WELDS

Single-pass welds were deposited over 100 s by a submerged arc process. As irreproducible conditions, including cracks, were obtained, however, a further five welds were made in AISI type 316 bar, using the following welding procedure and consumables:

welding process	tungsten inert gas
filler wire	1·2 mm diameter Oerlikon 18/12/3 ELC, batch 5801
current	280 A
voltage	$16\frac{1}{2}$–17 V
rotation speed	847 μm s^{-1} at electrode tip
electrode	W/2% TH, 3·175 mm diameter
included angle at tip	45°
shielding gas	argon (99·99%) at 50 750 mm^3 s^{-1}.

The filler wire had the same composition as the bar. The weld preparation groove was 15·875 mm wide with 2·38 mm radiused internal corners. Details of the welds are given in Table 2. The temperature measurement technique is summarised in Appendix 2.

TABLE 2

Weld	Groove depth, d (mm)	Number of weld passes	Wire feed speed (mm s^{-1})	Comments
W1864	7·9	4	25	Porous area; high crown
W1759	2·4	1	25	Generally flat contour
W1807	4·8	2	25	Good bead shape
W1808	7·9	2	50	High crown; wire feed too fast
W1829	7·9	4	25	Good bead shape

THERMAL CALCULATIONS

To assess the feasibility of weld calculations, preliminary investigations were made with C. A. Parsons programs initially on the Reyrolle Parsons IBM 360-50 computer and subsequently on an IBM 370-145. The first case considered was that of two slabs at different temperatures suddenly brought into contact [1]. The initial temperature at the interface was chosen to provide the appropriate heat content in the surrounding elements. For compatibility with subsequent runs, a triangular mesh was used, with a linear temperature variation in each element. Fifty elements were used, each over the complete thickness. Normal to the interface the elements were taken to be similar in size to those used by Hibbitt and Marçal [1] and no difficulty was found in reproducing their results.

Calculations were then carried out for the axisymmetric triangular mesh shown in Fig. 1, with linear temperature variation in each element. This mesh had 72 nodes at vertices and represented one-quarter of the section of a bar containing an axisymmetric weld 2·36 mm deep and 15·875 mm wide. The initial temperature of the welding was taken as 1800 K, that of the parent metal as 300 K and that of the nodes at the interface to be such that the heat content of the surrounding elements was correct. Heat transfer on the curved surface was taken to be partly proportional to the difference between the temperature and 300 K and partly proportional to the difference of the fourth powers of these temperatures, the coefficients being $5·9 \text{ kg s}^{-3} \text{ K}^{-1}$ and $15·4 \,\mu\text{g s}^{-3} \text{ K}^{-4}$, respectively. The time steps were chosen so that the difference between values after each step agreed to within 1% with those obtained after two steps of half the size. The CPU time taken by the computer for a run of 512 s was about 500 s (at a cost of under £10) using a Crank–Nicholson integration procedure with solution at each step by a sparse-banded elimination technique. Little difference was found on changing the value of the linear heat transfer coefficient, and the results did not drastically deteriorate when a coarser mesh (32 nodes) was used. The addition of elements to the weld over a period of 2 s instead of instantaneously had little effect after the first few seconds.

Attempts were made to determine the temperature variation using program FLHE on the CEGB computer, with the previous axisymmetric triangular mesh and also with a 144-element brick mesh over a quarter of the bar (Fig. 2), with or without a small central bore. The inclusion of the bore made negligible difference to the results, but significantly cheapened the analysis. However, it was found that oscillatory temperatures were calculated by FLHE. The essential difference between the axisymmetric form of the FLHE and C. A. Parsons programs was that, to reduce rounding errors, the C. A. Parsons program had lumped the heat capacity terms at the nodes,

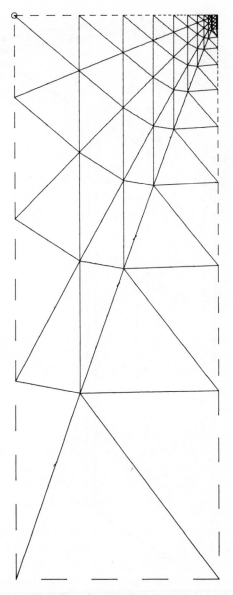

FIG. 1.   Mesh for axisymmetric finite-element analyses.

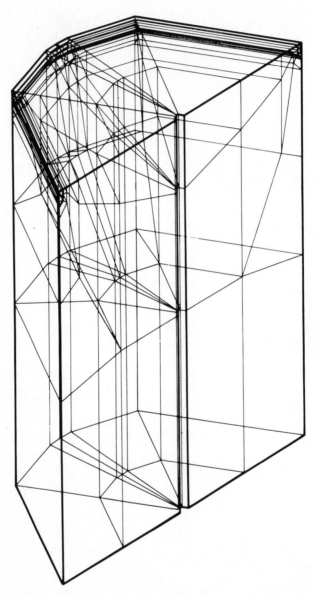

FIG. 2.   Mesh for non-axisymmetric finite-element analyses.

which is equivalent to a finite-difference representation of these terms. An analysis of the relative accuracy of the two techniques for a uniform rectangular mesh with linear temperature variation in each element for an insulated homogeneous rectangle indicated that their contributions to the errors arising as a result of finite mesh size were comparable. Numerical investigations using the mesh of Fig. 1 showed that the number of time steps required to ensure that the additional errors due to the finite time-step size remained within 1% over each step was also comparable for lumped and distributed heat capacities for triangles with linear temperature variations. Meanwhile, a numerical investigation of the effect of lumping on linear and quadratic elements had been carried out by Fullard [2] for both the mesh of Fig. 1 and a simplified version of part of it. Present conclusions are that:

(1) Reduced integration procedures should not be used for the heat capacity terms.
(2) Lumping should not be used for quadratic elements.
(3) Lumping underestimates changes and distributing overestimates them, the errors being comparable for linear rectangular elements except for localised changes, where rates of decay are obtained more accurately by distributing.

The results of the current temperature investigation are to be used in the BERSAFE stress analysis program with linear variations of stress and quadratic variations of displacement within an element. As temperatures are associated with stress, the use of linear temperature variations in this and in the thermal analysis also seemed appropriate. Lumping with linear elements also inhibits spurious oscillations, which would have particularly damaging effects if they caused plastic loading and unloading cycles. Hence, lumped linear brick elements (Fig. 2) were used for a three-dimensional transient temperature analysis.

Comparison between the calculated transient temperatures and temperatures measured by thermocouples 6, 13, 25 and 50 mm from the weld showed that a significant proportion of the heat input to the weld occurs after the weld has been laid but while it is still molten. Energy is produced at about 4700 kg m^2 s^{-3} from the arc, i.e. at a rate 23 400 000 kg s^{-3} per unit area for a groove 15 mm wide if the length of the molten zone subtends 10°, i.e. is slightly longer than the width. Of this, 900 000 kg s^{-3} is required to raise the temperature of the steel

from 300 to 1800 K. A heat inflow of 18 000 000 kg s^{-3} per unit area
and time (estimated from the axisymmetric calculations) was therefore taken to act over this area. Thermocouple reading confirmed
these values.

The mesh in Fig. 2 represents one-quarter of the bar and has 36
nodes on each of seven meridional planes at angles of ±90°, ±30°,
±5° and 0 to the plane of symmetry, 10 of these nodes being within
the weld, 4 on the interface and 22 within the bar. The nodes adjacent
to the interface are 317·5 μm at either side.

The analysis was carried out in three parts, each simulating one
phase of the welding cycle. In the first the initial temperature of the
weld nodes in the first three layers was taken to be 1800 K and at the
corresponding interface nodes 1300 K. After 8 s, the excess temperature over 300 K was scaled to allow for additional heat input. In the
second part the temperatures at the weld nodes in the third, fourth
and fifth planes were raised to 1800 K (in the third layer the temperature remained near the melting point) and those at the corresponding
interface nodes adjusted so that their heat contents were the averages of
those of the nodes at either side. Heat was applied over the weld surface
between these layers for 16 s (corresponding approximately to 13·3 mm
at 847 μm s^{-1}). In the third part heat was applied at a linearly decreasing
rate over 134 s on the weld surface between the fifth and seventh planes
at an average rate one-eighth of that previously applied on the other parts
of the weld surface.

The temperatures of interest are those in the plane of symmetry
(the fourth plane), which will have undergone preheating in the first
part, weld deposition in the second and cooling in the third. Least
attention was paid to the preheating phase, as any residual stresses
set up at points in the fourth plane during preheating will be annealed
out during the deposition of the weld. Thirty steps were taken over
the second part and eight over the third. This corresponds to an error
due to step size of about 1% over a step in the two-dimensional case.

STRESS ANALYSIS

The stress–strain–temperature surface denoting the yield stress of the
material was presented to BERSAFE [3] in the form shown in Table 3
obtained from the isothermal adjustments to the Gleeble machine
results (Appendix 1). The BERSAFE program allows seven tempera-

TABLE 3
Stress–strain–temperature data

Material	Plastic strain (%) Temperature (K)	0	0·01	0·02	0·03	0·04
			Stress (MN m^{-2})			
Weld	300	92	148	234	303	379
	470	83	148	207	231	241
	1 070	69	124	138	141	145
	1 700	69	124	138	141	145
	1 750	0·69	0·69	0·69	0·69	0·69
	1 800	0·69	0·69	0·69	0·69	0·69
Parent	300	77	427	434	441	448
	470	77	179	207	234	248
	1 070	77	159	186	200	207
	1 170	77	110	124	128	131
	1 700	77	110	124	128	131
	1 750	0·69	0·69	0·69	0·69	0·69
	1 800	0·69	0·69	0·69	0·69	0·69
HAZ	370	77	207	221	228	234
	970	77	276	276	276	276
	1 700	77	276	276	276	276
	1 750	0·69	0·69	0·69	0·69	0·69
	1 800	0·69	0·69	0·69	0·69	0·69

tures per material but the numbers used were considered adequate, as most of the intermediate temperature results could be simulated by linear interpolation.

Preliminary tests with a single square element with mid-side nodes under uniform stress or uniform strain produced the correct answer when the temperature was raised or lowered so that the material followed different stress–strain curves. The analysis was then carried out on a slice of Fig. 2 treated as an axisymmetric body.

In carrying out a transient thermo–elastic–plastic analysis with a specific mesh using BERSAFE, there are three choices to be made: (i) the tolerance to be achieved at each step, (ii) the number of steps per restart and (iii) the number of restarts. At the beginning of each restart, the stress–strain curve for the current temperature at each gauss point is located and used for the remainder of the restart. The tolerance is defined as the ratio of the sum of the squares of the residuals in the finite-element equations to the sum of the squares of the loads. The first submission was for the preheating and weld deposition stage. This was analysed with tolerances of 0·5, 1, 3 and 5% and between 5 and 20 steps. In some cases the analysis was not completed, as over 60 iterations were predicted for a single step.

It was concluded that a tolerance of 3% was acceptable. The errors in the average hoop stress in an element for tolerances of 1 and 3% were not more than 5 and 25% of the maximum hoop stress and, provided the number of steps was sufficient to ensure rapid convergence, little additional accuracy resulted from using more than ten steps; the difference for runs with five and ten steps, with a tolerance of 1%, was 10%.

For the 16 s heating period, a tolerance of 5% was used with six steps for either a single restart, or four restarts 0 to 4, 4 to 8, 8 to 12 and 12 to 16 s, or four restarts 0 to 1, 1 to 2, 2 to 3 and 3 to 4 s. While there were significant discrepancies in the results after 16 s, the difference in the results after 4 s was within 5% of the maximum hoop stress. (A coarser tolerance than for the initial stage was used, as the temperatures did not vary so much.) It was therefore decided to follow the initial preheating run by four restarts during heating, six restarts during cooling from 16 to 17 s, one restart during cooling from 17 to 150 s and one restart during subsequent cooling to room temperature. The times 17 and 150 s were those at which the surface nodes next to, and next but one to, the weld attained their maximum temperatures. Each restart had a tolerance of 5% and six steps as before.

Another analysis was carried out with a tolerance of 3% and six steps, but with the four, six, one and one restarts in the different stages replaced by eight, eight, two and two. The calculated residual stresses in the surface elements after cooling are given in Table 4. (The figures in parentheses correspond to the lower tolerance.)

TABLE 4

Direction of stress	Stress, MN m^{-2}		
	Centre of weld	HAZ	19 mm from weld
hoop	420 (410)	420 (290)	−30 (−160)
axial	150 (90)	90 (−70)	−50 (−200)

STRAIN MEASUREMENTS

Strains were already measured by a 'pillar isolation' technique but anomalous results were obtained for the single-pass weld, attributed to the erosion of the material to a greater depth than the weld. Therefore, a hole-drilling technique was finally used. The particular method employed is known as 'airbrasive drilling' and was developed by the CEGB Berkeley Laboratories [4]. It consists of attaching a rectangular rosette strain gauge to the metal surface on which the residual stresses are to be measured and drilling a circular hole by directing 53 μm diameter alumina particles through a rotating nozzle at high velocity on to the metal surface. Drilling to a depth of one diameter releases all the radial stress at the edge of the hole and causes local redistribution of the stress. The changes in strain measured by the three gauges before and after the hole-drilling are used with the measured final hole diameter to calculate the magnitude and direction of the major and minor principal stresses.

For the survey across the weld, rectangular rosettes were attached at selected positions around the circumference, viz. (i) on the weld centre, (ii) on the heat-affected zone and (iii) on the parent metal. These rosettes were located at positions unaffected by the previous trepanned holes. The metal surface was prepared by polishing with various grades of emery paper from medium to 400 grade.

The experimentally determined residual stress values are tabulated in Table 5. The results using the hole-drilling method on the weld agreed qualitatively with those obtained by the pillar isolation method, although the peak values were smaller.

DISCUSSION

Despite the improved experimental and theoretical techniques now available, adequate correlation between calculated and measured stresses has not been achieved. Possible reasons for the wide variations in the measured values and the large differences between those and the calculated values are

(1) inadequate knowledge of the material properties,
(2) phase changes within the metal,
(3) stress relaxation during the interval between welding and measuring the stresses, and

TABLE 5
Residual stresses

Location	Hole drilling technique [4]			Pillar isolation technique	
	Rosette	Stress MN m^{-2}	Direction (approx.)	Stress MN m^{-2}	Direction
Centre of weld	1	+ 19 − 46	hoop axial	31–93 tensile	hoop
	2	+ 19 − 56	hoop axial	0–108 compressive	axial
Heat-affected zone	3	+ 12 − 56	hoop axial	not measured	
	4	+ 58 − 37	hoop axial		
Parent metal (19 mm from weld)	5	+ 92 + 39	hoop axial	not measured	
	6	+ 42 + 4½	hoop axial		

Modulus of elasticity used to calculate the above figures = 189·6 GN m^{-2}; Poisson's ratio = 0·3.

(4)   the refinement level of the finite-element model in time and space.

In particular, the experimental techniques in no case located stresses anywhere near the yield stress, but the calculated residual stresses were at the yield stress in and near the weld, as would be expected for a material which is being plastically strained in tension after previous yield in compression.

The lower measured stresses (particularly the tensile stresses in the parent metal) cannot be understood on the basis of isotropic work-hardening; the data for more complex work-hardening theories would be prohibitively expensive to obtain and there is no real evidence to justify any attempt to do this. Creep relaxation of the stresses seems unlikely without exposure to heat, although mechanical relaxation by surface damage is just conceivable; the time between laying the weld and the final measurements was about two years, during which the weld was kept at room temperature with no other special precautions. The parent metal was austenitic, and therefore no solid state phase changes should have occurred, but the HAZ specimens used for the determination of material properties did exhibit some ferromagnetism.

In the finite-element analysis, there are five important forms of approximation deriving from the discrete nature of the analysis:

(1) mesh refinement,
(2) load-step sizes,
(3) time-step sizes,
(4) tolerance on the equation satisfaction and
(5) representation of material properties.

No one of these appears to be seriously out of line. The results show that the mesh refinement and load-step sizes are, on the whole, satisfactory; the time-step sizes are based on past experience with the program and are likely to be very reasonable. The equation tolerance is largely self-stabilising. It is less easy to be confident about the material properties representation but, except at the highest temperatures, there is no indication in the results that these properties need to be more accurately represented. The cumulative effect of so many tolerances, however, appears to be causing relatively poor convergence, but nevertheless, this should be sufficient for a far higher accuracy than would account for the discrepancy with the experimental results, especially with respect to the attainment or otherwise of residual stresses comparable with yield.

ACKNOWLEDGEMENTS

The authors wish to express their appreciation of the excellent cooperation shown throughout by the many groups concerned at C. A. Parsons, IRD and CEGB (BNL) and in particular of the personal interest and excellent liaison shown by the group leaders responsible for these undertakings. They wish to thank particularly Mr M. C. Murphy for the specimens and the load–displacement records, Mr R. Hill for manufacture of the welds, Mr G. E. White for recording temperatures during welding, Mr C. B. Jolly for strain measurements on the welds, Mr L. C. McCalvey for the pillar isolation technique measurements and Mr K. H. Fullard for modifications to FLHE and investigations of their effect. They also wish to thank the CEGB and C. A. Parsons for permission to publish the results. The work formed part of a collaborative programme between these organisations and GEC Ltd.

REFERENCES

1. HIBBITT, A. D. and MARÇAL, P. V., A numerical thermomechanical model for the welding and subsequent loading of fabricated structure, *Int. J. Comp. Struct.*, **3**, 1972, 1145–1174.
2. FULLARD, K. H., 'Thermal Transient Analysis Involving Large Temperature Gradients Solved by Means of FLHE, CEGB. Note RD/B/N3831, 1976.
3. HELLEN, T. K. and PROTHEROE, S. J., BERSAFE, *Computer Aided Design*, **6**, 1974, 15–24.
4. BEANEY, E. M., 'Accurate Measurement of Residual Stress on Any Steel Using the Centre Hole Method,' CEGB Report RD/B/N3568, 1976.

APPENDIX 1: USE OF BSC GLEEBLE MACHINE FOR OBTAINING MECHANICAL PROPERTIES DATA

The Gleeble machine enables specimens of material to be given an accurate thermal history and to be stressed at a known fixed temperature, or during a known temperature variation. Stress is determined from the imposed load and the specimen geometry; crosshead movement is also monitored, thereby enabling strain to be determined.

The choice of specimen geometry depends on the particular application. In this case, on the basis of initial tests, a long cylindrical form with a machined neck at the mid-point (Fig. 3) was used. Previous calibrations provided the axial and radial temperature variation in the specimens excluding the neck. Since the radial variation is only appreciable at high temperatures where there is negligible work-hardening and most of the heat is conducted axially, the axial variation should not depend appreciably on the outer radius. The temperature variations on the necked specimens were therefore taken to be as calibrated, the drop being up to 60°C for a maximum temperature of 900°C.

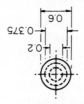

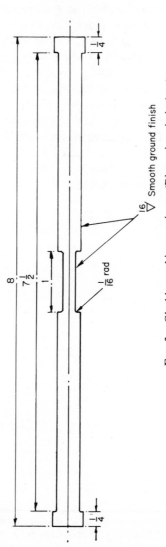

Fig. 3. Gleeble machine specimen. (Dimensions in inches.)

Specimens were machined from untreated parent metal and parts of a batch that had been heated to 1300°C to simulate the heat-affected zone. Weld metal specimens were taken from deposits made with the filler wire used for the test welds (Oerlikon 18/12/3 ELC, 1·2 mm diameter, batch 5801). They were tested at maximum temperatures from room temperature up to 1000°C, at 100°C intervals. Some delay was incurred owing to the largest batch of specimens being tested at BSC on incorrectly calibrated machines. Load–displacement records (Fig. 4) were obtained in each case. To convert these to stress–strain curves, data were analysed in ascending sequence of temperature and for each temperature in ascending sequence of load. By considering in turn each load increment and linearly interpolating with respect to both strain and temperature from the known values, the estimate of the strain at the current stress and temperature was systematically refined until satisfactory agreement was obtained between the displacement predicted and that measured for the corresponding maximum temperature and load. Hence, a series of stress–strain curves was produced for each temperature.

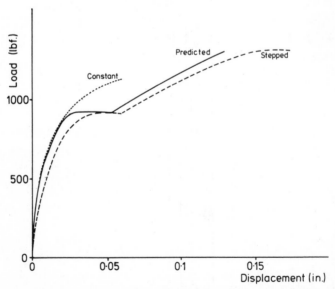

FIG. 4.   Load–displacement records for parent metal at 391°C and at 400°C stepped to 700°C, the predicted values also being shown.

To ensure that this process was satisfactory, one specimen was strained at 400°C, heated to 700°C and further strained. The deviations of the predictions based on the isothermal stress–strain curves from the measurements of deformation in the hotter part were similar to those for the cooler part (Fig. 4). Between 1700 and 1750 K, melting occurs and the yield stress was assumed to reduce by a factor of 100 from the room temperature value for the weld metal.

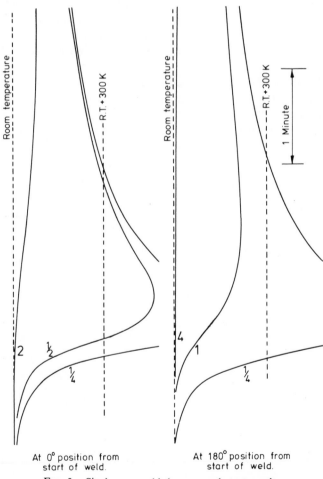

At 0° position from
start of weld.

At 180° position from
start of weld.

FIG. 5. Single pass weld thermocouple test results.

APPENDIX 2: TEMPERATURE MEASUREMENT DURING TEST WELDS

During the welding process, a sequence of temperature measurements were made in order to confirm the accuracy of the thermal fields calculated by the finite-element method.

Temperatures were measured by 30 SWG chromel/alumel thermocouples, spot-welded on to the mild steel bar, at measured distances from the groove that was to be welded. The hot-junction wires were approximately 1·6 mm apart and were insulated from the bar with glass silk tape. This tape was also wrapped on top of the wires to reduce errors due to air temperature or splashes from the weld. The outputs from these thermocouples were monitored on a three-channel 0–500°C Chassel recorder.

The distances from the thermocouples to the weld groove were chosen to give easy and informative comparison with the calculated temperatures. Little variation in temperature history resulted from their positions relative to the start (or even with the number of passes). Typical results are shown in Fig. 5 for different positions on the weld.

21

High-Temperature Plasticity Analysis of Diesel Engine Components

J. S. HOLT AND B. PARSONS
University of Leeds

SUMMARY

The stress analysis of components associated with the combustion chamber of a typical medium-speed Diesel engine is extremely complex and yet becomes increasingly important as the demands made on these components are extended and modified. The high gas temperature within the combustion chamber causes a significant reduction in the yield stress of the surrounding material and, hence, the combined effect of high gas load and thermal load is to frequently cause plastic deformation. The local stress peaks predicted by linear elastic analysis are much reduced and the corresponding strains increased, so that in order to satisfactorily predict failure it is essential to consider a full temperature-dependent plasticity analysis of these components.

In this paper the authors consider the demands made on a finite-element program for the analysis of combustion chamber components. Methods are considered by reference to ideal structures, progressing with experience to the analysis of an exhaust valve and piston. These analyses are performed by the finite-element method, using high-order isoparametric elements, the 'initial stress' method of non-linear analysis being applied in the plastic regime. The numerically integrated elements allow for a continuous variation of mechanical properties with temperature and offer a consistent treatment of structural analysis from elastic, through plastic, to residual stress. In this way a realistic modelling is possible in which due account is taken of temperature and load-cycling in a plastically deforming structure.

437

NOTATION

$[D]$	elastic stress–strain matrix
$[D]^*_{ep}$ or $[D(\sigma)]^*_{ep}$	elasto-plastic stress–strain matrix (a function of stress state $\{\sigma\}$ and H')
E	Young's modulus
$[H]$	strain–displacement matrix
H'	input plastic slope (the slope of the plot of uniaxial stress versus plastic strain)
$\bar{\epsilon}$	equivalent (uniaxial) strain
$\bar{\sigma}$	von Mises equivalent stress or effective stress
$\{\ \}$	denotes a vector
$[\ ]$	denotes a matrix

The following terms apply to the initial stress method of non-linear analysis used in the paper and are appropriate to a particular load iteration (defined in the text):

$\{\Delta\epsilon'\}, \{\Delta\sigma'\}$	increments of elastic strain and stress (*note*: $\{\Delta\sigma'\} = [D]\{\Delta\epsilon'\}$)
$\{\lambda\}_{init}$	initial loads which are equivalent to (or in equilibrium with) the initial stresses
$\{\sigma_0\}, \{\sigma\}$	stress state at beginning and end of iteration
$\{\sigma'\}$	total elastic stress
$\{\Delta\sigma\}$	stress increment predicted by $[D]^*_{ep}$ (*note*: $\{\Delta\sigma\} = [D]^*_{ep}\{D\}^*_{ep}$)
$\{\Delta\sigma''\}$	initial stresses (*note*: $\{\Delta\sigma''\} = \{\sigma'\} - \{\sigma\}$)

INTRODUCTION

The Diesel engine industry has a definite requirement for finite-element analysis and, since the exposition of the method in the 1950s, it has been used to good effect in the heat and stress analysis of a variety of components. Following early work by Fiska, Iversen and Sarsten [1] in elaborating the application of the method to thermally loaded Diesel engine components, there has been a general acceptance of the method.

As the power output of engines increases, however, it becomes increasingly difficult to predict with confidence the life of components using linear (i.e. elastic) theory only. In recent years the preoccupation with accurate geometric modelling has been put into perspective by a growing awareness of the implications of material non-linearity. The high thermal gradients within combustion chamber components and the correspondingly reduced yield stress of the material on the gas side of the combustion chamber walls require a full temperature-dependent plasticity analysis to satisfactorily predict life [2].

LOADING IN THE DIESEL ENGINE

Fatigue damage during the life of an engine is assumed to be due to a combination of the low-cycle thermal fatigue of the stop–start cycles and the high-cycle gas load fluctuation. From cold the new engine with an assumed unstressed material will undergo a running-in schedule of small load increments, building up gradually to full load. The transient stresses are thus kept to a minimum, the 'raw' material being allowed to gradually work-harden from its initial condition.

If the load is repeatedly applied and then removed from a tensile test specimen, the stress–strain curve will initially change with each cycle but will become stable after a number of cycles. The curve for the first cycle is called the monotonic curve; the stable curve which eventually develops is called the cyclic curve. The analysis of engine components should consider the initial warm-up in terms of monotonic material data, with suitable transient temperature distributions. Subsequent warm-ups will be more rapid, the higher transient stresses being accommodated to some extent by the higher yield of the work-hardened material. Following warm-up, the medium-speed Diesel engine is assumed to attain steady running conditions characterised by a steady mean cycle gas temperature and pressure. After the initial running-in period the material behaviour may then be represented by cyclic data.

THE FINITE-ELEMENT PROGRAM

The work of this paper was carried out using a general-purpose two- and three-dimensional finite-element program written in the Department of Mechanical Engineering at the University of Leeds. Quadratic displacement isoparametric elements are used exclusively (see Fig. 1), element stiffnesses being usually integrated by a three-point gauss

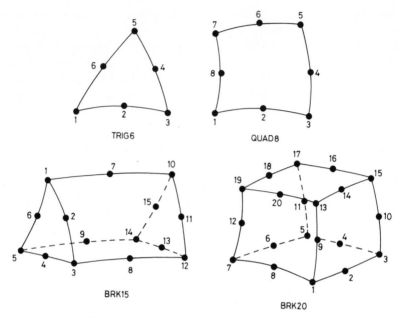

FIG. 1.   Quadratic isoparametric finite elements.

rule. The displacement method of finite-element analysis and the concept of isoparametric elements are well known [3, 4] and will not be discussed further here.

The Initial Stress Method of Non-Linear Stress Analysis

The progression up the stress–strain curve is based on the 'initial stress' approach of Zienkiewicz, Valliappan and King [5]. The calculation of initial loads is performed by a two-point gauss rule, using the expression

$$\{\lambda\}_{init} = \int [H]^{T}\{\Delta\sigma''\}\,dv \tag{1}$$

where $\{\lambda\}_{init}$ are the initial loads, $[H]$ is the strain–displacement matrix and $\{\Delta\sigma''\}$ are the initial stresses. This takes advantage of the accuracy of the stresses and strains at these gauss points [6].

With regard to load, the following definitions are used throughout. 'Load step' is taken to mean a new application of load which the program is to consider; this may be the addition or removal of a pressure load, thermal load, etc. In the plastic regime a load step is applied in stages termed 'load increments', within each of which the problem is tackled in an iterative manner, as defined by the initial stress approach. This requires the use of the term 'load iteration'.

The Elasto-Plastic Stress–Strain Matrix

The initial stresses $\{\Delta\sigma''\}$ are obtained via the elasto-plastic stress–strain matrix $[D]^*_{ep}$, as shown in Fig. 2. As normally defined, $[D]^*_{ep}$ relates an infinitesimally small change of strain to stress, i.e.

$$\{d\sigma\} = [D(\sigma)]^*_{ep}\{d\epsilon\} \qquad (2)$$

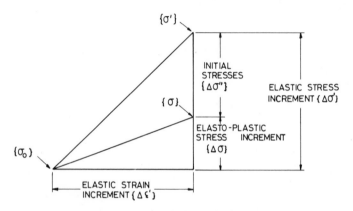

FIG. 2. The determination of initial stresses.

Its use in the determination of initial stresses requires that eqn (2) is considered valid for finite changes in stress and strain. It will be noted that $[D]^*_{ep}$ is evaluated at a particular stress state and the success of its use during a finite change of strain can be judged by comparing the effective plastic slope H'_e during the iteration with the input plastic slope H', where

$$H'_e = \frac{\overline{\{\sigma_0 + \Delta\sigma\}} - \bar{\sigma}_0}{\bar{\epsilon}_p} \qquad (3)$$

and is thus the value of plastic slope determined by the program during the iteration. Successively better approximations of H'_e to H' are obtained as $[D]^*_{ep}$ is evaluated closer to the average stress during the current load iteration (see Fig. 3). In particular, the calculation of

ITERATION		$H'_e (\times 10^{-6})$
1	$D^*_{ep}(\bar{\sigma}_0)$	2·0096
2	$D^*_{ep}\left(\frac{\bar{\sigma}_1+\bar{\sigma}_0}{2}\right)$	0·6643
3	$D^*_{ep}\left(\frac{\bar{\sigma}_2+\bar{\sigma}_0}{2}\right)$	0·7007
4	$D^*_{ep}\left(\frac{\bar{\sigma}_3+\bar{\sigma}_0}{2}\right)$	0·6996
5	$D^*_{ep}\left(\frac{\bar{\sigma}_4+\bar{\sigma}_0}{2}\right)$	0·6996

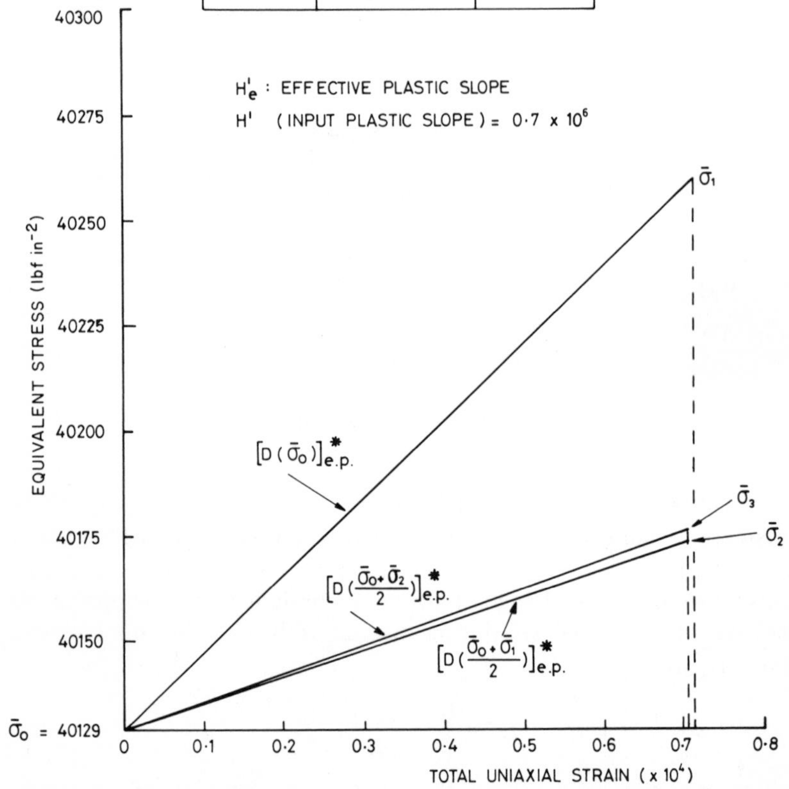

FIG. 3. Iterative improvement of the elasto-plastic stress–strain matrix $[D]^*_{ep}$.

$[D]_{ep}^*$ at $\{\sigma'\}$ as given in the computational scheme of Ref. [5] can give rise to reduced plastic strain and a reduction of equivalent stress, corresponding to an apparent strain-softening.

The effects of iteration in $[D]_{ep}^*$ and of the number of load increments above yield were investigated in a thermal context by taking a thick cylinder with a radial temperature distribution from 600°C on the inside to 0°C on the outside, and comparing the equivalent plastic strain at the innermost gauss points (Fig. 4). Regardless of whether $[D]_{ep}^*$ is calculated on the basis of $\{\sigma_0\}$ or any of the subsequent approximations to the average stress state, the plastic strain $\bar{\epsilon}_p$ shows a uniformly monotonic convergence from below with load increment. This was also the case for the equivalent total uniaxial strain, $\bar{\epsilon}_T$, defined by

$$\bar{\epsilon}_T = \frac{\bar{\sigma}}{E} + \bar{\epsilon}_p \qquad (4)$$

By contrast, the equivalent stress converges monotonically, but does so either from above or from below, depending on how $[D]_{ep}^*$ is calculated.

Convergence Criteria

A load iteration is judged to be complete by reference to a convergence criterion, for which various alternatives are possible [7]. Convergence may be assumed, for instance, when a change in plastic work or plastic strain is sufficiently small, or when the initial loads are less than a specified fraction of, say, the vector norm of the total applied loads in the current load step.

Thermal and mechanical load problems, however, differ in their rates of convergence and in their sensitivity to convergence criteria (see Fig. 5). The pressure load case has a noticeably slower rate of convergence than an equivalent temperature load case, and, if a criterion using the norm of the applied loads is set at a reasonable level for the pressure situation, then this would cause the program to terminate the temperature load case prematurely.

Some compromise is necessary, and so, bearing in mind that the pressure load is typically only 30% of the temperature load in combustion chamber components, a criterion based on plastic work has been adopted in all cases. The program used in the present work assumes convergence when the change in plastic work summed over all gauss points is less than 1% of the current total. Referring to Fig.

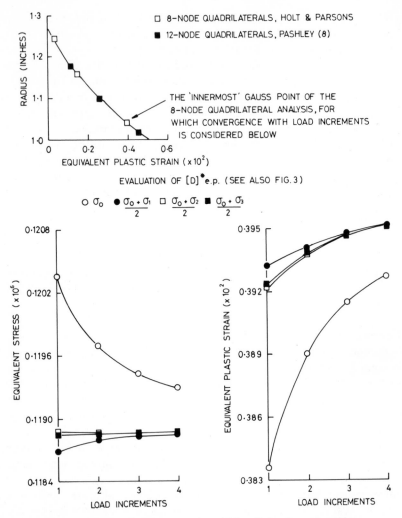

FIG. 4.   Aspects of convergence in a thick-cylinder thermal problem.

5, it will be noticed that the slower rate of convergence of the mechanically loaded structure implies that, for a given cut-off, the percentage error may be greater than for the equivalent thermal problem. For this reason the option to select a criterion of 0·5% of plastic work is given, although this is at the expense of greater computer time.

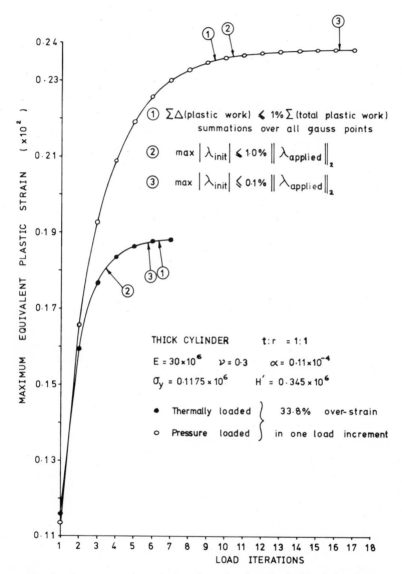

FIG. 5.   Convergence in equivalent thermal and mechanical load situations.

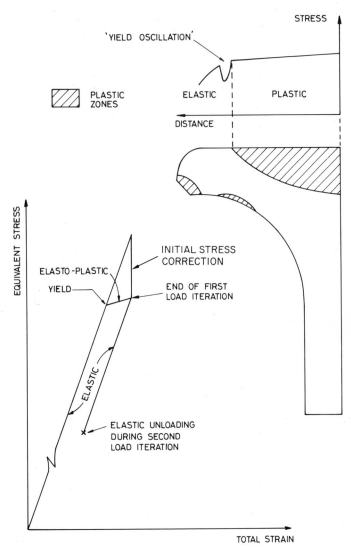

FIG. 6.   The effect of 'yield oscillation' in the analysis of an exhaust valve.

A further point in connection with demonstrating convergence with load increments, as in Fig. 4, is worth noting. The selection of a particular criterion, say 1% of total plastic work, will not necessarily show a monotonic convergence, since the cut-off is insensitive to the margin within the tolerance. Thus, in Fig. 4 it was necessary to condition all load increments to have seven iterations, although for some load increments convergence as defined by the criterion used occurred after only six iterations.

Yield Oscillation

In the initial stress method the adjustment brought about by a re-solution with the initial loads may cause already yielded elements to unload and return to the elastic region during the load increment. It is therefore desirable that this problem should be recognised and allowed for if a program is used in combustion chamber component analysis where high thermal and gas loads occur.

In such cases, however, some finite-element programs allow the elastic unloading to take place, although this can give rise to anomalies in the final stress output. Following a reduction in stress due to plasticity, the unloading into the elastic region means that the stresses are then considerably lower than if yielding had not occurred. This situation and its effect, as noted in an analysis carried out using a commercially available program [9], is shown in Fig. 6, in which the elastic unloading in an increment in which yielding had already occurred is called 'yield oscillation'.

A paper describing a correction scheme to take account of yield oscillation is being prepared by the first author. Such oscillation occurs frequently in combustion chamber components, and the scheme has been applied in all the plasticity analyses which are described in this paper.

MATERIAL PROPERTIES AND TEMPERATURE-DEPENDENCE

The variation in temperature of the combustion chamber components is sufficiently great to make changes in the material properties very significant. Accordingly, the temperature-dependence within a finite-element program intended for combustion chamber component analysis is of fundamental importance.

Temperature-Dependence in the Finite-Element Program

The temperature-dependence of material properties may be communicated to the finite-element program in a number of ways. Of the two methods considered, the first, called here the 'zone method', can be used in programs where a true temperature-dependence is not inbuilt, but where individual elements may have properties assigned directly to them. Material properties which remain constant are specified for all elements within a particular region or zone defined by a temperature band. The structure is divided into zones by choosing those elements which lie within a particular temperature range. In this way the properties varying with temperature are modelled by step functions over the structure.

The second method is to input tables of material property values and the corresponding temperatures; the local element properties, whether at gauss point or centroid, are determined when needed by interpolation. The temperature at the sampling point is first interpolated from the input nodal values, this temperature being the interpolating point for the various material properties.

Of the two methods described, the former has very severe limitations. To be successful it requires the structure to be idealised with a great many elements in order to allow for an adequate variation of properties and, for this reason, it is usually only possible with constant stress elements. When sufficient elements are used in the idealisation, however, then this method using constant stress triangles produces results comparable with those obtained with the temperature-interpolated data either in the constant stress elements themselves or in higher-order isoparametric elements [10].

The zone method, however, becomes unmanageable once load-cycling is considered; it then becomes desirable to have some means of relating each sampling point in the structure to its own stress–strain curve. For this reason, therefore, the zone method is not offered as an option in thermal problems in the program used by the authors. Instead data are input in tabular form, and for each sampling point in the structure (i.e. the gauss points) a vector of local material properties is set up. In this way, as loading and unloading proceeds, the yield stress and plastic work, for instance, are continually updated and so may act as a steering mechanism within the element plasticity calculations.

Stress–Strain Input

Following earlier experiences with the specification of material stress–strain curves [10], it was concluded that the advantages of restricting these to bilinear form far outweighed any small resulting loss of accuracy. In practice it was found that the high sensitivity of the resulting stresses and strains to changes in the initial yield stress and their relative insensitivity to strain-hardening rate made the detailed modelling of strain-hardening disadvantageous.

Accuracy of Material Data

Temperature-dependent material data of the combustion chamber components are not only sparse, but also subject to statistical uncertainty, and therefore the calculation of temperature distributions in these components is prone to error. A combined heat conduction and stress analysis of a Diesel engine exhaust valve has been carried out by the authors to determine the magnitude of these effects.

Temperature measurements on a sectioned valve, following thermal soak in an engine under particular loading conditions, were obtained by the hardness relaxation technique. Using a known mean cycle gas temperature, these measurements allowed equivalent heat transfer film coefficients to be determined in order to assess the possible behaviour of an alternative valve material. The ambiguity inherent in choosing a representative set of film coefficients was investigated by retaining three separate sets of data, any one of which could have been used in the normal course of events by the analyst in subsequent stress analysis. The temperature distributions produced by these three sets of film coefficients, using the mesh of Fig. 7, are summarised in Table 1.

TABLE 1
Summary of heat conduction analyses

Film coefficients group	Sample temperatures[a] (°C)					
	A	B	C	D	E	F
1	731·2	624·9	581·1	579·8	581·3	587·7
2	731·6	626·3	582·9	581·8	583·3	588·6
3	732·9	626·0	584·0	583·1	584·7	589·4

[a]Refer to Fig. 7.

Material properties of three representative samples of the austenitic valve steel were obtained. While the elastic properties were in excellent agreement, the temperature-dependent proof stresses (from

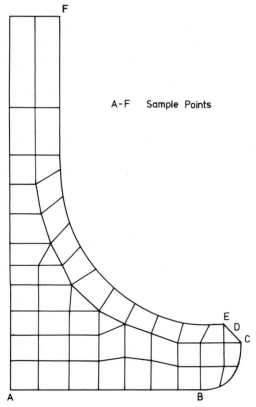

FIG. 7.   Finite-element idealisation of an exhaust valve.

which yield is determined by extrapolation) showed some variation (Table 2).

Accordingly, the three sets of yield stresses covering the temperature range of the valve were retained and combined with the three temperature distributions previously obtained. Stress analyses for temperature only and temperature with pressure were compared on the basis of 'yield potential', i.e. the ratio of the von Mises equivalent stress to the current temperature-dependent yield stress (Table 3). The value of yield potential in comparing structures under different loads where the temperature variation of material properties is significant can be seen in Fig. 8. The maximum equivalent stress occurs in the seat region, while the highest yield potential occurs at the centre of the valve face, where the high temperature has caused a reduction in yield stress.

TABLE 2
Material properties of an austenitic valve steel

Material property sample	Temp.(°C)	Yield stress (tonf in^{-2})	Proof stress (tonf in^{-2})		
			0·1%	0·2%	0·5%
1		19·0	23·0	25·6	29·7
2	500	17·4	20·0	22·0	25·0
3		19·6	22·5	24·5	27·7
1		15·3	19·0	21·4	24·5
2	600	16·6	18·8	20·4	23·2
3		14·4	17·4	19·6	22·8
1		14·6	17·4	19·4	21·8
2	700	14·0	16·5	18·0	20·0
3		14·3	17·2	19·2	22·0
1		11·4	13·6	15·1	17·0
2	800	11·1	13·1	14·4	15·7
3		13·1	14·2	15·2	17·6

TABLE 3
Maximum gauss point yield potentials in the analyses of an exhaust valve

Temperature distribution (Table 1)	Material property sample					
	1		2		3	
	T	$P + T$	T	$P + T$	T	$P + T$
1	0·651 8	0·788 5	0·640 2	0·813 1	0·667 5	0·807 8
2	0·649 5	0·783 6	0·636 0	0·808 6	0·666 1	0·802 5
3	0·655 0	0·787 7	0·641 8	0·814 3	0·673 7	0·806 4

T: Temperature only.
$P + T$: Combined temperature and pressure.

The largest variation (Table 3) is seen to be due to the scatter in material data and serves as a reminder that any numerical analysis is only as good as the data it uses. In particular, the yield stress of the valve material played a most important rôle in the analyses, virtually determining the final stress levels.

J. S. Holt and B. Parsons

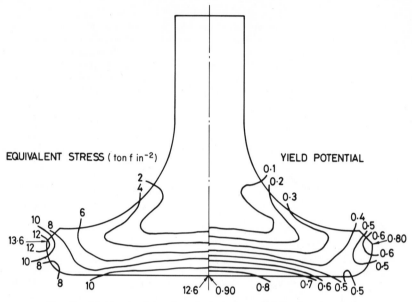

EQUIVALENT STRESS (ton f in⁻²)   YIELD POTENTIAL

FIG. 8.   Yield potential and equivalent stress distributions in a valve.

COMPONENT PLASTICITY ANALYSIS

In this section the thermo-elasto-plastic loading and subsequent residual stress calculations for a Diesel engine piston crown are considered in some detail.

The Piston Crown

The steel crown of a two-part piston (Fig. 9) is considered under the action of thermal loads at steady running conditions and the gas load due to the peak combustion pressure. The gas pressure is assumed to be constant over the crown down to, and including, the top ring groove. The loads are reacted at the bolted inner ring-type seat by a full axial restraint, and at the outer seat by a partial restraint which allows some rotation (see Fig. 10). The maintenance of contact (or even some interference) between the crown and piston body at the outer seat is essential, although the modelling does represent a simplification of the actual boundary conditions. At this stage of the work it is not intended to make an exhaustive analysis of the crown, but rather to consider the crown in the context of plastic deformation.

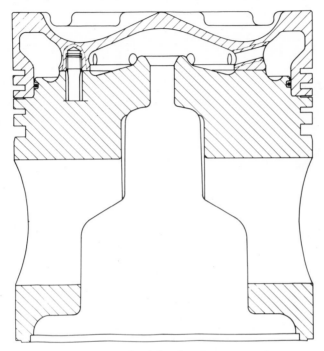

FIG. 9.   Cross-sectional drawing of a two-part piston.

The crown is analysed by a mesh of 69 quadratic isoparametric elements, with a total of 274 nodes, as shown in Fig. 11. Particularly noteworthy is the thin layer of elements on the surface of the crown, where plastic flow is expected. Material properties are input in tabular form at ten discrete temperatures spanning the temperature range of the crown, values at the element gauss points being interpolated as required. The large variation of yield stress of the piston crown material with temperature (Fig. 12) is notable; a 50% reduction is seen in the hottest regions.

Finite-Element Analysis
In a previous paper [10] consideration was given to the general problems of modelling the thermo-elasto-plastic behaviour of the piston crown under combined loading only, i.e. temperature with pressure. The gas load during steady running is, however, a cyclic loading and unloading, superimposed on the assumed steady thermal

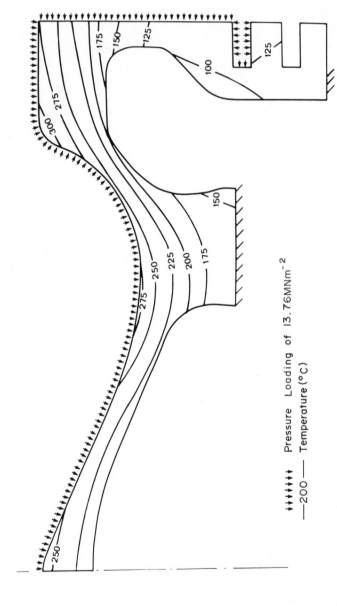

++++++   Pressure Loading of 13.76MNm^{-2}

—200—   Temperature (°C)

FIG. 10.   Loading and restraint conditions in the piston crown.

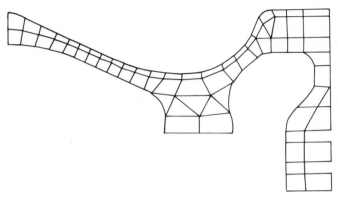

FIG. 11.   Piston crown finite-element idealisation.

load. The effects of the order and method of application during this stage of the engine cycle are therefore considered here by analysing various load cases as defined below (the constitutive load steps are in parentheses), isotropic hardening being assumed:

Load case 1: $(T) + (P) - (P) - (T)$
Load case 2: $(T + \frac{1}{2}P) + (\frac{1}{2}P) - (P) - (T)$
Load case 3: $(T + P) - (P) - (T)$

T: thermal load
P: pressure load

In load cases 1 and 3 the load reaches a maximum by either the separate or combined application of the thermal and pressure components, while in case 2 the mean of the high-cycle fatigue load is

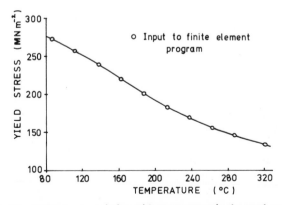

FIG. 12.   Yield stress variation with temperature in the steel crown.

first applied. Following the attainment of maximum load in each case, the pressure and then the thermal load are removed in order to determine residual stress.

TABLE 4
Summary of piston crown analyses

Plastic zone (Fig. 13)	Maximum gauss point equivalent plastic strain, $\bar{\epsilon}_p \times 10^3$				Maximum gauss point yield potential	
	(T)	$+$	(P)	$-$	(P)	$-(T)^*$
A	$-$		0·942 8	$\rightarrow$	0·942 8	0·421 5
B	$-$		0·194 3	$\rightarrow$	0·194 3	0·117 0
C	0·327 1	$\rightarrow$	0·327 1		0·329 8	0·404 7
	$(T + \frac{1}{2}P)$	$+$	$(\frac{1}{2}P)$	$-$	(P)	$-(T)^*$
A	0·181 5		0·960 6	$\rightarrow$	0·960 6	0·420 2
B	0·041 7		0·192 9	$\rightarrow$	0·192 9	0·115 9
C	0·295 6	$\rightarrow$	0·295 6		0·329 4	0·404 9
	$(T + P)$			$-$	P	$-(T)^*$
A	0·967 4			$\rightarrow$	0·967 4	0·414 4
B	0·190 4			$\rightarrow$	0·190 4	0·115 3
C	0·266 5				0·329 4	0·405 0

*Elastic unloading.
$\rightarrow$, No further yielding in this load step.

Results

The results are summarised in Table 4, which gives the plastic strain at each load step of the three load cases, and in Fig. 13, which shows the development of the plastic enclaves A, B and C at three stages during the load cases (each load step is analysed in three equal load increments). In Table 4 the maximum gauss point plastic strain within each of the three plastic zones A, B and C is given. Zone C is due essentially to the thermal load; the combined analysis of temperature and pressure causes a reduction in its size (compare Figs 13a and 13b). In each load case there is a further amount of work-hardening in this zone when the pressure is taken off; in fatigue

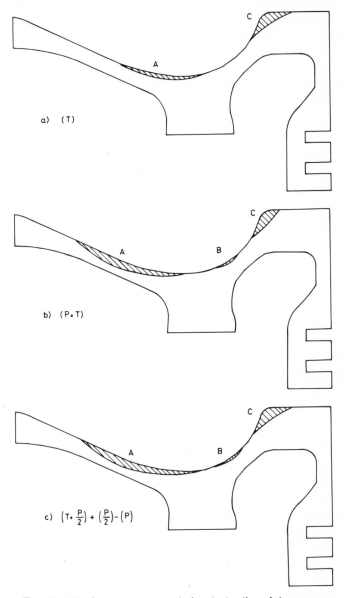

FIG. 13.   Plastic zones at stages during the loading of the crown.

calculations this is usually considered to be an elastic unloading. Zone B is due primarily to the superposition of pressure in a region which is already close to yield under the thermal loading. Zone A is a region where yielding due initially to the thermal load is considerably extended by the gas pressure cycle. It is also in this zone where a significant amount of yield oscillation occurs, as shown in Table 5.

TABLE 5

Occurrences of gauss point yield oscillations during each increment of loading

Yield oscillations in load increments				
(T) $(0 + 0 + 3)$	+	(P) $(6 + 15 + 38)$	−	(P) $(0 + 0 + 0)$
$(T + \frac{1}{2}P)$ $(0 + 1 + 0)$	+	$(\frac{1}{2}P)$ $(12 + 22 + 36)$	−	(P) $(0 + 0 + 0)$
	$(T + P)$ $(7 + 18 + 28)$	−	(P) $(0 + 0 + 0)$	

From the results of the various analyses it was found that the order of application of the loads was not as important as was first thought. Differences did occur in the amount of plastic strain developed, although, as expected from the foregoing discussion, the plastic strain in region C remained essentially unchanged, since here the plastic flow is due to only one load, the thermal load. Even so, as more of the pressure component is included in the initial load step, then it becomes more important to unload the pressure to realise the full plastic strain.

On complete unloading, the results of each of the load conditions gives rise, with only minor variations, to the residual stress distribution of Fig. 14, given in the form of yield potential contours. Subsequent reapplication of load with the assumption of isotropic hardening does not cause any further work-hardening.

CONCLUDING REMARKS

The methods described in this paper have been carefully developed to meet the special requirements of combustion chamber component analysis. By reference to ideal structures it has been possible to consider aspects of convergence in the initial stress method, and due regard has been paid to the question of temperature-dependence. Of particular interest has been the attention focused on yield oscillation, which, far from being unimportant, has been shown to occur in everyday analysis to a quite significant extent.

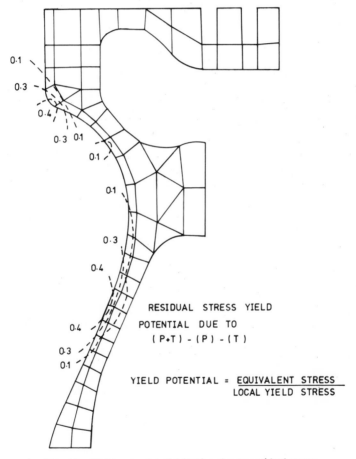

FIG. 14. Yield potential distribution due to residual stress.

ACKNOWLEDGEMENTS

In presenting this paper the authors would like to thank Mr B. B. Miatt of the Department of Mechanical Engineering for the use of his plotting program in Figs 7 and 12. Figure 3 was produced with the help of Mr M. Hawkes, also of the Department of Mechanical Engineering. The extensive use of the University of Leeds ICL 1906A computer and the cooperation of the computing staff are gratefully acknowledged. The authors would also like to thank Ruston Diesels Ltd and Mirrlees Blackstone Ltd for their kind assistance in many aspects of the work.

REFERENCES

1. FISKA, G., IVERSEN, P. A. and SARSTEN, A., Computer calculation of stresses in axi-symmetric thermally loaded components, *Proc. I. Mech. E.*, **182**, 1967–1968, 152–168.
2. LUTON, P. and SINHA, S. K., 'Development of analytical aspects of Diesel engine design', 12th Int. Congress on Combustion Engines, Tokyo, 1977.
3. ZIENKIEWICZ, O. C., *The Finite Element Method in Engineering Science*, McGraw-Hill, 1971.
4. ERGATOUDIS, J. G., IRONS, B. M. and ZIENKIEWICZ, O. C., Curved, isoparametric, quadrilateral elements for finite element analysis, *Int. J. Solids Struct.*, **4**, 1968, 31–42.
5. ZIENKIEWICZ, O. C., VALLIAPPAN, S. and KING, I. P., Elasto-plastic solutions of engineering problems; initial stress finite element approach, *Int. J. Num. Meth. Eng.*, **1**, 1969, 75–100.
6. BARLOW, J., Optimal stress locations in finite element models, *Int. J. Num. Meth. Eng.*, **10**, 1976, 243–251.
7. NAYAK, G. C. and ZIENKIEWICZ, O. C., Elasto-plastic stress analysis. A generalisation for various constitutive relations including strain softening, *Int. J. Num. Meth. Eng.*, **5**, 1972, 113–135.
8. PASHLEY, D., pers. comm.
9. LUTON, P., pers. comm.
10. HOLT, J. S. and HOPE, P. S., 'Thermo-elasto-plastic analysis of a Diesel engine piston crown by finite elements', 12th Int. Congress on Combustion Engines, Tokyo, 1977.

22

Non-Linear Stress Analysis: Reflections from a Remote Past

SIR DERMAN CHRISTOPHERSON
University of Durham

I think a good many of those attending this conference must have seen my name on the programme with a certain amount of surprise. In recent years, when I have had the opportunity of being present at occasions when progress in engineering research has been under discussion, I have become accustomed to identifying a certain startled look in the eyes of the people I encounter there, indicating that, although my name was perhaps vaguely familiar, they had supposed me to have been dead these fifteen years. And, of course, from the point of engineering research, they were not far wrong.

However, it seems to have occurred to the organisers of the conference that it might be of interest on this occasion to ask me to say something different from the kind of welcoming and generally benevolent remarks which people like myself are often invited to deliver to those attending conferences. Instead, it was thought that this conference might be entertained for a short time to hear, as it were, from Rip van Winkle, and that he might consider a number of questions: how far the situation revealed in the conference papers is just what would have been expected with reasonable foresight by those working in the field twenty years ago; how far they contain material which could not have been anticipated at that time; and, contrariwise, whether there are themes which would have been confidently expected to appear, and which are more or less absent. It is perhaps also worth posing another question: Does the thrust of research in the last twenty years lend support to the view that what is done in research laboratories, perhaps particularly in university laboratories in the UK, is a self-generating and self-sustaining activity, with little relation to the real needs of industry, or, on the other hand, does it suggest that what everyone has been doing has been strongly influenced by the need to find answers to practical questions, whether of a short-term or a long-term character?

Of course, to attempt to answer these questions is a tall order. I recognise that whatever I say is bound to be inadequate, and no doubt to seem biased to many, perhaps most, of the audience. I confess that I undertook the job in part because it gave me a cast-iron excuse, which I seldom get, for actually attending the working sessions of a conference, and getting a first-hand picture of what is going on.

Let me start with one very general point, which applies not just to the field of stress analysis, but to every field of research. I do not think that twenty years ago in the middle 1950s anyone would have foreseen the very large increase in the resources available for research, in the universities and no doubt in all the other contexts in which research takes place. We looked forward, of course, to rising standards brought about by improved technology and management. We hoped that research, along with everything else, would be a beneficiary. But most of us would, I think, have been reasonably satisfied if we had been assured that the real resources devoted to research in 1975 were going to be, say, three or four times what they were in 1955. In fact, funds provided to universities specifically for research increased in these twenty years by a factor of 30. The proportion devoted to engineering and applied science, as against pure science, is not easy to estimate, but certainly grew by an even greater factor. Of course, we have to take account of the decline in the value of money, but when that is done, the increase in the real resources provided specially for research cannot have been less than about seven or eight times. Moreover, what evidence there is suggests that the extent to which the universities support research from their general income, not provided specifically for that purpose, has increased in roughly the same proportion. Of course, the university system is much larger, but real resources provided for each individual are more than double what they were. So that in comparing what can be done now with what was being done twenty years ago, we have to remember that we are now living in a very different situation.

However, that is by the way. In the field of non-linear stress analysis what were the problems which were then receiving the most attention, and in which we should have expected and hoped for a lot of progress in the next twenty years? I think, without the exercise of too much wisdom after the event, we should have been able to identify at least some of the themes which figure prominently in our present agenda. The first session of all, for example, contained papers about crack propagation. That subject was first brought very prominently to everyone's mind by the remarkable failures in ship structures which occurred during the war years, in part because the widespread use of welding eliminated the rivet-holes which had no doubt always acted as 'crack stoppers' without anyone fully appreciating their function in that connection, and in part because some of the steel produced in wartime conditions had unusual ductile-brittle transition temperatures. There is a reference to room temperature in Radon's paper, but for his work to be relevant to these materials would have required an exceedingly warm room. These catastrophies meant that a very big experimental effort, by the standards of the time, went into the experimental study of notch brittleness by a great variety of experimental techniques. But the theoretical concepts which some of the conference papers develop were still some way in the future, and for my own part I do not think I would have foreseen their arrival.

Rather similar remarks apply to papers 7 to 12 dealing with problems in which the departures from linearity are due to creep. Creep testing of metallic materials has, of course, been going on for a long time and indeed is, and will continue to be, a very significant and time-consuming study; the paper by Fessler and Hyde suggests a significant means of economy in this field. But the basic properties of the important engineering materials have been reasonably well-known for a long time. Twenty years ago not much progress had been made in making use of all this information in actual design in anything more than a very unsophisticated way. To do so in real engineering problems is, of course, a formidable task, and I was particularly interested to see the contemporary state of the art—perhaps it is not unfair to use the word 'art'—as revealed in papers such as those by Ewing and by Patterson and Hitchings. As a simple-minded person, it seemed to me that the design approach recommended by Danks and Lomax represents a most promising line, which is worth examining in a wider range of contexts.

Again, it was obvious in the 1950s that numerical analysis based on finite-difference and finite-element methods was due for a very good innings. A good deal of the spadework had already been done—indeed I did a small part of it myself in the remote past when I was a research student—and it had been shown that even when the computing had to be done on a mechanical hand calculator, remarkable results could be achieved within a reasonable time. Large modern computers accelerate this work by I don't know what factor, certainly by thousands, perhaps by millions, of times. Much of the work which we have heard about during this conference would have been impossible without them, and we have reached a stage when their use is so commonly assumed that we make hardly more than a passing reference in research papers to the systems and programs actually employed. What has been created is not just a new specialisation and a new profession, but a new climate in which the whole of quantitative applied science (and much of pure science) now operates. It is the provision of computing services which has taken up a large proportion of the increased resources which I referred to a few minutes ago. In the University of Durham, which may be a bit untypical because in our joint enterprise with Newcastle we are nett providers of computing power for other people and not nett importers, taking account of both capital and recurrent costs, total expenditure on computer services is about one-third of the total of all expenditure on equipment and services for research, and not very different from the total of expenditure on all library and information services. I suppose I show the extent of my out-of-dateness by regarding computing as one kind of activity and the business of scientific communication—which is what library and information services are about—as another. I ought to use terms such as 'information retrieval' and regard them both as the same thing, or at least as alternative ways of doing the same thing. Now, of course, it looks as though we are in for another revolution in thinking, and the day of the mammoth computer used for every conceivable purpose may be coming to an end, to be replaced by a host of intercommunicating minicomputers. Perhaps there is no area in which people of my age, and perhaps even some people rather younger than me, feel less familiar with the landscape, and less capable of taking even a layman's part in the formulation of policy.

In the precomputer age, we thought we were being rather clever if we managed to deal (only in cases for which the geometry was straightforward, such as a flat plate, or a symmetrical shell) with one source of non-linearity at a time, e.g. with a non-linear stress–strain relationship only, with large-deflection theory only, or with cases in which thermal stresses and mechanical loadings were simultaneously present only. An analytical method which produced a result within a few weeks of computation in such cases was a reasonably satisfactory one. But only one of our authors (Turner, with his interesting solution of a classical problem of elastic contact between solids) is content to deal with one source of non-linearity at a time. Turvey has complex geometry and large deflections at the same time. Holt and Parsons have complex geometry, thermal stresses and mechanical loading under dynamic conditions all together. Kråkeland and Mo combine inelastic materials, large deflections and the need to take account of developing instability, all at the same time. These authors are, I think, the only ones to tell us exactly how long the job takes; with 12 successive load increments and 22 iteration cycles for updating the stiffness matrix for each increment, it takes just under twelve minutes.

Of course, I am not suggesting that the computers make everything easy. The very fact that the computational part of the work can be passed over in a few lines proves the opposite. What have to be recognised are first the form in which the physics of the problem can best be formulated and then, having established the appropriate equations, the iterative processes which will make the most efficient use possible of the software (and hardware) available. This is obvious. What is not obvious is the way in which people acquire the ability to do this, and the kind of preparation which is most likely to produce these skills.

Let me turn now briefly to one or two subjects which in 1955 we might have thought would have figured in the conference papers, but which are absent. I hope I do not sound too depressing if I introduce the word 'catastrophe'. After all, the mathematicians are now engaged in developing something which they call 'catastrophe theory'. Of course, many of the papers in the programme are about averting catastrophes of various more-or-less predictable kinds. But it is not the predictable that I have in mind, but the unpredictable—earthquakes, floods or man-made happenings, accidents in the broadest sense. I do not think it can be argued that these things have become any less common in the intervening years. What was foreseeable was that the scale of the resulting catastrophes would increase. (After all, in the field of nuclear power generation we are in the middle of a debate in which one of the principal matters in dispute is how far catastrophe can be foreseen and provided for.)

We may well think that that is not a problem for the stress analyst. How is he to design for a situation, the nature of which by definition is unknown. But I think that is to accept too limited a view of our rôle. It is quite possible to think out what will happen in all the various forms of catastrophe, and strive to limit the effect to what is acceptable. For example, one can design structures so that the removal of any member or section has no effect elsewhere. I have not much doubt that a good deal of this kind of thinking goes on already in many contexts. For understandable reasons, we in this field are perhaps rather reluctant to talk about our ideas in public, even among our colleagues at academic conferences. But it must be recognised that without public discussion ideas will circulate only slowly and many opportunities will be missed.

Another subject I mention with some hesitation, because it used to be very much an interest of my own, is the centuries-old technology of the metal-forming processes—rolling, drawing, forging, cutting, and so on. In the 1950s what appeared to be rather rapid progress was being made in relating these processes, usually involving very large strains, to plasticity theory, and, of course, some work continues to appear on the subject. But, considering the enormous investment and effort which go into manufacture, I would have thought not very much. What is the explanation? Is it that theoretical understanding in this field contributes not much to practice? Is it that the processes have evolved in a purely empirical way to such a high degree of perfection that there is not much more to be done? Or is it, as I suspect, that what is new inevitably attracts a lot more attention from everybody than what has been going on for a long time?

There is just one other structure which operates sometimes in the elastic range, sometimes plastically, on which I would like to have seen some work reported, and that is the one on which we all depend—the skeleton. The surgeons are now busily engaged in providing and fitting spare parts. Do the spare parts have the same elastic and plastic properties as the parts they replace? And how far does it matter, from the point of view of the stress on what remains of the original structure, if they are different? No doubt we are finding out the answers to these questions rather rapidly by experience. But I would expect that theoretical contributions at an early stage would make the process a good deal more comfortable.

I have exhausted, or rather more than exhausted, the allotted time. I should apologise to the authors who in this brief survey I have not found the opportunity to mention. I have very much enjoyed my two days of freedom from being a Vice-Chancellor, and I am very grateful to the Conference for giving me the opportunity.

Index